Zero-Dimensional Schemes

Zero-Dimensional Schemes

Proceedings of the International Conference
held in Ravello, June 8–13, 1992

Edited by

Ferruccio Orecchia
Luca Chiantini

Walter de Gruyter · Berlin · New York 1994

Editors

Ferruccio Orecchia
Dipartimento di Matematica
e Applicazioni
Università degli Studi di Napoli
Via Cintia
I-80126 Napoli
Italia

Luca Chiantini
Dipartimento di Metodi
e Modelli Matematici
Università di Roma "La Sapienza"
Via Scarpa 10
I-00185 Roma
Italia

1991 Mathematics Subject Classification: 14-06; 14 Pxx

♾ Printed on acid-free paper which falls within the guidelines of the ANSI to ensure permanence and durability.

Library of Congress Cataloging-in-Publication Data

Zero-dimensional schemes : proceedings of the international conference held in Ravello, June 8–13, 1992 / editors, Ferruccio Orecchia, Luca Chiantini.
p. cm.
"The International Conference on 'Zero-Dimensional Schemes' was held in Ravello (Salerno) from June 8 to 13, 1992" – Pref.
ISBN 3-11-013934-0 (acid-free) : $ 98.95 (U.S.)
1. Schemes (Algebraic geometry) – Congresses.
I. Orecchia, F. (Ferruccio), 1948– , II. Chiantini, Luca, 1957– III. International Conference on "Zero-Dimensional Schemes" (1992 : Ravello, Italy)
QA664.Z46 1994
516.3'5–dc20 94-20654
CIP

Die Deutsche Bibliothek – Cataloging-in-Publication Data

Zero-dimensional schemes : proceedings of the international conference held in Ravello, June 8–13, 1992 / ed. by Feruccio Orecchia ; Luca Chiantini. – Berlin ; New York : de Gruyter, 1994
ISBN 3-11-013934-0
NE: Orecchia, Ferruccio [Hrsg.]

Printed in Germany. Printing: Arthur Collignon GmbH, Berlin. Binding: Lüderitz & Bauer GmbH, Berlin. Cover design: Thomas Bonnie, Hamburg.

Preface

The International Conference on "Zero-Dimensional Schemes" was held in Ravello (Salerno) from June 8 to 13, 1992. This volume contains most of the lectures given at the Conference as well as papers grown up from discussions among participants. In addition we have inserted at the end of the volume a list of problems and questions; most of them were discussed during the conference. We are very grateful to the:

Università di Napoli "Federico II",
Consiglio Nazionale delle Ricerche,
Dipartimento di Matematica e Applicazioni "R. Caccioppoli" di Napoli,
Istituto Nazionale per gli Studi Filosofici,

for their financial support,

Ente Provinciale per il Turismo di Napoli,

for having allowed the use of the wonderful "Villa Rufolo" where the Conference took place.

We would like to thank also all participants for their contribution to the success of the Conference and the staff and graduate students of the Department for their precious help in the organization. Finally we are deeply grateful to the contributors and referees of this volume.

Ferruccio Orecchia

Luca Chiantini

Contents

Zero-dimensional Schemes: Singular Curves and Rational Surfaces

Anthony V. Geramita

Abstract. In this expository talk I discuss two very important influences on the current study of zero-dmensional schemes.

It is an honour and a pleasure to be giving the opening lecture in this first international conference on "Zero-dimensional Schemes". The beauty of the Sorrentino peninsula and the clarity of the sky and water here form a magnificent benchmark (in terms of both beauty and clarity) for our own talks at this gathering.

We are all extremely grateful for the many hours that Professor F. Orecchia and his colleagues at the Universitá di Napoli (especially Prof. L. Chiantini) have put into the preparations for this week. On behalf of all the participants I want to thank them, and also the graduate students and staff from the Dipartimento di Matematica, for their kindness and ingenuity in organizing a 20th century conference in a 14th century villa.

1. Introduction

The study of zero-dimensional schemes, in one form or another, has interested mathematicians for hundreds of years. The reasons for this interest have varied over time but I have observed three major reasons why this subject has flourished in our time.

The first, and probably most important, reason for the study of schemes of dimension zero has come from the, so-called, *Castelnuovo Theory.* This theory is concerned with the study of algebraic curves embedded in particular projective spaces. It was G. Castelnuovo who showed how to discover non-trivial facts about an embedded curve from a careful study of the zero-dimensional schemes arising as hyperplane sections of the curve. A wonderful introduction to Castelnuovo Theory can be found in the Montreal Lecture Notes of D. Eisenbud and J. Harris [6].

Attempts to expand and extend Castelnuovo's original ideas continue to influence the study of zero-dimensional schemes. This is attested to by the fact that several of the lectures at the conference are devoted to explaining and exploiting

* The author gratefully acknowledges support he has received during the preparation of this work from the Natural Sciences and Research Council of Canada and the Consiglio Nazionale delle Ricerche of Italia.

this point of view. Consequently, I will not discuss it further in this lecture but, instead, will discuss two other major influences on the modern development of the subject.

The second place where I have observed a substantial discussion of schemes of dimension zero is in the study of certain interpolation problems. In $\mathbb{P}^2$, for example, these problems ask: given a fixed finite set of points, when is there an algebraic curve of some fixed degree passing through these points and having, at each point, a singularity of some prescribed multiplicity?

Put another way, if $\mathbb{X} = \{P_1, \ldots, P_s\} \subseteq \mathbb{P}^2$ and $\alpha_1, \ldots, \alpha_s$ are non-negative integers, is there a form of degree d in $k[x_0, x_1, x_2]$ all of whose partial derivatives (up to degree $\alpha_i - 1$) vanish at P_i? These problems can be refined to questions about the Hilbert function of some ideal and those questions can, in turn, be translated to calculations of cohomology groups of invertible sheaves on certain rational surfaces. I will discuss this more precisely, but rather briefly, at the end of this talk since there are two expository articles about this point of view which are currently available (Gimigliano [15], and Geramita [8]). Both papers contain extensive bibliographies on this particular aspect of zero-dimensional schemes.

But, I was first attracted to the study of zero-dimensional schemes because of their appearance in yet a third context. Prof. F. Orecchia introduced me to this context in 1979, during his year long visit to Queen's University. (I will always remain very indebted to him for that favour.) Applications of ideas from this point of view provide, among other things, a simple and elegant way of looking at Macaulay's famous examples of affine curves whose defining ideals have arbitrarily large numbers of generators.

Since this third source of problems about zero-dimensional schemes appears to be less well-known, I will devote the major part of this lecture to an exposition of this point of view. This is the content of the next section.

2. Ordinary Singularity of Curves

Let C be a reduced and irreducible curve, $C \subseteq \mathbb{A}^{n+1}(k)$, where k is an algebraically closed field of characteristic zero. Assume further, with no loss of generality, that C passes through the origin of $\mathbb{A}^{n+1}$. Then C corresponds to a prime ideal $\wp \subseteq k[x_0, \ldots, x_n]$, where $\wp$ has height n and $\wp \subseteq (x_0, \ldots, x_n)$.

Let $A = k[x_0, \ldots, x_n]/\wp$ denote the (affine) coordinate ring of C and $m = (x_0, \ldots, x_n)$. Then the *associated graded ring* is,

$$gr_m(A) = (A/m) \oplus (m/m^2) \oplus \ldots = k[x_0, \ldots, x_n]/J$$

where J is a homogeneous ideal of height n.

Consequently, we can write the primary decomposition of J as:

$$J = \mathfrak{q}_1 \cap \ldots \cap \mathfrak{q}_s \cap \mathfrak{q}$$

where $\text{rad}(\mathfrak{q}_i) = (L_{i1}, \dots L_{in}) = \wp_i$, and the L_{ij} $(j = 1, \dots, n)$ are linearly independent linear forms. Moreover, $\text{rad}(\mathfrak{q}) = (x_0, \dots, x_n)$.

Notice that $\wp_i$ describes a line, ℓ_i, through the origin of $\mathbb{A}^{n+1}$ or, alternatively, a point in $\mathbb{P}^n$. (It can happen that J is unmixed, i.e. $\mathfrak{q} = R$.) The lines ℓ_i are usually referred to as the *tangent lines* to C at the origin.

Definition 2.1.

i) $\text{Proj}(gr_m(A))$ – a zero-dimensional subscheme of $\mathbb{P}^n$ – is called the (projectivized) *tangent cone* to C at the origin.

ii) The origin is an *ordinary singularity* of C if $\text{Proj}(gr_m(A))$ is a reduced ring, i.e. if

$$J = \wp_1 \cap \dots \cap \wp_s \cap \mathfrak{q}.$$

Examples 2.2.

i) If $C \subseteq \mathbb{A}^2$ is given by the equation $y^2 = x^2(x+1)$ then $gr_m(A) = k[x,y]/(y-x)(y+x)$. The line $\ell_1 : y = x$ and the line $\ell_2 : y = -x$, are the tangent lines to C at $(0,0)$ and hence $(0,0)$ is an ordinary singularity of C.

ii) If $C \subseteq \mathbb{A}^2$ is given by the equation $y^2 = x^3$ then $gr_m(A) = k[x,y]/(y^2)$. The line $\ell : y = 0$ is the only tangent line to C at $(0,0)$ and so $(0,0)$ is NOT an ordinary singularity.

Orecchia had been interested in schemes like those defined by J, especially from the point of view of K-theory. He, along with my colleague at Queen's, L. Roberts (among others), had been interested in the Picard group of lines through the origin in $\mathbb{A}^{n+1}$. H. Bass and M.P. Murthy [4] had given a description of the conductor of rings, like A, which was crucial to a calculation of the Picard group and Orecchia [21] had found a very nice geometric interpretation of the Bass-Murthy result which I will now explain.

This time let $\mathcal{A} = k[x_0, \dots, x_n]/(\wp_1 \cap \dots \cap \wp_s)$, where the $\wp_i$ are as above. Thus $\mathcal{A}$ is the affine coordinate ring of s lines through the origin of $\mathbb{A}^{n+1}$ OR (alternatively) $\mathcal{A}$ is the homogeneous coordinate ring of s points, $P_1, \dots, P_s$ in $\mathbb{P}^n$.

Since $\mathcal{A}$ is a reduced ring and $\mathcal{A}/\wp_i \cong k[T_i]$ is integrally closed, if we let $\bar{\mathcal{A}}$ denote the integral closure of $\mathcal{A}$ in its total ring of fractions we have $\bar{\mathcal{A}} = \oplus k[T_i]$ and:

$$\begin{array}{ccccccc} \bar{\mathcal{A}} & = & (\bar{\mathcal{A}})_0 & \oplus & (\bar{\mathcal{A}})_1 & \oplus & \cdots \\ \cup\!\mid & & \cup\!\mid & & \cup\!\mid & & \\ \mathcal{A} & = & \mathcal{A}_0 & \oplus & \mathcal{A}_1 & \oplus & \cdots \end{array}$$

where $\dim(\bar{\mathcal{A}})_i = s$ for all i. Thus, for $t \gg 0, \mathcal{A}_t = (\bar{\mathcal{A}})_t$.

Notice that ideals of $\bar{\mathcal{A}}$ are easy to describe. Every ideal is given by a sequence of s positive integers, $m_1, \dots, m_s$ which corresponds to the ideal $(T_1^{m_1}, \dots, T_s^{m_s})$.

Theorem 2.3 (Orecchia). *Let $\mathfrak{C} =$ the conductor of $\mathcal{A}$ in $\bar{\mathcal{A}}$. Then, as an ideal of $\bar{\mathcal{A}}$,*

$$\mathfrak{C} = (T_1^{m_1}, \ldots, T_s^{m_s})$$

where m_i is the least degree of a hypersurface in $\mathbb{P}^n$ which does not vanish at P_i but does vanish at the points $P_1, \ldots, P_{i-1}, P_{i+1}, \ldots, P_s$.

Remark 2.4.
i) It is more complicated to describe $\mathfrak{C}$ as an ideal of $\mathcal{A}$. If, however, all the m_i are equal (say to r) then $\mathfrak{C} = m^r$ as an ideal of $\mathcal{A}$.
ii) A set of points $\{P_1, \ldots, P_s\}$ for which all m_i above are equal, is said to have the *Cayley-Bacharach property.* That name comes from a theorem (due to Cayley and Bacharach) which says that complete intersection sets of points have the Cayley-Bacharach property (see [9] for a more complete discussion.)

This sort of "uniformity" on the m_i led us, in a natural way, to a codification and extension of this notion. We came to consider the following situation in some detail:

let $\mathbb{X} = \{P_1, \ldots, P_s\} \subseteq \mathbb{P}^n$ be a set of s distinct points and let $I = \wp_1 \cap \ldots \cap \wp_s \subseteq k[x_0, \ldots, x_n] = R$ and let $\mathcal{A} = R/I$. The Hilbert function of $\mathcal{A}$ (or of $\mathbb{X}$) is *maximal* if:

$$H(\mathbb{X}, t) = H(\mathcal{A}, t) = \dim_k(R_t/I_t) = \min\{s, \dim R_t\}.$$

If $\mathbb{X}$ has maximal Hilbert function then $\mathbb{X}$ is said to be in *generic position.* It is not hard to show that almost every set of s distinct points in $\mathbb{P}^n$ is in generic position.

The importance of these notions is illuminated by another theorem of Orecchia [22].

Theorem 2.5 (Orecchia). *If* $\mathrm{Proj}(gr_m(A))$ *(A the affine coordinate ring of a curve) is reduced and consists of points in generic position then*

$$gr_m(A) \textit{ is arithmetically Cohen-Macaulay.}$$

(i.e., in our notation, there is no "q" in the primary decomposition of J).

Orecchia and I realized in 1982 [14] that we could use the Cohen-Macaulay property to "lift" information from $gr_m(A)$ to A_m. For example:

Theorem 2.6 (Geramita-Orecchia). *Let $A = k[x_0, \ldots, x_n]/\wp$ be the affine coordinate ring of a curve C through the origin of $\mathbb{A}^{n+1}$ and let m correspond to the origin on C. Write $gr_m(A) = k[x_0, \ldots, x_n]/J$.*

Suppose $J = \wp_1 \cap \ldots \cap \wp_s$ ($\wp_i \leftrightarrow P_i \in \mathbb{P}^n$) and $\mathbb{X} = \{P_1, \ldots, P_s\}$ is in generic position.

Then,

$$\textit{the minimal number of generators of the ideal } J =$$

$$\textit{minimal number of generators of the ideal } \wp \in k[x_0, \ldots, x_n]_m.$$

Since J is a homogeneous ideal one can easily manufacture instances in which the determination of the number of generators of J is rather easy. As a consequence one reconstructs Macaulay's examples of space curves requiring large numbers of generators. Orecchia and I did that in [14].

Notice, however, that "numbers of generators" can be viewed as giving information about the first Betti number in a minimal free resolution of J (or of $\wp$ in the local ring $S = k[x_0, \ldots, x_n]_m$). It is natural to ask about the other Betti numbers.

Since J (as above) is a perfect homogeneous ideal, one knows that J has a minimal free resolution of the form

$$0 \to F_n \to \ldots \to F_2 \to F_1 \to J \to 0$$

where $F_n \neq 0$. The rank of F_n is also known as the *Cohen-Macaulay type* (or *CM-type*) of the ring $A = k[x_0, \ldots, x_n]/J$.

In other works, with Orecchia we investigated this last Betti number and showed it was the same as that of $\wp$ in the local ring S. These comparison theorems were generalized, by R. Froberg [7], to include all the Betti numbers.

It was these investigations which led us to make several conjectures about the *expected* value of the Betti numbers in a minimal free resolution like the one above, for a general enough set of points.

The first conjecture was about the numbers of generators of a generic enough set of distinct points in $\mathbb{P}^n$. This conjecture was made in [14] (where several instances of it were proved) and is known as the *Ideal Generation Conjecture.* It was first proved completely for points in $\mathbb{P}^2$ by P. Maroscia and me [11] and then by E. Ballico in $\mathbb{P}^3$ [2] (see also L. Roberts' addenda [24]). In another paper with Gregory and Roberts [10] we gave a much simpler proof for $\mathbb{P}^2$ as well as scattered results for $\mathbb{P}^n$ for $n \geq 3$. In [12] we found yet another proof for $\mathbb{P}^2$.

L. Roberts was led, at this time, to a natural conjecture about the last Betti number of the ideal of a generic set of distinct points in $\mathbb{P}^n$ [23]. This is known as the *Cohen-Macaulay type conjecture.* The results of Maroscia and me in $\mathbb{P}^2$ solved Roberts' conjecture there, and earlier results with Orecchia gave many other instances of the correctness of this conjecture in $\mathbb{P}^n$ for $n > 2$. In [23], Roberts also gave several new instances where this conjecture was correct. In [3], Ballico and I solved Roberts' conjecture in $\mathbb{P}^3$ and then Trung and Valla [25] gave a proof affirming Roberts' conjecture in $\mathbb{P}^n$ for every $n \geq 3$.

Emboldened by these results, A. Lorenzini in her Queen's Ph.D. dissertation [20], formulated a conjecture about all the Betti numbers in a free resolution of the ideal of a generic set of points in $\mathbb{P}^n$. This conjecture is known as the *Minimal Resolution Conjecture (MRC).*

Again, the results of Maroscia and me affirm the MRC in $\mathbb{P}^2$ and the results of Ballico and me (referred to above) affirm the MRC for $\mathbb{P}^3$. In her dissertation Lorenzini gave many other cases where the MRC is correct.

It is perhaps pertinent to mention that at this meeting I've heard rumblings about the MRC – with the hope of a counter-example for few points in $\mathbb{P}^n$ for $n \geq 4$ AND the hope of a proof for all large enough sets of generic points in $\mathbb{P}^n$ for $n \geq 4$. If both "rumblings" turn into facts this will be a wonderful state of affairs and we will have a great deal to understand about why the conjecture fails for few points. It is worth mentioning, also, that the hoped-for counterexamples to MRC are about "central" Betti numbers and so there is no indication yet about any counter-example to the Ideal Generation Conjecture.

Before I turn back to the fork in the road which led us down this interesting path, I want to mention that these conjectures all have Artinian counterparts - the so-called "zero-dimensional versions" (where this time "dimension" is Krull dimension). See [10] for a discussion of this. The point is that the homogeneous coordinate ring of a set of points is Cohen-Macaulay and, since our fields are all infinite, it is possible to mod out a linear non-zero divisor and get a zero-dimensional quotient (by some homogeneous ideal) of a polynomial ring in one fewer variable. The Betti numbers of this homogeneous ideal are the same as those of the original ideal and all the conjectures about the Betti numbers of ideals of points have a purely algebraic formulation for Artinian quotients of polynomial rings. There are not even rumoured counterexamples to these conjectures as of this writing. Unfortunately, one knows how to "lift" Artinian examples to points in only a few situations and so a proof of these zero-dimensional conjectures does not necessarily settle the corresponding conjectures for points.

In spite of the fact that so much has been learned about points in $\mathbb{P}^n$, there has not been a corresponding increase in our knowledge of the precise nature of the rings that arise as tangent cones of ordinary singularities. It would be interesting to know the full range of possibilities. More explicitly, one can consider the following:

Let C be a reduced and irreducible curve in $\mathbb{A}^{n+1}$ which passes through the origin of $\mathbb{A}^{n+1}$ and has an ordinary singularity there. Let $\wp$ be the defining ideal of C and let $A = k[x_0, \ldots, x_n]/\wp$. Denote by m the maximal ideal of A corresponding to the origin.

Characterize geometrically, the set of points of $\mathrm{Proj}(gr_m(A))$ for which:

i) $gr_m(A)$ is reduced (and hence Cohen-Macaulay); AND

ii) $gr_m(A)$ has a prescribed set of Betti numbers (compatible with what is permissible for the set of points of $\mathrm{Proj}(gr_m(A))$.

Remark 2.7.

a) Orecchia and I [14] showed how to construct $C's$ as above with tangent cone consisting of points in generic position, i.e. having maximal Hilbert function.

b) S. Gupta and L. Roberts [17] showed how to construct such $C's$ as above whose tangent cones had many possible Hilbert functions (including maximal ones).

However, none of these results ever considered even the 1st Betti number of the ideal defining the tangent cone.

It would be very interesting if there were theorems relating particularities of the Betti numbers of the tangent cone to particularities of the ordinary singularity of the curve. There is clearly lots left to be done in this area.

3. Interpolation

For the remainder of this lecture I want to explain (by way of introduction) a little bit about an aspect of zero-dimensional schemes that has fascinated me for several years and may not be very familiar to some of you. In his lecture, Gimigliano will relate the results of some of our recent investigations in this area, but I will give, in this lecture, some of the foundational ideas.

The usual interpolation problem (in $\mathbb{P}^n$) asks:

given $\mathbb{X} = \{P_1, \ldots, P_s\}$,a set of s distinct points, is there a hypersurfacee of degree d which contains these points? If yes, how many?

This is clearly a question about the Hilbert function of $\mathbb{X}$. Moreover, if $\mathbb{X}$ is a general set of s points in $\mathbb{P}^n$, we've already noted that:

$$H(\mathbb{X}, t) = \min\{s, \binom{t+n}{n}\}$$

This, then, is a "general" answer to the usual interpolation problem.

The *differential interpolation problem* is more involved. Given $\mathbb{X}$ as above and also given integers, $\alpha_1, \ldots, \alpha_s (\alpha_i \geq 1)$, this problem asks:

Is there a hypersurface of degree d containing the points of $\mathbb{X}$ such that P_i is a singularity of the hypersurface with multiplicity at least α_i.

If, as in §2, we let $\wp_i \leftrightarrow P_i$ then this is also a problem about Hilbert functions, but this time for the ideal $J = \wp_1^{\alpha_1} \cap \ldots \cap \wp_s^{\alpha_s}$ (affectionately referred to as an ideal of *fat points*).

Is there a "general" answer to this interpolation problem?

The obvious first guess is that the answer is the same as for the usual interpolation problem, i.e. the Hilbert function of such an ideal is the maximal Hilbert function for an ideal of that multiplicity. But, this is easily seen to not be the case.

Example 3.1.

Let P_1, P_2, P_3 be three non-collinear points in $\mathbb{P}^2$ and let $\alpha_1 = 3, \alpha_2 = 2$ and $\alpha_3 = 1$, i.e. $J = \wp_1^3 \cap \wp_2^2 \cap \wp_3$. It is easy to see, e.g. by Nagata's addition formula for multiplicities, that J is an ideal of multiplicity 10.

If $A = k[x_0, x_1, x_2]/J$ had maximal Hilbert function, its Hilbert function would have to be: 1 3 6 10 10 .... Consequently, there would be no cubic curves in J.

But, there is always a cubic curve in J: double the line between P_1 and P_2 and adjoin the line from P_1 to P_3. Since a cubic curve with a triple point is always a union of lines through that point we see that J contains a unique cubic. So, the "general" Hilbert function for ideals of this type is: 1 3 6 9 10 10 .... This is definitely NOT the maximal Hilbert function.

Even in $\mathbb{P}^2$ the general answer to the differential interpolation problem is a matter for conjecture. An excellent reference for the conjectures related to this problem can be found in the comprehensive survey done by Gimigliano [15]. There have been noteworthy recent contributions to this problem (in case all $m_i = 2$ but for all $\mathbb{P}^n$) by J. Alexander [1] and A. Hirschowitz [19]. Also, M. Catalisano [5] gives the best results about the state of this problem in $\mathbb{P}^2$. I also wrote a more elaborate introduction to this aspect of zero-dimensional schemes for the Europroj Workshop on Points, organized in Genova, Italia in February of 1992, [8], as well as compiling a bibliography of all the relevant papers about interpolation (I knew) that had appeared to that date. Both the talk and the bibliography will appear in the next volume (IX) of the Curves Seminar at Queen's.

Let us restrict our attention to $\mathbb{P}^2$. Then the graded pieces of the ideals of fat points, like J above, give incomplete linear systems of plane curves. Thanks to B. Harbourne, my attention was drawn to the fact that these ideals give *complete* linear systems on the blow-up of $\mathbb{P}^2$ at the support of J. So, one can move from $\mathbb{P}^2$ to these new surfaces (more complicated than $\mathbb{P}^2$, to be sure) and try to use other methods to discover information about complete linear systems on these new surfaces. This is exactly the Riemann-Roch problem for these surfaces.

For example, one can use classical results about the non-singular cubic surface, $\mathcal{S}$, in $\mathbb{P}^3$ to completely solve the Riemann-Roch problem for that surface (i.e. calculate $h^0(D)$, for D any divisor on $\mathcal{S}$). One can then translate these results to make an analysis of the Hilbert function of any ideal of the type $J = \wp_1^{\alpha_1} \cap \ldots \cap \wp_6^{\alpha_6}$ (where the P_i correspond to 6 general points of $\mathbb{P}^2$). This is the case because $\mathbb{P}^2$ blown up at 6 general points can be embedded in $\mathbb{P}^3$ as a non-singular cubic surface.

This example has provoked interest in the nature of other embeddings of these simple blow-ups of $\mathbb{P}^2$. There is a considerable (and growing) literature on this subject which is in the bibliography I referred to above.

Returning, for a moment, to the cubic surface, I want to emphasize that, by and large, the calculation of $h^0(D)$ depends on an algorithm. That algorithm was implemented by a (then) undergraduate at Queen's (now Dr.Robert McCann) and is freely available on request.

Important generalizations of these calculations have been made by Giuffrida and Maggioni [16] (in the case of the cubic surface) and Harbourne [18] (in the case of other del Pezzo surfaces).

Finally, ideals of the type $\wp_1^{\alpha_1} \cap \ldots \cap \wp_s^{\alpha}$ are among the simplest examples of what Zariski called *complete ideals* (and, in the modern commutative algebra literature, are called *integrally closed ideals)* i.e. ideals whose graded pieces "become" complete linear systems on blow-ups of $\mathbb{P}^2$. One wonders to what extent the recent results on fat points extend to other complete ideals.

4. Concluding Remarks

I hope that in this opening lecture of the conference, I've given some idea of the scope of research I know about concerning zero-dimensional schemes. I see from the titles announced for many of the remaining talks that by the time this week is over there will be many other interesting aspects of zero-dimensional schemes that you and I will learn about.

Zero-dimensional schemes sounds like a very "small" topic, but

"*Anche le possente quercie crebbero da piccole ghiande .....*"

References

[1] Alexander, J., Singularites imposables en position generale a une hypersurface projective, Compositio Math. 68 (1988), 305-354.

[2] Ballico, E., Generators for the Homogeneous Ideal of s general points in $\mathbb{P}^3$, J. Algebra 106 (1987), 46-52.

[3] Ballico, E., Geramita, A.V., On the Ideal of a General Set of Points in $\mathbb{P}^3$, Proc. of the 1984 Vancouver Conference in Algebraic Geometry, CMS Conference Proceedings, Vol. 6, 1-10.

[4] Bass, H., Murty, P., Grothendieck Groups and Picard Groups of Abelian Group Rings, Ann. of Math. 86 (1967), 16-73.

[5] Catalisano, M.V., Linear Systems of Plane Curves Through Fixed "Fat" Points of $\mathbb{P}^2$, preprint.

[6] Eisenbud, D., Harris, J., Curves in Projective Space, Les Presses de L'Universite de Montreal, 1982.

[7] Froberg, R., Relations Between a Local Ring and its Associated Graded Ring, J. Algebra 111 (1987), 300-305.

[8] Geramita, A.V., Some Problems on Rational Surfaces, preprint, to appear in Queen's Papers in Pure and Applied Mathematics, The Curves Seminar, Vol. IX.

[9] Geramita, A.V., Kreuzer, M., Robbiano, L., Cayley-Bacharach Schemes and their Canonical Modules, to appear in Trans. AMS.

[10] Geramita, A.V., Gregory, D., Roberts, L., Monomial Ideals and Points in Projective Space, J. Pure Appl. Algebra 40 (1986) , 33-62.

[11] Geramita, A.V., Maroscia, P., The Ideal of Forms Vanishing at a Finite Set of Points in $\mathbb{P}^n$, J. Algebra 90 (1984), 528-555.

[12] Geramita, A.V., Migliore, J.C., Liaison Addition, to appear in J. Algebra.

[13] Geramita, A.V., Orecchia, F., On the Cohen-Macaulay type of s-lines in $\mathbb{A}^{n+1}$, J. Algebra 70 (1981), 116-140.

[14] Geramita, A.V., Orecchia, F., Minimally Generating Ideals Defining Certain Tangent Cones, J. Algebra 78 (1982), 36-57.

[15] Gimigliano, A., Our Thin Knowledge of Fat Points, The Curves Seminar at Queen's, Vol.VI, Queen's Papers in Pure and Applied Math., no. 83, (1989).

[16] Giuffrida, S., Maggioni, R., On the Resolution of a Curve Lying on a Smooth Cubic Surface in $\mathbb{P}^3$, Trans. AMS 311 (1992), 181-201.

[17] Gupta, S.K., Roberts, L., Cartesian Squares and Ordinary Singularities of Curves, Comm. Algebra 11 (1983), 127-182.

[18] Harbourne, B., Hilbert Function of Good Points in $\mathbb{P}^2$, The Curves Seminar at Queen's, Vol. VI, Queen's Papers in Pure and Applied Math., no.83, Queen's University, Kingston, Ontario (Canada).

[19] Hirschowitz, A., La Methode d'Horace pour l'Interpolation a Plusieurs Variables, Manuscripta Math. 50 (1985), 337-388.

[20] Lorenzini, A., On the Betti Numbers of Points in Projective Space, Ph.D. Dissertation, Queen's University, Kingston, Ontario (Canada) (1987).

[21] Orecchia, F., Sui gruppi di Picard di certe algebra finite non integre, Ann. Univ. Ferrara, Sez. VII, vol. XXI, 25-36 (1975).

[22] Orecchia, F., Points in Generic Position and the Conductor of Curves with Ordinary Singularities, J. Lond. Math. Soc. 24 (1981), 85-96.

[23] Roberts, L., A Conjecture on Cohen-Macaulay type, C.R. Math. Rep. Acad. Sci. Canada, vol. 3, (1981), 43-48.

[24] Roberts, L., The Ideal Generation Conjecture for 28 Points in $\mathbb{P}^3$, Can. J. Math., vol. XXXVIII, No. 5, (1986), 1228-1238.

[25] Trung, N.V., Valla, G., The Cohen-Macaulay type of Points in Generic Position, J. Algebra 125 (1989), 110-119.

Duality, derivations and deformations of zero-dimensional singularities

A. G. Aleksandrov

Abstract. Several types of duality in the tangent cohomology of a multiple point are described. A relationship between Milnor and Tjurina numbers of a Gorenstein zero-dimensional singularity are obtained.

1991 Mathematics Subject Classification: 32S10, 14B05, 14M10, 13D10, 13N05.

0. Introduction

There are lots of papers devoted to the study of various aspects of the theory of Artinian algebras or zero-dimensional singularities (see e.g. the overview [Ia]). However most of authors investigate combinatorial or computational properties of such algebras. It seems till present there is no satisfactory progress in understanding the analytic structure of such singularities. In a certain sense this note is a sequel of the paper [AM] where we analyzed some problems that arise in the deformation theory of zero-dimensional singularities.

In the first Section we recall some basic notations and definitions. In the Section 2 we discuss some properties of the primitive ideal. In the next Section a duality between tangent and cotangent modules of zero-dimensional singularities are obtained. Numerical properites of the tangent cohomology of Gorenstein artinian algebras are considered in the Section 4. Our aim is to describe relations between the Tjurina and Milnor numbers of such singularities. Among other things we have proved a corrected version of Corollary 2.3 from [AM]. Then we describe the special case of complete intersection defined by the gradient of a holomorphic function with an isolated critical point. In particular, we obtain a natural nondegenerate pairing between the module of derivation and the first cotangent cohomology module. In the Section 6 we consider graded Artinian algebras. In the last Section we briefly discuss some properties of the module of holomorphic vector fields in the graded case.

I would like to acknowledge useful conversation with A.N. Kuznetsov, V.P. Palamodov, B. Martin and A.N. Shoshitaishvili. I am very grateful to the Chairman of the Conference in Ravello, Professor Ferruccio Orecchia, for the opportunity to attend this enjoyable meeting.

1. Differential forms, derivations and the tangent cohomology modules

Let κ be a valued field of characteristic zero and let A be a local analytic κ-algebra, i.e., a factor algebra of the convergent power series ring $H = \kappa\{z_1, \dots, z_m\}$ in m variables over κ. For instance, if the base value of κ is trivial then local analytic κ-algebras are complete local rings containing κ as the coefficient field. We shall denote the maximal ideal of A by $\mathcal{M}_A$ and the module of regular holomorphic differential 1-forms of H by $\Omega^1_{H/\kappa}$, that is, $\Omega^1_{H/\kappa} \cong H\{dz_1, \dots, dz_m\}$.

Let I be an ideal of H and let $A = H/I$ be a factor algebra of H. One can define the module $\Omega^1_{A/\kappa}$ of Kähler differentials of A over κ from the following standard exact sequence of A-modules

$$I/I^2 \xrightarrow{d} \Omega^1_{H/\kappa} \otimes_H A \longrightarrow \Omega^1_{A/\kappa} \longrightarrow 0\ , \qquad (*)$$

where d is given by $d(\bar{f}) = d_{H/\kappa}(\bar{f}) \otimes 1$ for $\bar{f} \in I/I^2$ and $d_{H/\kappa} : H \longrightarrow \Omega^1_{H/\kappa}$ is the universal differentiation.

Let us consider the functor $\mathrm{Der}_\kappa(A, -)$ of κ-derivations (from A-modules to A-modules). Then there is a fundamental functorial isomorphism $\mathrm{Der}_\kappa(A, -) \cong \mathrm{Hom}_A(\Omega^1_{A/\kappa}, -)$. We put $\mathrm{Der}(A) = \mathrm{Der}_\kappa(A, A)$ so that $\mathrm{Der}(A)$ is the A-module of κ-derivations of A.

Note. Suppose that the ideal I is generated by the sequence of functions $f_1, \dots, f_k \in H$. Then

$$\Omega^1_A \cong \Omega^1_H / \left(\sum_{j=1}^{k} f_j \cdot \Omega^1_H + H \cdot df_j \right) .$$

Let κ be the field $\mathbf{C}$ of complex numbers. Then the local $\mathbf{C}$-algebra $A = H/I$ corresponds to the germ $X \subset (\mathbf{C}^m, 0)$ with the dual analytic $\mathbf{C}$-algebra $\mathcal{O}_X \cong A$. The modules Ω^1_A and $\mathrm{Der}(A)$ are usually called the module of regular holomorphic (or, equivalently, Pfaffian) differential 1-forms and the module of holomorphic vector fields on the germ X, respectively. They are also denoted by Ω^1_X and $\mathrm{Der}(X)$.

We will also consider the tangent and cotangent functors of analytic algebras (see [P1]) denoted by T_i and T^i, respectively, $i \geq 0$. By definition, for any A-module M there exist the following isomorphisms

$$T_0(A/\kappa, M) \cong \Omega^1_{A/\kappa} \otimes_A M\ , \quad T^0(A/\kappa, M) \cong \mathrm{Der}_\kappa(A, M)$$

and exact sequences of A-modules

$$0 \longrightarrow T_1(A/\kappa, M) \longrightarrow I/I^2 \otimes_A M \xrightarrow{d \otimes 1_M} \Omega^1_{H/\kappa} \otimes_H M \longrightarrow T_0(A/\kappa, M) \longrightarrow 0\ ,$$

$$0 \longrightarrow T^0(A/\kappa, M) \longrightarrow \mathrm{Hom}_A(\Omega^1_{H/\kappa} \otimes_H A, M) \longrightarrow$$
$$\longrightarrow \mathrm{Hom}_A(I/I^2, M) \longrightarrow T^1(A/\kappa, M) \longrightarrow 0\ .$$

The first sequence contains $(*)$ tensored with M over A. Applying the functor $\mathrm{Hom}_A(-, M)$ to $(*)$, we get the second sequence.

For brevity we shall denote the tangent and cotangent modules $T_i(A/\kappa, A)$ and $T^i(A/\kappa, A)$ by $T_i(A)$ and $T^i(A)$ respectively.

Hence

$$T_0(A) \cong \Omega^1_{A/\kappa}, \quad T^0(A) \cong \mathrm{Der}(A), \quad T_1(A) \cong \mathrm{Ker}(d).$$

Moreover, if $A = A_{red}$ is *reduced* and an A-module M has *no embedded* associative primes then there is an isomorphism (see [KL], (1.4.3))

$$T^1(A/\kappa, M) \cong \mathrm{Ext}^1_A(\Omega^1_{A/\kappa}, M) .$$

This is also true if instead of the above condition on A we shall assume that the analytic **C**-algebra A corresponding to the germ $X \subset (\mathbf{C}^m, 0)$ has positive depth along its singular locus.

2. The primitive ideal

Now we will analyze the sequence $(*)$ from the preceding Section. Firstly recall the following definition (see [Sei], §2).

For any ideal $I \subseteq H$ one can consider the set $\mathcal{D}(I) \subset H$ of all $f \in I$, such that $\partial(f) \in I$ for all $\partial \in \mathrm{Der}\,(H)$. It is obvious that $\mathcal{D}(I)$ is an ideal of H. Following [Pel] we shall call $\mathcal{D}(I)$ the primitive ideal of I and denote it by $\int I$.

Proposition 2.1 (see [Sei, §2]). *The primitive ideal has the following properties:*

1) $I^2 \subseteq \int I \subseteq I$;

2) *for any ideals* $I_1, I_2 \subset H$ *we have* $\int (I_1 \cap I_2) = \int (I_1) \cap \int (I_2)$;

3) *if* I *is a primary ideal of* H, *then* $\int I$ *is primary too.*

Note. The first and second properties are also proved in [Pel, (1.3)].

Lemma 2.2. *Let* I *be an ideal of* H *and* $A = H/I$. *Then there is the following exact sequence of* A*-modules*

$$0 \longrightarrow \int I/I^2 \longrightarrow I/I^2 \xrightarrow{d} \Omega^1_{H/\kappa} \otimes_H A \longrightarrow \Omega^1_{A/\kappa} \longrightarrow 0 .$$

In particular, there are isomorphisms

$$T_1(A) \cong \mathrm{Ker}(d) \cong \int I/I^2$$

and the sequence $(*)$ *is exact from the left side if and only if* $\int I = I^2$ *or, equivalently,* $T_1(A) = 0$.

Proof. Follows from definitions.

It should be mentioned that in the terminology of [Sei] local analytic κ-algebras with trivial first cotangent module are called *L*-algebras.

Theorem 2.3 (see [Sei], (4.1)). *For any local analytic κ-algebra $A = H/I$ the first tangent module vanishes, i.e., $T_1(A) = 0$, if and only if A is reduced and the ideal $I^2 \subset H$ has no embedded primes.*

For example, if $I = (f)$ is a *principal* ideal of H generated by a function without multiple factors then $T_1(A) = 0$.

Definition 2.4. A local analytic κ-algebra $A = H/I$ (or, equivalently in the case $\kappa = \mathbf{C}$, the corresponding germ $X \subset (\mathbf{C}^m, 0)$) is called a *complete intersection* singularity if the ideal I is generated by a *regular* H-sequence of function germs $f_1, \dots, f_k$, that is, f_j is not a zero-divisor in $H/(f_1, \dots, f_{j-1})H$ for every $j = 1, \dots, k$. It is well-known that in this case $T_i(A) = T^i(A) = 0$ for all $i \geq 2$ (see [P2], (4.5)).

Claim 2.5. *Suppose that a local κ-algebra A corresponds to a reduced complete intersection singularity. Then $T_1(A) = 0$ or, equivalently, $\int I = I^2$.*

Proof. This is a direct consequence of the theorem above (see [Sei], §4, ex. (1)).

Remark. More generally, if A is an arbitrary complete intersection then the first tangent module $T_1(A) \cong \operatorname{Ker}(d)$ is a *projective* A-module (see [F]). In fact, $T_1(A)$ is a free A-module as a projective module over a local ring.

In particular, the above theorem implies that the sequence $(*)$ is not exact from the left in the following two cases:

1) A is a local analytic κ-algebra associated with a *nonreduced* hypersurface germ, i.e., I is generated by an element $f \in P$ having multiple factors;

2) A is a *zero-dimensional* local analytic algebra associated with a multiple point germ.

Both cases give us, in a certain sense, the simplest examples of singularities having non-trivial first tangent module $T_1(A)$ and $T_i(A) = T^i(A) = 0$ for all $i \geq 2$. Moreover, in the first case one can easily calculate the primitive ideal. Thus, suppose $I = (f) \cdot H$, $f = \prod f_i^{e_i}$, $e_i \geq 1$, $i = 1, \dots, h$, $(f_i) \cdot H$ are prime ideals and f_i, f_j have no factor in common for $i \neq j$. Then $\int (f) = \left(\prod f_i^{e_i+1}\right) \cdot H$ (see [Sei], (3.3) or [Pel], (1.7)). Hence $T_1(A)$ is a free A-module of rank one. In Section 5 we shall compute the primitive ideal explicitly in the special case of *gradient* zero-dimensional singularities.

3. Duality for zero-dimensional singularities

Let us consider the case of zero-dimensional singularities or, equivalently, *Artinian* algebras. It is well-known that such algebras are *Cohen-Macaulay* (see e.g. [HK]). It should be remarked that the tangent and cotangent modules T_i and T^i, $i \geq 0$ are finite dimensional over κ. This follows from the fact that the supports of these modules are concentrated in the point. Of course the dimension over κ of such module is equal to its length as a κ-module.

Proposition 3.1. *Suppose that A is an Artinian local analytic κ-algebra with residue field $\kappa \cong A/\mathcal{M}_A, \mathcal{K} = \mathcal{K}(A)$ is its canonical module (see* [HK], (5.6)*) and M is an A-module of finite type. Then there exists a natural perfect pairing*

$$M \times \mathrm{Hom}_A(M, \mathcal{K}(A)) \longrightarrow \kappa \ .$$

Proof. Let $\mathcal{F} = \mathcal{F}(\kappa)$ be the injective hull of the residue field κ. The Matlis duality (see [HK], (1.35)) implies an isomorphism

$$M \xrightarrow{\cong} \mathrm{Hom}_A(\mathrm{Hom}_A(M, \mathcal{F}), \mathcal{F}) \ ,$$

which induces the perfect pairing

$$M \times \mathrm{Hom}_A(M, \mathcal{F}) \longrightarrow \mathcal{F} \ .$$

By definition (see [Gr], ex. 1, p.63), the injective hull $\mathcal{F}$ is isomorphic to the module of *continuous* (in the $\mathcal{M}_A$-adic topology of the ring A) homomorphisms to the residue field κ:

$$\mathcal{F} \cong \mathrm{Hom}^c_\kappa(A, \kappa) = \lim_{n} \mathrm{Hom}_\kappa(A/\mathcal{M}^n_A, \kappa) \ .$$

Furthermore for every A-module N there is a canonical isomorphism

$$\mathrm{Hom}_A(N, \mathcal{F}) \cong \mathrm{Hom}^c_\kappa(N, \kappa) = \lim_{n} \mathrm{Hom}_\kappa(N/\mathcal{M}^n_A \cdot N, \kappa).$$

In case N is an Artinian A-module we also have

$$\mathrm{Hom}^c_\kappa(N, \kappa) \cong \mathrm{Hom}_\kappa(N, \kappa)$$

and as a result

$$\mathrm{Hom}_A(N, \mathcal{F}) \cong \mathrm{Hom}_\kappa(N, \kappa) \ .$$

It is well known (see [HK], (6.4)) that the canonical module of an Artinian ring A and the injective hull of the residue field κ of A are isomorphic: $\mathcal{K} \cong \mathcal{F}$. Because $\mathcal{F}$ is Artinian [Gr] and M is Noetherian by our assumption we see that $N := \mathrm{Hom}_A(M, \mathcal{F}) \cong \mathrm{Hom}_A(M, \mathcal{K})$ is an Artinian A-module. Hence

$$M \xrightarrow{\cong} \mathrm{Hom}_A(\mathrm{Hom}_A(M, \mathcal{K}), \mathcal{F}) \cong \mathrm{Hom}_\kappa(\mathrm{Hom}_A(M, \mathcal{K}), \kappa),$$

which implies the required result.

Theorem 3.2. *Let A be an Artinian local analytic κ-algebra. Then for all $i \geq 0$ there exist natural nondegenerate pairings of finite dimensional κ-vector spaces*

$$T_i(A/\kappa, A) \times T^i(A/\kappa, \mathcal{K}(A)) \longrightarrow \kappa,$$
$$T_i(A/\kappa, \mathcal{K}(A)) \times T^i(A/\kappa, A) \longrightarrow \kappa.$$

Proof. Let us consider the universal coefficient spectral sequence (see e.g. [P1], (1.4))

$$E_2^{p,q} = \mathrm{Ext}_A^p(T_q(A/\kappa, A), \mathcal{K}(A)) \Rightarrow T^{p+q}(A/\kappa, \mathcal{K}(A)).$$

Because A is a zero-dimensional Cohen-Macaulay ring it follows from properties of the canonical module ([HK], (6.1)) that

$$\mathrm{Ext}_A^p(\kappa, \mathcal{K}(A)) = 0 \quad \text{for all} \quad p \geq 1 \ .$$

Hence this spectral sequence is degenerated and we get the following canoncial isomorphisms

$$\mathrm{Hom}_A(T_i(A/\kappa, A), \mathcal{K}(A)) \xrightarrow{\cong} T^i(A/\kappa, \mathcal{K}(A)), \qquad i \geq 0 \ .$$

Using the properties of the injective hull as in the above proposition we get also

$$\mathrm{Hom}_A(T_i(A/\kappa, A), \mathcal{K}(A)) \xrightarrow{\cong} \mathrm{Hom}_\kappa(T_i(A/\kappa, A), \kappa),$$

and the desired statement for $i \geq 0$.

Remark. It should be mentioned that in the note [Ba] the existence of an analog of the second pairing has been established in the case of *compact* complex spaces of arbitrary dimension by a different technique.

4. Numerical properties of the tangent cohomology of zero-dimensional singularities

Recall that an analytic κ-algebra A is called Gorenstein if $\mathrm{Ann}(\mathcal{M}_A) \subset A$ is a principal ideal (see [HK], (1.44)). For a local ring or algebra A of Krull dimension $n \geq 0$ this condition is equivalent to the following

$$1)\ \mathrm{Ext}_A^i(\kappa, A) = 0 \quad \text{for all} \quad i \neq n; \qquad 2)\ \mathrm{Ext}_A^n(\kappa, A) \cong \kappa.$$

In particular, an Artinian (i.e. *zero-dimensional*) algebra A is Gorenstein if and only if $\mathrm{Ext}_A^i(\kappa, A) = 0$ for $i \geq 1$ and $\mathrm{Hom}_A(\kappa, A) \cong \kappa$.

Proposition 4.1 (cf. [A2]). *Let A be a zero-dimensional Gorenstein analytic κ-algebra. Then there exist natural nondegenerate pairings of finite-dimensional κ-vector spaces*

$$T_i(A/\kappa, A) \times T^i(A/\kappa, A) \longrightarrow \kappa \qquad \textit{for all} \quad i \geq 0.$$

In particular, we obtain

$$\dim_k T_i(A) = \dim_k T^i(A) \quad \textit{for all } i \geq 0.$$

Proof. As we mentioned in Section 3 ([HK], (6.4)) the canonical module of an Artinian ring A and the injective hull of the residue field κ of A are naturally isomorphic: $\mathcal{K} \cong \mathcal{F}$. Moreover, for a zero-dimensional Gorenstein ring A we have $A \cong \mathcal{F}(\kappa)$ by virtue of the Dieudonné assertion (see [HK], (1.43)), that is, such A is a self-injective ring. Hence our statement is a direct consequence of the duality Theorem from the previous Section.

Remark. In the special cases $i = 0, 1$ we have already established (see [AM], (2.1)) the equalities

$$\dim_\kappa \Omega^1_{A/\kappa} = \dim_\kappa \operatorname{Der}(A), \qquad \dim_\kappa T_1(A) = \dim_\kappa T^1(A)$$

by slightly different arguments (see also the Section 5).

Now we will consider the de Rham (Poincaré) complex for a local analytic κ-algebra A

$$A \xrightarrow{d_0} \Omega^1_{A/\kappa} \xrightarrow{d_1} \Omega^2_{A/\kappa} \xrightarrow{d_2} \cdots$$

where $\Omega^i_{A/\kappa} = \wedge^i \Omega^1_{A/\kappa}$, $d_0 = d$ is the universal differentiation, d_1 is the natural extension of d, and d_i is induced by the exterior power of d_1, $i \geq 2$.

Claim 4.2 (see [AM], §2). *Let $A = H/I$ be a zero-dimensional analytic κ-algebra and suppose $I \subset \mathcal{M}^2_H$. Then*

$$\dim_\kappa \Omega^1_{A/\kappa} \geq \dim_\kappa A - \dim_\kappa(\operatorname{Ker} d) + \dim_\kappa \mathcal{M}_A/\mathcal{M}^2_A - 1 .$$

Proof. This statement follows from the fact that $n_i = z_i dz_m$, $i = 1, \ldots, m-1$, are κ-independent non-trivial elements in $\Omega^1_{A/\kappa}$ not contained in $d(A)$.

Remark. Suppose $I \subseteq \mathcal{M}^s_H$, $s \geq 2$. Analogously we get

$$\dim_\kappa \Omega^1_{A/\kappa} \geq \dim_\kappa A - \dim_\kappa(\operatorname{Ker} d) + (s-1) \cdot \dim_\kappa \mathcal{M}_A/\mathcal{M}^2_A - 1 .$$

Denote by $r(I)$, $e(A)$ the minimim number of generators of the ideal I and the embedding dimension of A, respectively. It is easy to see that

$$r(I) = \dim_\kappa T_1(A/\kappa, \kappa) = \dim_\kappa T^1(A/\kappa, \kappa) ,$$

$$e(A) = \dim_\kappa \mathcal{M}_A/\mathcal{M}^2_A = \dim_\kappa T_0(A/\kappa, \kappa) = \dim_\kappa T^0(A/\kappa, \kappa).$$

The *complete intersection defect* of A is defined to be the integer $\Delta(A) = r(I) - e(A) + \dim A$, or, equivalently,

$$\Delta(A) = \dim_\kappa T_1(A/\kappa, \kappa) - \dim_\kappa T_0(A/\kappa, \kappa) + \dim A ,$$

where by $\dim A$ we mean the Krull dimension of the local ring A. It is well known that $\Delta(A) \geq 0$, and equality holds if and only if A is a complete intersection (see [Av] or [P2], (4.7)). Let h be the height of the ideal I. Recall also that usually the difference $d(I) := r(I) - h(I)$ is called the deviation of the ideal I. In our case where $I \subseteq H$ defines an Artinian algebra $A = H/I$ it is clear that $h(I) = e(A)$ and $d(I) = \Delta(A)$.

Below we shall examine in more detail the following statement (cf. [AM], (2.3)).

Theorem 4.3. *Let $A = H/I$ be a zero-dimensional analytic κ-algebra and suppose $I \subseteq \mathcal{M}_H^2$. Then*

$$\dim_\kappa T_1(A) - \dim_\kappa T_0(A) \geq \dim_\kappa T_2(A/H, A) \quad for \quad \Delta(A) \geq 0$$

and equality holds if $\Delta(A) = 0, 1, 2$. In addition, if we suppose that A is Gorenstein then

$$\dim_\kappa T^1(A) - \dim_\kappa T^0(A) \geq \dim_\kappa T_2(A/H, A) \quad for \quad \Delta(A) \geq 0$$

with equality in cases $\Delta(A) = 0,\ 1,\ 2$.

Proof. From the exact sequence $(*)$ of the Section 1 we deduce

$$\dim_\kappa T_1(A) - \dim_\kappa T_0(A) = \dim_\kappa I/I^2 - \dim_\kappa \Omega^1_{H/\kappa} \otimes_H A = \dim_\kappa I/I^2 - e(A) \cdot \dim_\kappa A\ .$$

Suppose that $r := r(I) > 0$. By definition of the tangent cohomology (see [LS], 3.2) there is the following exact sequence:

$$0 \longrightarrow T_2(A/H, A) \longrightarrow \mathcal{H}_1(K.(I)) \longrightarrow (H/I)^r \longrightarrow I/I^2 \longrightarrow 0,$$

where $T^2(A/H, A)$ is the second lower tangent cohomology module of A/H and $\mathcal{H}_1(K.(I)) := \mathcal{H}_1(K.(I), H)$ is the first homology of the usual Koszul complex $K.(I)$ with the coefficients in H associated to a system of generators of the ideal I. Hence

$$\dim_\kappa I/I^2 = r \cdot \dim_\kappa A - \dim_\kappa \mathcal{H}_1(K.(I)) + \dim_\kappa T_2(A/H, A)\ .$$

So the proof of the first inequality will be completed if we can obtain the following estimate:

$$\dim_\kappa \mathcal{H}_1(K.(I)) \leq \Delta(A) \cdot \dim_\kappa A\ .$$

If A is *Gorenstein* we can replace $\dim_\kappa T_0(A)$ by $\dim_\kappa T^0(A) = \dim_\kappa \Omega^1_{A/\kappa}$ in virtue of the above proposition. So we obtain the desired result for the upper cotangent modules.

Lemma 4.4. *With the above notation we have*

$$\dim_\kappa \mathcal{H}_1(K.(I)) \leq \Delta(A) \cdot \dim_\kappa A$$

and equality holds if $\Delta(A) = 0, 1, 2$.

Proof. If $A = H/I$ is a complete intersection then $\mathcal{H}_1(K.(I)) = 0$ and I/I^2 is a *free* A-module. Thus, we can suppose that A is not a complete intersection.

Recall that the maximal length of a regular H-sequence contained in the ideal I is called *I-depth* of H and denoted by depth (I, H). The following basic property of a Cohen-Macaulay local ring

$$\dim H = \text{depth } (I, H) + \dim\ H/I$$

implies that under our assumptions depth $(I/H) = \dim\ H = m = e(A)$.

It is clear that $\Delta(A) = r - m = k \geq 0$.

We can choose the minimal set of generators of I in the following way: $I = (f, g) = (f_1, \ldots, f_m, g_1, \ldots, g_k)$, where $(f) = (f_1, \ldots, f_m)$ is a regular H-sequence of maximal length contained in I.

Let us consider the standard spectral sequence (see [Bou], §9, Exercise 7,b)):

$${}'E^2_{p,q} = \mathcal{H}_p(K.(g), \mathcal{H}_q(K.(f), \mathcal{M})) \Rightarrow \mathcal{H}_{p+q}(K.(f,g), \mathcal{M}).$$

Since (f) is H-regular then $\mathcal{H}_q(K.(f), H) = 0$ for all $q > 0$. Hence ${}'E^2_{p,q}$ degenerates and $\mathcal{H}_p(K.(I), H) \cong \mathcal{H}_p(K.(g), H/(f))$, $p \geq 0$. In particular, $\mathcal{H}_1(K.(I), H) \cong \mathcal{H}_1(K.(g), H/(f))$.

We shall denote the ideal $(f_1, \ldots, f_m, g_1, \ldots, g_s)$ by $I_s, s = 1, \ldots, k$.

Consider the case of an *almost* complete intersection, that is, $r = m + 1$. Set $B \cong H/(f)$, $A \cong B/(g_1)$. Then B is a zero-dimensional complete intersection, i.e., B is *self-injective.* By definition of the Koszul homology we obtain

$$\mathcal{H}_1(K.(g_1), B) \cong \mathrm{Hom}_B(B/(g_1), B) \cong \mathrm{Hom}_\kappa(B/(g_1), \kappa)\ .$$

As a result $\dim_\kappa \mathcal{H}_1(K.(I_1)) = \dim_\kappa \mathrm{Hom}_\kappa(A, \kappa) = \dim_\kappa A$.

Let us consider the case $r = m + 2$. The above spectral sequence yields the following exact sequence (cf. [Bou], §9, n.5):

$$\begin{aligned} 0 \longrightarrow B/(g_2) \otimes_B \mathcal{H}_1(K.(g_1), B) \longrightarrow \mathcal{H}_1(K.(I_2)) \longrightarrow \\ \longrightarrow \mathrm{Hom}_B(B/(g_2), \mathcal{H}_0(K.(g_1), B)) \longrightarrow 0 \end{aligned}$$

or, equivalently,

$$\begin{aligned} 0 \longrightarrow B/(g_2) \otimes_B \mathrm{Hom}_B(B/(g_1), B) \longrightarrow \mathcal{H}_1(K.(I_2)) \longrightarrow \\ \longrightarrow \mathrm{Hom}_B(B/(g_2), B/(g_1)) \longrightarrow 0. \end{aligned}$$

Under our assumptions there exists ([CE], VI, (5.3)) the isomorphism

$$\sigma : B/(g_2) \otimes_B \mathrm{Hom}_B(B/(g_1), B) \cong \mathrm{Hom}_B(\mathrm{Hom}_B(B/(g_2), B/(g_1)), B)$$

because B is self-injective. It is also clear that $\mathrm{Hom}_B(B/(g_2), B/(g_1))$ is Artinian. Hence

$$\mathrm{Hom}_B(\mathrm{Hom}_B(B/(g_2), B/(g_1)), B) \cong \mathrm{Hom}_\kappa(\mathrm{Hom}_B(B/(g_2), B/(g_1)), \kappa)$$

and their dimensions are equal to $\dim_\kappa \mathrm{Hom}_B(B/(g_2), B/(g_1))$. This implies

$$\dim_\kappa \mathcal{H}_1(K.(I_2)) = 2 \cdot \dim_\kappa \mathrm{Hom}_B(B/(g_2), B/(g_1))\ .$$

Finally, by definition,

$$\mathrm{Hom}_B(B/(g_1), B/(g_2)) \cong \{u \in B/(g_2) : u \cdot g_1 \in (g_2)B\} .$$

Hence there exists the exact sequence

$$0 \longrightarrow \mathrm{Hom}_B(B/(g_2), B/(g_1)) \longrightarrow B/(g_1) \xrightarrow{g_2} B/(g_1) \longrightarrow B/(g_1, g_2) \longrightarrow 0$$

where the middle map is induced by the usual multiplication by g_2. As a result we get

$$\dim_\kappa \mathrm{Hom}_B(B/(g_2), B/(g_1)) = \dim_\kappa B/(g_1, g_2) = \dim_\kappa A$$

and the desired equality

$$\dim_\kappa \mathcal{H}_1(K.(I_2)) = 2 \cdot \dim_\kappa A .$$

Let $r = m + 3$, i.e., $\Delta(A) = 3$. As before we get the following exact sequence (cf. [Bou], §9, n.5)

$$\begin{aligned} 0 \longrightarrow B/(g_3) \otimes_B \mathcal{H}_1(K.(g_1, g_2), B) \longrightarrow \mathcal{H}_1(K.(I_3)) \longrightarrow \\ \longrightarrow \mathrm{Hom}_B(B/(g_3), B/(g_1, g_2)) \longrightarrow 0. \end{aligned}$$

It does mean that $\dim_\kappa \mathcal{H}_1(K.(I_3)) =$

$$\dim_\kappa \mathrm{Hom}_B(B/(g_3), B/(g_1, g_2)) + \dim_\kappa (B/g_3) \otimes_B \mathcal{H}_1(K.(g_1, g_2), B)) .$$

In order to calculate the latter term we tensor with $B/(g_3)$ over B the base exact sequence in case $\Delta(A) = 2$:

$$\begin{aligned} \ldots \longrightarrow \mathrm{Tor}_1^B(B/(g_3), \mathrm{Hom}_B(B/(g_2), B/(g_1)) \longrightarrow \\ \longrightarrow B/(g_3) \otimes_B B/(g_2) \otimes_B \mathrm{Hom}_B(B/(g_1), B) \longrightarrow B/(g_3) \otimes_B \mathcal{H}_1(K.(I_2)) \longrightarrow \\ \longrightarrow B/(g_3) \otimes_B \mathrm{Hom}_B(B/(g_2), B/(g_1)) \longrightarrow 0 . \end{aligned}$$

So $\dim_\kappa (B/(g_3) \otimes_B \mathcal{H}_1(K.(g_1, g_2), B)) \leq$

$$\begin{aligned} \dim_\kappa (B/(g_3) \otimes_B B/(g_2) \otimes_B \mathrm{Hom}_B(B/(g_1), B)) + \\ \dim_\kappa B/(g_3) \otimes_B \mathrm{Hom}_B(B/(g_2), B/(g_1)). \end{aligned}$$

It remains to show that the values of each term from the right side of the inequality are equal to $\dim_\kappa (B/(g_1, g_2, g_3))$. As above we have isomorphisms

$$\begin{aligned} \sigma : B/(g_3) \otimes_B B/(g_2) \otimes_B \mathrm{Hom}_B(B/(g_1), B) \cong \\ \mathrm{Hom}_B(\mathrm{Hom}_B(B/(g_3), \mathrm{Hom}_B(B/(g_2), B/(g_1))), B) \cong \\ \mathrm{Hom}_B(\mathrm{Hom}_B(B/(g_2, g_3), B/(g_1)), B) \end{aligned}$$

and the equality

$$\begin{aligned} \dim_\kappa (B/(g_3) \otimes_B B/(g_2) \otimes_B \mathrm{Hom}_B(B/(g_1), B)) = \\ \dim_\kappa \mathrm{Hom}_B(B/(g_2, g_3), B/(g_1)) = \\ \dim_\kappa \mathrm{Hom}_B(B/(g_3), \mathrm{Hom}_B(B/(g_2), B/(g_1))) = \\ \dim_\kappa B/(g_3) \otimes_B \mathrm{Hom}_B(B/(g_2), B/(g_1)). \end{aligned}$$

The latter equality follows from the exact sequence:

$$
\begin{aligned}
0 \longrightarrow \operatorname{Hom}_B(B/(g_3), \operatorname{Hom}_B(B/(g_2), B/(g_1))) \longrightarrow \\
\longrightarrow \operatorname{Hom}_B(B/(g_2), B/(g_1)) \xrightarrow{g_1} \operatorname{Hom}_B(B/(g_2), B/(g_1)) \\
\longrightarrow B/(g_3) \otimes_B \operatorname{Hom}_B(B/(g_2), B/(g_1)) \longrightarrow 0 .
\end{aligned}
$$

Moreover by [CE], II, 5.2) there exists a natural isomorphism

$$\operatorname{Hom}_B(B/(g_3), \operatorname{Hom}_B(B/(g_2), B/(g_1))) \cong \operatorname{Hom}_B(B/(g_2, g_3), B/(g_1)).$$

As a result we get

$$\dim_\kappa(B/(g_3) \otimes_B \mathcal{H}_1(K.(g_1, g_2), B)) \leq 2 \cdot \dim_\kappa \operatorname{Hom}_B(B/(g_2, g_3), B/(g_1)).$$

By our assumptions B is self-injective. Due to Matlis duality (see the Section 3) we have the natural *B-module* isomorphism

$$B/(g_1) \xrightarrow{\cong} \operatorname{Hom}_B(\operatorname{Hom}_B(B/(g_1), B), B).$$

Hence $\operatorname{Hom}_B(B/(g_2, g_3), B/(g_1)) \cong$

$$
\begin{aligned}
\operatorname{Hom}_B(B/(g_2, g_3), \operatorname{Hom}_B(\operatorname{Hom}_B(B/(g_1), B), B)) \cong \\
\operatorname{Hom}_B(\operatorname{Hom}_B(B/(g_1), B), \operatorname{Hom}_B(B/(g_2, g_3), B))
\end{aligned}
$$

where the latter isomorphism is also well known (see [CE], II, exercise 4). In order to calculate

$$\dim_\kappa \operatorname{Hom}_B(\operatorname{Hom}_B(B/(g_1), B), \operatorname{Hom}_B(B/(g_2, g_3), B))$$

we split the long exact sequence of B-modules

$$0 \longrightarrow \operatorname{Hom}_B(B/(g_1), B) \longrightarrow B \xrightarrow{g_1} B \longrightarrow B/(g_1) \longrightarrow 0$$

into two short sequences

$$0 \longrightarrow \operatorname{Hom}_B(B/(g_1), B) \longrightarrow B \longrightarrow g_1 B \longrightarrow 0,$$

$$0 \longrightarrow g_1 B \longrightarrow B \longrightarrow B/(g_1) \longrightarrow 0.$$

Denote by $H := \operatorname{Hom}_B(B/(g_2, g_3), B)$ and apply the functor $\operatorname{Hom}_B(-, H)$ to both sequences. Since $\operatorname{Ext}^1_B(B, H) = 0$ we get the corresponding two long exact sequences of B-modules:

$$
\begin{aligned}
0 \longrightarrow \operatorname{Hom}_B(g_1 B, H) \longrightarrow \operatorname{Hom}_B(B, H) \longrightarrow \\
\longrightarrow \operatorname{Hom}_B(\operatorname{Hom}_B(B/(g_1), B), H) \longrightarrow \operatorname{Ext}^1_B(g_1 B, H) \longrightarrow 0, \\
0 \longrightarrow \operatorname{Hom}_B(B/(g_1), H) \longrightarrow \operatorname{Hom}_B(B, H) \longrightarrow \\
\longrightarrow \operatorname{Hom}_B(g_1 B, H) \longrightarrow \operatorname{Ext}^1_B(B/(g_1), H) \longrightarrow 0.
\end{aligned}
$$

It is clear that H is Artinian. This implies that all these B-modules are Artinian too. In particular this means that they are finite-dimensional vector spaces over

κ. It is easy to verify that

$$\dim_\kappa \mathrm{Hom}_B(\mathrm{Hom}_B(B/(g_3), B), H) =$$

$$\dim_\kappa \mathrm{Hom}_B(B/(g_1), H) + \dim_\kappa \mathrm{Ext}^1_B(g_1 B, H) - \dim_\kappa \mathrm{Ext}^1_B(B/(g_1), H).$$

Because B is self-injective there are natural isomorphisms (see [CE], VI, 5.1)

$$\mathrm{Ext}^1_B(g_1 B, \mathrm{Hom}_B(B/(g_2, g_3), B)) \cong \mathrm{Hom}_B(\mathrm{Tor}^B_1(g_1 B, B/(g_2, g_3)), B),$$

$$\mathrm{Ext}^1_B(B/(g_1), \mathrm{Hom}_B(B/(g_2, g_3), B)) \cong \mathrm{Hom}_B(\mathrm{Tor}^B_1(B/(g_1), B/(g_2, g_3)), B).$$

To compute the dimensions over κ of torsion modules we first remark that from the short exact sequence

$$0 \longrightarrow g_1 B \longrightarrow B \longrightarrow B/(g_1) \longrightarrow 0$$

we get an isomorphism

$$\mathrm{Tor}^B_1(g_1 B, B/(g_2, g_3)) \cong \mathrm{Tor}^B_2(B/(g_1), B/(g_2, g_3)).$$

Moreover there exist the following two isomorphisms (see [CE], VI, exercise 19):

$$\mathrm{Tor}^B_1(B/(g_1), B/(g_2, g_3)) \cong g_1 B \cap (g_2, g_3)B / g_1 B \cdot (g_2, g_3)B,$$

$$\mathrm{Tor}^B_2(B/(g_1), B/(g_2, g_3)) \cong \mathrm{Ker}\{g_1 B \otimes_B (g_2, g_3)B \xrightarrow{\pi} g_1 B \cdot (g_2, g_3)B\},$$

where π is induced by the usual multiplication of B. At last let us consider the long exact sequences of B-modules

$$0 \longrightarrow \mathrm{Ker}(\pi) \longrightarrow g_1 B \otimes_B (g_2, g_3)B \xrightarrow{\pi} g_1 B \cap (g_2, g_3)B \longrightarrow$$
$$\longrightarrow g_1 B \cap (g_2, g_3)B / g_1 B \cdot (g_2, g_3)B \longrightarrow 0.$$

By definition of the tensor product for arbitrary ideals $\mathcal{F}$ and $\mathcal{G}$ of B we have $\mathcal{F} \otimes_B \mathcal{G} \subseteq B \otimes_B B$ and B-module isomorphisms

$$\mathcal{F} \otimes_B \mathcal{G} \cong \mathcal{F} \otimes_B B \cap B \otimes_B \mathcal{G} \cong \mathcal{F} \cap \mathcal{G} \subseteq B \cong B \otimes_B B\ .$$

Hence in our case we get the equality

$$\dim_\kappa (g_1 B \otimes_B (g_2, g_3)B) = \dim_\kappa (g_1 B \cap (g_2, g_3)B)$$

which yields

$$\dim_\kappa \mathrm{Ker}(\pi) = \dim_\kappa (g_1 B \cap (g_2, g_3)B / g_1 B \cdot (g_2, g_3)B) =$$

$$\dim_\kappa \mathrm{Tor}^B_1(g_1 B, B/(g_2, g_3)) = \dim_\kappa \mathrm{Tor}^B_1(B/(g_1), B/(g_2, g_3))\ .$$

This means that

$$\dim_\kappa \mathrm{Hom}_B(\mathrm{Hom}_B(B/(g_1), B), H) = \dim_\kappa \mathrm{Hom}_B(B/(g_1), H)\ .$$

But the *right* term is equal to $\dim_\kappa(B/(g_1,g_2,g_3))$ in virtue of the standard exact sequences

$$0 \longrightarrow \mathrm{Hom}_B(B/(g_1),H) \longrightarrow H \xrightarrow{g_1} H \longrightarrow B/(g_1)\otimes_B H \longrightarrow 0\ ,$$
$$0 \longrightarrow \mathrm{Hom}_B(B/(g_1),B/(g_2,g_3)) \longrightarrow$$
$$\longrightarrow B/(g_2,g_3) \xrightarrow{g_1} B/(g_2,g_3) \longrightarrow B/(g_1,g_2,g_3) \longrightarrow 0$$

and the isomorphism

$$\sigma : B/(g_1)\otimes_B \mathrm{Hom}_B(B/(g_2,g_3),B) \cong \mathrm{Hom}_B(\mathrm{Hom}_B(B/(g_1),B/(g_2,g_3)),B)\ .$$

On the whole $\dim_\kappa \mathcal{H}_1(K.(I_3)) \leq 3\cdot \dim_\kappa A$.

Now we shall finish the proof by induction. Let $r = m+k$, i.e., $\Delta(A) = k \geq 3$. As before from the above spectral sequence we get the following exact sequence (cf. [Bou], §9, n.5):

$$0 \longrightarrow B/(g_k)\otimes_B \mathcal{H}_1(K.(g_1,\dots,g_{k-1}),B) \longrightarrow \mathcal{H}_1(K.(I_k)) \longrightarrow$$
$$\longrightarrow \mathrm{Hom}_B(B/(g_k),B/(g_1,\dots,g_{k-1})) \longrightarrow 0$$
$$\dim_\kappa \mathcal{H}_1(K.(I_k)) =$$
$$\dim_\kappa \mathrm{Hom}_B(B/(g_k),B/(g_1,\dots,g_{k-1})) + \dim_\kappa(B/(g_k)\otimes_B \mathcal{H}_1(K.(I_{k-1}))).$$

It is evident that the first term on the right side of the equality is equal to $\dim_\kappa(B/(g_1,\dots,g_k))$. The second one can be computed in the following way. Tensoring with $B/(g_k)$ over B the exact sequence

$$0 \longrightarrow B/(g_{k-1})\otimes_B \mathcal{H}_1(K.(g_1,\dots,g_{k-2}),B) \longrightarrow \mathcal{H}_1(K.(I_{k-1})) \longrightarrow$$
$$\longrightarrow \mathrm{Hom}_B(B/(g_{k-1}),B/(g_1,\dots,g_{k-2})) \longrightarrow 0\ ,$$

we get the long exact sequence

$$\dots \longrightarrow \mathrm{Tor}_1^B(B/(g_k),\mathrm{Hom}_B(B/(g_{k-1}),B/(g_1,\dots,g_{k-2}))) \longrightarrow$$
$$\longrightarrow B/(g_k)\otimes_B B/(g_{k-1})\otimes_B \mathcal{H}_1(K.(I_{k-2})) \longrightarrow$$
$$\longrightarrow B/(g_k)\otimes_B \mathcal{H}_1(K.(I_{k-1})) \longrightarrow$$
$$\longrightarrow B/(g_k)\otimes_B \mathrm{Hom}_B(B/(g_{k-1}),B/(g_1,\dots,g_{k-2})) \longrightarrow 0\ .$$

So $\dim_\kappa(B/(g_k)\otimes_B \mathcal{H}_1(K.(I_{k-1}))) \leq$

$$\dim_\kappa(B/(g_k)\otimes_B \mathrm{Hom}_B(B/(g_{k-1}),B/(g_1,\dots,g_{k-2})))+$$
$$\dim_\kappa(B/(g_k)\otimes_B B/(g_{k-1})\otimes_B \mathcal{H}_1(K.(I_{k-2}))) \leq$$
$$\dim_\kappa(B/(g_k)\otimes_B \mathrm{Hom}_B(B/(g_{k-1}),B/(g_1,\dots,g_{k-2})))+$$
$$\dim_\kappa(B/(g_k)\otimes_B B/(g_{k-1})\otimes_B \mathrm{Hom}_B(B/(g_{k-2}),B/(g_1,\dots,g_{k-3})))+$$
$$\dim_\kappa(B/(g_k)\otimes_B B/(g_{k-1})\otimes_B B/(g_{k-2})\otimes_B \mathcal{H}_1(K.(I_{k-3}))) \leq \dots \leq$$
$$\dim_\kappa(B/(g_k)\otimes_B \mathrm{Hom}_B(B/(g_{k-1}),B/(g_1,\dots,g_{k-2})))+$$
$$\dim_\kappa(B/(g_k)\otimes_B B/(g_{k-1})\otimes_B \mathrm{Hom}_B(B/(g_{k-2}),B/(g_1,\dots,g_{k-3}))) + \dots$$
$$\dim_\kappa(B/(g_k)\otimes_B \dots \otimes_B B/(g_2)\otimes_B \mathcal{H}_1(K.(g_1),B)).$$

By the same manner as in the case $\Delta(A) = 3$ one can verify that all terms are equal to $\dim_\kappa(B/(g_1, \dots, g_K)) = \dim_\kappa A$. Hence

$$\dim_k \mathcal{H}_1(K.(I_k)) \leq$$

$$\dim_\kappa(B/(g_1, \dots, g_k)) + (k-1) \cdot \dim_\kappa(B/(g_1, \dots, g_k)) = k \cdot \dim_\kappa A.$$

Remark. It should be mentioned that in case $\Delta(A) = 2$ the equality

$$\dim_\kappa \mathcal{H}_1(K.(I)) = 2 \cdot \dim_\kappa A$$

and its direct consequence

$$\dim_\kappa I/I^2 = m \cdot \dim_\kappa A + \dim_\kappa T_2(A/H, A)$$

has been established in ([SV], 2.4) by slightly different arguments.

Corollary 4.5. *Let A be a zero-dimensional Gorenstein analytic κ-algebra. Then*

$$\dim_\kappa T^1(A) \geq \dim_\kappa A + \dim_\kappa T_2(A/H, A) - \dim_\kappa \mathrm{Ker}(d) + e(A) - 1 \ .$$

Proof. This follows immediately from Theorem and Claim above.

Remark. We note that $T_2(A/H, A) = 0$ in the complete intersection case (see [LS], (3.2.1)). However there exist many examples of Gorenstein singularities with $T_2(A/H, A) = 0$ that are not complete intersections (see e.g. [U]).

5. The gradient case

For convenience we also recall the following assertion (see [AM], (2.2)).

Proposition 5.1 (see [AM]). *Let $A = H/I$ be an Artinian complete intersection, that is, I is generated by a regular H-sequence $f_1, \dots, f_m \in H$. Then the tangent and cotangent modules $T_i(A)$ and $T^j(A)$ have the same dimension for $i, j = 0, 1$, i.e.,*

$$\dim_\kappa \Omega^1_{A/\kappa} = \dim_\kappa T_1(A/\kappa, A) = \dim_\kappa \mathrm{Der}(A, A) = \dim_\kappa T^1(A/\kappa, A).$$

Remark. This has led to the thought that besides pairings from the preceding Section between the tangent and cotangent modules with equal indices there exist natural cross pairings. In fact, we will show that it is true in the following special case.

Definition 5.2. Let $F \in H$ be an analytic function with isolated critical point at the origin. The ideal I generated by partial derivatives $f_i = \partial F/\partial z_i$, $i = 1, \dots, m$, is called the *gradient* ideal. In this case the H-sequence $f_1, \dots, f_m$ is regular so that $A = H/I$ is an Artinian complete intersection.

Theorem 5.3 (see [A2]). *Let I be the gradient ideal defined by an analytic function $F \in H$ with isolated critical point at the origin. Then there exist two canonical nondegenerate pairings*

$$T^0(A) \times T^1(A) \longrightarrow \kappa,$$
$$T_0(A) \times T_1(A) \longrightarrow \kappa.$$

Proof. For brevity we shall use the following convention. We shall denote the vector-column $\mathrm{grad}(F) = (f_1, \ldots, f_m)$ by dF or, equivalently, by $\mathbf{f}$. Then we can write an element $h \in I$, $h = g_1 f_1 + \ldots + g_m f_m$, in the following way: $h = \mathbf{g} \cdot \mathbf{f} = \mathbf{g} \cdot dF$. If $\mathbf{u} = (u_1, \ldots, u_m)$ and $u_i \in H$, then $d\mathbf{u} = (du_1, \ldots, du_m)$ is an $m \times m$ square matrix . Hence $d^2F = \mathrm{hess}(F)$ is the Hesse matrix of the function F.

In order to prove the theorem we shall describe the factor-module $\int I/I^2$ where $\int I$ is the primitive ideal from Section 2. So let $\mathbf{f} = dF$ and $h \in \int I$. By definition of the primitive ideal we have $dh = \mathbf{u} \cdot f$ where $\mathbf{u} = (u_1, \ldots, u_m)$ and $u_i \in H$. On the other hand, by Proposition 2.1 (1), $h \in I$, $h = \mathbf{v} \cdot \mathbf{f}$. Hence

$$dh = d\mathbf{v} \cdot \mathbf{f} + \mathbf{v} \cdot d\mathbf{f},$$

besides

$$\mathbf{v} \cdot d^2F = \mathbf{v} \cdot d\mathbf{f} = dh - d\mathbf{v} \cdot \mathbf{f} = \mathbf{u} \cdot \mathbf{f} - d\mathbf{v} \cdot \mathbf{f} \equiv 0 \pmod{\mathbf{f}}.$$

Let us consider a vector field $\mathcal{V} = v_1\partial/\partial z_1 + \ldots + v_m\partial/\partial z_m \in \mathrm{Der}(H)$. Then $\mathcal{V}(f_j) = \sum_i v_i \partial^2 F/\partial z_i \partial z_j = \mathbf{v} \cdot df_j = \mathbf{v} \cdot d^2F \equiv 0 \pmod{\mathbf{f}}$. Hence $\mathcal{V}$ defines the element $v \in \mathrm{Der}(H/I) = \mathrm{Der}(A)$. In this manner the correspondence $h \longrightarrow v$ defines a natural map

$$\int I \longrightarrow T^0(A).$$

Suppose that $h \mapsto 0$, i.e., $\mathcal{V}(\varphi) \in I$ for every $\varphi \in H$. This means that $v_i \in I$, $i = 1, \ldots, m$, (put $\varphi = z_i$), that is, $h \in I^2$. Thus we get an injection

$$\int I/I^2 \longrightarrow T^0(A) \quad \text{or, equivalently,} \quad T_1(A) \longrightarrow T^0(A).$$

The above proposition gives us the equality of dimensions of these spaces. Clearly this implies the isomorphism $T_1(A) \xrightarrow{\cong} T^0(A)$. It is well known that complete intersections are Gorenstein germs. Hence under our assumptions there exist the pairings of the theorem from the Section 3. This completes the proof.

Remark. Using elementary properties of tangent cohomology one may easy calculate an explicit representation of the pairings from our theorem:

$$\mathrm{Der}(A) \times A^m/\mathrm{hess}(F) \cdot A^m \longrightarrow A \longrightarrow \kappa$$
$$(v_1, \ldots, v_m) \times (a_1, \ldots, a_m) \mapsto \sum_i v_i a_i \mapsto \kappa,$$

where the last pointer is the projection on the one-dimensional component of the algebra A that corresponds to the element $\det(\mathrm{hess}(F)) = df_1 \wedge \ldots \wedge df_m/dz_1 \wedge \ldots \wedge dz_m$.

6. Graded singularities

A function $f \in H$ is called *quasihomogeneous* with weights $w_1, \ldots, w_m \in Z$ of degree $d \in Z$ if $f = \sum a_{\underline{i}} z_1^{i_1} \ldots z_m^{i_m}$ and $a_{\underline{i}} = 0$ for all $\underline{i} = (i_1, \ldots, i_m)$ such that $i_1 w_1 + \ldots i_m w_m \neq d$.

We say that A is a *graded* analytic κ-algebra if $A = \oplus_{n \in Z} A(n)$, where $A(n) \cdot A(m) \subseteq A(n+m)$ and $\dim_\kappa A(n) < \infty$, $n \in Z$ (see [SW]). The vector space $A(n)$ consists of *quasihomogeneous* elements of degree n. In the same manner one can define a graded A-module M. As usual, for any graded A-module M with finite dimensional quasihomogeneous components $M(n)$, $n \in Z$, the Poincaré series

$$P(M;t) = \sum_{n \in Z} \dim_\kappa M(n) \cdot t^n$$

is defined.

Choosing quasihomogeneous generators of A over κ, we put $A = H/I$, where $H = \kappa\{z_1, \ldots, z_m\} = \oplus_{n \geq Z} H_n$ is a graded convergent power series ring over κ, I is an ideal of H. Denote the weight z_i by w_i. One may assume $\{w_i \in Z, i = 1, \ldots, m\}$ are *relatively prime*, and $I = (f_1, \ldots, f_p)$, where $f_j \in H(d_j)$ is a *quasihomogeneous* element of degree d_j, $j = 1, \ldots, p$.

A graded analytic κ-algebra A is called *positive* graded (or, equivalently, Z_+-graded) if $A = \oplus_{n \geq 0} A_n$. In this case all weights of z_i are positive, i.e., $w_1, \ldots, w_m \in Z_+$. Such algebras are called also *weighted homogeneous.* It is obvious that any grading on A induces the natural grading on $\mathrm{Der}(A)$ and, more generally, on $T_i(A), T^i(A)$, $i \geq 0$.

We recall that in the case of a *complete intersection* germ of positive dimension with isolated singularities the grading on the dual analytic algebra is defined uniquely (excluding the trivial case of multiplicity-two hypersurfaces). As a matter of fact such *quasihomogeneous* algebras are *weighted* homogeneous (see [A1], §6). The zero-dimensional complete intersection case is essentially different. In fact, there exist many such algebras admitting several gradings. Moreover, in any grading there are negative, positive and even zero weights at the same time.

Example (cf. [SW], (7.3), [Sh4], Remark 3). Put $f_1 = x^5 + y^5 + x^3y^3z$, $f_2 = y^8$, $f_3 = z^3$, $I = (f_1, f_2, f_3)$, $A = \kappa\{x, y, z\}/I$. Then A is Z-graded with weights of x, y, z equal to 1, 1, -1 respectively.

However in case of Artinian algebras with the embedding dimension 2, one can prove the following statement.

Proposition 6.1 ([Sh4], Th.2). *Let A be an Artinian algebra with $e(A) = 2$. If $A = \kappa\{x, y\}/I$ is quasihomogeneous and the ideal I is not monomial then the grading on A is defined uniquely up to a multiple factor. Moreover, in any case one can choose the weights $w(x)$, $w(y)$ as positive rational.*

Therefore, every Z-graded Artinian algebra of embedding dimension two admits a Z_+-grading.

Now we can improve the estimate from the Section 4 using the following well-known assertion due to H.J. Reiffen.

Theorem 6.2 ([SW], (4.2)). *Suppose char $\kappa = 0$ and A is positive graded. Then the de Rham (Poincaré) complex $\Omega^\bullet_{A/\kappa}$ is acyclic in positive dimensions and $H^*_{DR}(\Omega^\bullet_{A/\kappa}) \cong H^0_{DR}(\Omega^\bullet_{A/\kappa}) \cong \kappa$.*

Corollary 6.3. *Let A be a weighted homogeneous Artinian algebra. Then*

$$\dim_\kappa T_0 \geq \dim_\kappa A + e(A) - 2,$$

$$\dim_\kappa T_1(A) \geq \dim_\kappa A + \dim_\kappa T_2(A/H, A) + e(A) - 2\ .$$

Proof. The inequalities follow from the Claim and Theorem of Section 4 respectively.

Note. As usual we shall denote by $\tau(A) = \dim_\kappa T^1(A)$ the so-called Tjurina number of A. In the case of an Artinian algebra the Milnor number $\mu(A)$ is given by $\mu = \dim_\kappa A - 1$. In the case of a *quasihomogeneous* zero-dimensional Gorenstein algebra of embedding dimension $e(A) \geq 2$, as a consequence of the second inequality we obtain

$$\dim_\kappa T^1(A) \geq \dim_\kappa A + \dim_\kappa T_2(A/H, A) + e(A) - 2\ .$$

If, for instance, $\dim_\kappa T_2(A/H, A) = 0$ and A is a complete intersection, the inequality $\tau \geq \mu + 1$ holds in contrast to the case of a complete intersection of positive dimension, where always $\tau(A) \leq \mu(A)$ with equality for quasihomogeneous singularities (see [GMP]). If $e(A) = 2$ the inequality $\tau \geq \mu + 1$ is reached in the case $I = (xy,\ x^p + y^q)$, $p, q \geq 2$, $\tau = p + q$. For completeness we also remark that for a zero-dimensional *complete intersection* germ the following inequality is known (see [Sh3], (3.1)):

$$\dim_{\mathbf{C}} \mathrm{Der}(A) \geq \dim_{\mathbf{C}} A - \dim_{\mathbf{C}} C_1$$

where C_1 is the image of $\{g \in H : \partial g/\partial z_i \in I\}$ in $A = H/I$. It is weaker than our first one because $\dim_{\mathbf{C}} C_1 \geq 1$.

We also mentioned that, in general, the conditions on A upon which

$$H^0_{DR}(\Omega^\bullet_{A/\kappa}) \cong \kappa$$

are *unclear*. But in some special cases one can state more.

For instance, most probably $H^0_{DR}(\Omega^\bullet_{A/\kappa}) \cong \kappa$ in the case of an Artinian algebra A with *non-solvable* Lie algebra of κ-derivations $T^0(A) \cong \mathrm{Der}(A)$. This is verified at least in case $e(A) \leq 3$ [see ([SW], §6)], where the authors also mentioned without proof that it holds for any embedding dimension of A.

However, it seems that the condition of non-solvability is very restrictive. For example, in case $e(A) = 2$ any Artinian algebra has *solvable* Lie algebra of κ-derivations (see [Sh4]). Moreover the same is also true for every local *gradient* algebra of a function with isolated critical points (see [Y]).

At the end of this Section we describe some considerations due to A. Shoshitaishvili that enable us to get an additional useful information on the structure $H^0_{DR}(\Omega^\bullet_{A/\kappa}) = \mathrm{Ker}(d)$.

He has considered (see [Sh3], (1.1)) the following two subalgebras C_1 and C_2 of A. The algebra C_1 has been defined above and $C_2 = \{a \in A : \partial a = 0 \quad \forall \partial \in \mathrm{Der}(A)\}$ is the algebra of "invariants" of A. It is evident that $\kappa \subseteq C_1 \subseteq C_2$.

Proposition 6.3. *Let* $d : A \longrightarrow \Omega^1_{A/\kappa}$ *be the universal differentiation. Then* $\kappa \subseteq C_1 \subseteq \mathrm{Ker}(d) \subseteq C_2$.

Proof. Given $\bar{g} \in C_1 \subseteq A$ representing by function $g \in H$. By definition, $dg \in I \cdot \Omega^1_{H/\kappa}$, i.e., $d\bar{g} = 0$ in $\Omega^1_{A/\kappa}$. Hence $\bar{g} \in \mathrm{Der}(d)$. On the other hand, an arbitrary $\partial \in \mathrm{Der}(A)$ admits a unique factorization $\partial = \iota_\partial \circ d$, with $\iota_\partial \in \mathrm{Hom}_A(\Omega^1_{A/\kappa}, A)$. This gives the inclusion $\mathrm{Ker}(d) \subseteq C_2$.

Corollary 5.8. (cf. [Sh3], (1.3)). *Suppose in addition that* A *is a complete intersection. Then* $C_1 = \mathrm{Ker}(d) = C_2$.

7. Derivations in the graded case

In conclusion we recall (see also [A2], [AM]) the following important problem closely related with many questions that arise in the deformation theory of singularities (see e.g. [A1], (6.3)).

Problem. Does there exist a grading on A such that A has no derivation of negative weight in this grading?

Comment. In the notations of the above Example let us consider

$$\mathcal{V} = z^2 x\partial/\partial x + z^2 y\partial/\partial y.$$

It is easy to see that $\mathcal{V}$ defines the derivation v of the Artinian algebra A and $\deg(v) = -2 < 0$!

Proposition 7.1. *Let A be an Artinian homogeneous κ-algebra, i.e., all weights $w_i = 1$, $i = 1, \ldots, m$. Then there are no derivations on A of negative weight.*

Proof. In [AM] we have proved this assertion using a variant (see [AM], §4, Claim) of the well-known Rossi theorem [R].

Conjecture (see [A2]). Let **A** be an Artinian Z_+-graded *complete intersection*, i.e. all weights w_i, $i = 1, \ldots, m$, are positive. Then there are no derivations on A of negative weight: $T^0(A)_- = \operatorname{der}(A)_- = 0$.

It should be mentioned that in [AM, §4] we have proved this conjecture in case of two varibales. In fact it is easy to verify that our conjecture is also true in cases of three and four variables.

Remark. Let $F \in \mathbf{C}\{z_1, \ldots, z_m\}$ be a *quasihomogeneous* function with an isolated critical point at the origin and I be the *gradient* ideal generated by partial derivatives $f_i = \partial F/\partial z_i$, $i = 1, \ldots, m$. In this case there exists the canonical Saito's grading on F that induces Z_+-grading on $A = \mathbf{C}\{z_1, \ldots, z_m\}/I$. The analogue of the above Conjecture in the case of gradient ideal has been formulated in [Sh1] as Theorem (iii). This Theorem was reformulated in slightly different terms in [Sh2], Proposition 1. Moreover, the author represented the proof of the latter. Unfortunately, the calculation of the integral II_t in the proof is not correct. Thus the problem in the gradient case remains open!

Further applications are concerned with problems of calculating the Poincaré series of the tangent cohomology modules.

Proposition 7.2. *Let $A = H/I$ be an Artinian quasihomogenous complete intersection, i.e., I is generated by a regular H-sequence of quasihomogeneous functions $f_1, \ldots, f_m \in H$ of degrees $d_1, \ldots, d_m$ with respect to weights $w_1, \ldots, w_m \in Z$. Then there exist natural nondegenerate pairings of finite dimensional κ-vector spaces*

$$T_i(A) \times T^i(A) \longrightarrow \kappa[-c] \qquad \textit{for all } \ i \geq 0 \ ,$$

where $c = \sum d_j - \sum w_i$ and in the square brackets we point out the value of moving of trivial grading of base field κ. Moreover, if A is a gradient Artinian algebra associated with a quasihomogeneous function F of degree $d \in Z_+$ there exist the following natural nondegenerate pairings

$$T^0(A) \times T^1(A) \longrightarrow \kappa[-c]$$
$$T_0(A) \times T_1(A) \longrightarrow \kappa[-c]$$

where $c = m \cdot d - 2 \cdot \sum w_i$.

Proof. It is direct consequence of our above statements.

Corollary 7.3. *Under the above assumption we have the following identities for Poincaré series:*

$$P(T_i(A);t) = t^c \cdot P(T^i(A);t^{-1}) \qquad \text{for all} \quad i \geq 0$$

and in the gradient case

$$P(\mathrm{Der}(A);t) = t^c \cdot P(T^1(A);t^{-1}), \quad P(\Omega^1_A;t) = t^c \cdot P(T_1(A);t^{-1}).$$

References

[A1] Aleksandrov, A.G., *Cohomology of a quasihomogeneous complete intersection.* (Russian) Izv. Akad. Nauk SSSR, ser. Matem. **49** (no 3) (1985), 467–510.

[A2] Aleksandrov, A.G., *Duality and deformations of Artinian algebras.* In: Proceedings of the International Conference on the Theory of Rings, Algebras and Modules in Memory of A.I. Shirshov, Barnaul, USSR, 1991.

[AM] Aleksandrov, A.G., Martin, B., *Derivations and deformations of Artinian algebras.* Beiträge zur Algebra und Geometrie **33** (1992), 115–130.

[Av] Avramov, L.L., *Homology of local flat extensions and complete intersection defects.* Math. Ann. **228** (1977), 27–37.

[Ba] Baran, A., *Dualité pour l'hyperext sur les espaces complexes,* C.R. Acad. Sci. Paris Ser. I **302** (1986), 627–630.

[Bou] Bourbaki, N., *Algèbre homologique.* Paris, Masson, 1980.

[CE] Cartan, H., Eilenberg, S., *Homological algebra.* Princeton Math. Ser. 19, Princeton, Univ. Press, Princeton, NJ, 1956.

[F] Ferrand, D., *Suite régulière et intersection complète.* C.R. Acad. Sci. Paris Ser. I **264** (1967), 427–428.

[GMP] Greuel, G.-M., Martin, B., Pfister, G., *Numerische Charakterisierung quasihomogener Kurvensingularitäten.* Math. Nachr. **124** (1985), 123–131.

[Gr] Grothendieck, A., *Local cohomology.* Lecture Notes in Math. 41, Springer-Verlag, 1967.

[HK] Herzog, J., Kunz, E., *Der kanonische Modul eines Cohen-Macaulay-Rings.* Lecture Notes in Math. 238, Springer-Verlag, 1971.

[Ia] Iarrobino, A., *Hilbert scheme of points: Overview of last ten years.* In: Algebraic Geometry Bowdoin 1985. Proc. Symp. in Pure Math. 46, Part 2, Amer. Math. Soc., Providence, Rhode Island, 1987, 297–320.

[KL] Kleiman, S.L., Landolfi, J., *Geometry and deformations of special Schubert varieties.* In: Algebraic Geometry, Oslo 1970. Wolters-Noordhoff Publ., 1972, 97–124.

[LS] Lichtenbaum, S., Schlessinger, M., *The cotangent complex of a morphism.* Trans. Amer. Math. Soc. **128** (1967), 41–70.

[P1] Palamodov, V.P., *Deformations of complex spaces.* (Russian) Uspeki Mat. Nauk, **31** (no 3) (1976), 129–194.

[P2] Palamodov, V.P., *Cohomology of analytic algebras.* (Russian) Trudy Mosc. Matem. ob. **44** (1982), 3–61.

[Pel] Pellikaan, R., *Hypersurface singularities and resolution of Jacobi modules,* Dissertation, Rijks-Universiteit te Utrecht, Utrecht, 1985.

[R] Rossi, H., *Vector fields on analytic spaces.* Ann. Math. (2) **78** (1983), 455–467.

[Sei] Seibt, P., *Infinitesimal extensions of commutative algebras.* Journal of Pure and Applied Algebra **16** (1980), 197–206.

[Sh1] Shoshitaishvili, A.N., *Reconstruction of the analytic type of a quasihomogenous function with respect to the local ring of a mapping given by its gradient.* (Russian) Trudy Mosk. Matem. Ob-va **34** (1977), 253–259.

[Sh2] Shoshitaishvili, A.N., *Functions with isomorphic Jacobian ideals.* (Russian) Funktsional. Anal. i Prilozh. **10** (no.2) (1976),, 57–62.

[Sh3] Shoshitaishvili, A.N., *Subalgebra of invariants of the local ring of a zero-dimensional complete intersection.* (Russian) Uspeki Mat. Nauk **35** (no.4) (1980),, 213–214.

[Sh4] Shoshitaishvili, A.N., *Analytic sets that are biholomorphically equivalent to quasihomogenous sets.* (Russian) Funktsional. Anal. i Prilozh. **17** (no.2) (1983), 92–93.

[SV] Simis, A., Vasconselos, W.V., *The syzygies of the conormal module.* Amer. J. Math. **103** (1981), 203–224.

[SW] Sheja, G. Wiebe, H., *Über Derivationen von lokalen analytischen Algebren.* Symposia Math., Roma, Vol. XI (1973), 161–192.

[U] Ulrich, B., *Vanishing of cotangent functors.* Math. Z. **196** (1987), 463–484.

[Y] Yau, Stephen S.-T., *Solvability of Lie algebras arising from isolated singularities and nonisolatedness of singularities defined by* $sl(2, \mathbf{C})$ *invariant polynomials.* Amer. J. Math. **113** (1991), 773–778.

On Higher Order Embeddings and Vector Bundles

*Edoardo Ballico**

Abstract. If E is a vector bundle over the variety X, we say that E is "k-very ample" if the restriction map $r_{z,E} : H^0E \to H^0E_{|z}$ surject for every subscheme Z of length $\leq k+1$. We show that when E is k-very ample, the map $\text{Hilb}^b(X) \to$ Grassmannian, induced by E, injects. Also we study the Chern classes of k-very ample bundle.

0. Introduction

In [BFS] the authors discussed several notions of spannedness for a line bundle L on a projective variety related to the higher order geometry of the map into a projective space induced by $H^0(X, L)$. The first aim of this paper is to introduce and discuss briefly the corresponding notions in the case of a vector bundle E and the map into a Grassmannian associated to $H^0(X, E)$. The discussion will be very short, but in our opinion it is sufficient to show how natural these notions seem to be.

The second aim of this note is to show how the computations given in [BS] not only can be extended to the vector bundle case (which is quite obvious), but from a slightly different point of view can be improved in a substantial way even in the line bundle on smooth surface case considered in [BS] (compare the bound in Theorem 0.1 below with the bound in [BS], (3.1.2)). We work over an algebraically closed base field **K**. The length of a **K**-vector space is its dimension over **K** (if finite). Let X be a projective variety and E a rank r vector bundle over X. Fix a vector space $W \subseteq H^0(X, E)$ (although usually we will take $W = H^0(X, E)$). For any closed subscheme Z of X, let $r_{Z,W} : W \to H^0(Z, E|Z)$ be the restriction map; if $W = H^0(X, E)$, write $r_{Z,E}$ (or r_Z if there is no danger of misunderstanding) instead of $r_{Z,W}$. Fix an integer $k \geq 0$. E is called k-very ample if for every subscheme $Z \subset X$ with length$(Z) \leq k+1$, the restriction map $r_{Z,E}$ is surjective; if we consider $W \subseteq H^0(X, E)$ we will say that E is k-very ample with respect to W if for all such Z we have the surjectivity of $r_{Z,W}$. Note that "0-very ample" is equivalent to "spanned". We will check in 1.3 that for $k > 0$ a k-very ample bundle is ample.

* The author was supported by MURST and by GNSAGA of CNR (Italy).

In the first section we will introduce and discuss briefly (following both [BFS] and the appendix to [ABP]) many other related notions.

Fix a complete scheme X and an integer t; let U be an open subset of the Hilbert scheme $\mathrm{Hilb}^t(X)$ of length t 0-dimensional subschemes of X. Fix a rank r vector bundle E on X and a vector space $W \subseteq H^0(X,E)$. Assume that the rank of the restriction map $r_{Z,W}$ is the same, say g, for all $Z \in U$ (e.g. assume E $(t-1)$-very ample and take $U = \mathrm{Hilb}^t(X)$ and $g = tr$). We denote by ϕ (or ϕ_U or ϕ^t or ϕ_{U^t}) the morphism from U to the Grassmannian G of codimension g subspaces of W defined by $\phi(Z) = \mathrm{Ker}(r_{Z,W})$ for $Z \in U$; we will say that ϕ^t is the t-th morphism associated to W. Now we can state the following result.

Theorem 0.1. *Let X be a projective variety, E a rank r vector bundle on X, k a positive integer. Assume that E is k-very ample. Let $\phi := \phi^k : \mathrm{Hilb}^k(X) \to G$ (with G Grassmannian of codimension kr subspaces of $H^0(X,E)$) be the k-th morphism associated to $H^0(X,E)$. Then ϕ is an embedding.*

Theorem 0.1 is a particular case of Theorem 1.10 which will be stated in Section 1. For the case $r = 1$ of Theorem 0.1, see [CG].

The second (and last) section is devoted to "examples". In the first part of this section we show how the notion of k-very ampleness fits in the classification of pairs (X,E) with X smooth projective surface and E rank 2 vector bundle. In the second part of the section we consider the stratification according to the order of k-very ampleness for "natural" families of rank 2 vector bundles on a smooth surface (e.g. the moduli schemes of stable vector bundles with fixed Chern classes). However the only result we prove here (Theorem 2.4) gives a complete picture in the case of unstable rank 2 vector bundles on $\mathbf{P}^2$. At the end of the section we discuss a class of bundles a class of bundles (the π-uniform bundles in the sense of [I]) on relatively minimal ruled surfaces.

1. Elementary properties

In this section we introduce and discuss briefly some notions related to geometric properties of higher order of morphisms to Grassmannians (i.e. the morphisms associated to spanned vector bundles). The remarks and elementary lemmas contained in this section seem to be the "minimal prerequisites" for any discussion and use (e.g. for classification of pairs (X,E) with low invariants) of these notions.

Definition 1.1. Let E be a rank r vector bundle on an algebraic scheme X and $W \subseteq H^0(X,E)$ a finite dimensional vector space. Fix non negative integers a, k. We will say that E is (k,a)-very ample with respect to W or (k,a)-spanned with respect to W if for every 0-dimensional subscheme Z of X with $\mathrm{length}(Z) \leq k+1$ and $\dim_x(T_xZ) \leq a$ for all $x \in Z_{red}$ (i.e. with embedding dimension at most a) the restriction map $r_{Z,W}$ from W to $H^0(Z, E|Z)$ is surjective. We will omit "with respect to W" if $W = H^0(X,E)$. If E is (k,a)-very ample with respect to W for an integer a which is at least the embedding dimension of X, we will say that E is k-very ample with respect to W.

Remark 1.2. It is obvious that a bundle is 0-very ample (or $(k,0)$-very ample) (with respect to W) if and only if it is spanned by its global sections (resp. by W).

Remark 1.3. Let E be a k-very ample with $k > 0$. Here we will check that E is ample. Since E is spanned, by [Gi], prop. 2.1, it is sufficient to check the following claim: for every integral curve $T \subseteq X$ (with, say, $\pi : A \to T$ as the normalization map) every quotient line bundle B of of $\pi^*(E|T)$ has $\deg(B) > 0$. It is easy to check that B is spanned and that for any two points P, Q with $\pi(P) \neq \pi(Q)$, the restriction map $r_{\{P,Q\},B}$ is surjective. Hence $\deg(B) \geq 1$ (and $\deg(B) > 1$ if A is not rational, $\deg(B) > 2$ if A is not hyperelliptic, and so on).

For related matter, see 1.4, 1.9, Definition 1.8 and Remark 1.15.

Lemma 1.4. *Let X be a smooth projective variety and E a $(1,1)$-very ample vector bundle on X. Then* $\det(E)$ *is very ample.*

Proof. Fix a length 2 zero dimensional subscheme Z of X (hence with embedding dimension at most 1). Since $r_{Z,E}$ is surjective, $r_{Z,\det(E)}$ is surjective. □

Definition 1.5. A bundle E over X is called *generically k-very ample* if there is an open dense subset of the symmetric product $X^{(k+1)}$ such that $r_{Z,E}$ is surjective for every $Z \in U$.

Remark 1.6. Fix a bundle E on a projective variety X; set $r := \operatorname{rank}(E)$. It is easy to check that if E is k-very ample then $\det(E)$ is k-very ample (and that all tensor powers, exterior powers, symmetric powers, and so on (Schur functors, ...) have the same property). But much more should be true for many such functors.

The next easy lemmas 1.7 and 1.9 (which are a tool needed for the classification problems considered in §2) summarize the elementary numerical consequences of k-very ampleness for bundles on a curve.

Lemma 1.7. *Let F be a rank r generically k-very ample bundle on an integral complete curve T. Let $\pi : A \to T$ be the normalization. Set $E := \pi^*(F)$. Then* $\deg(\det(E)) = \deg(E) \geq kr$.

Proof. Fix $k+1$ general points P_i, $0 \leq i \leq k$, on A; set $S := \{P_0, \ldots, P_k\}$ and $S' = S \backslash \{P_0\}$. Note that E is generically k-very ample. Since $r_{S,E}$ is surjective, it is easy to find $s \in H^0(A, E)$ vanishing on S' but not at P_0. Hence the non zero map $\mathbf{O}_A \to E$ induced by s, induces an inclusion $\mathbf{i} : \mathbf{O}_A(S') \to E$; let $R \subseteq E$ be the saturation of the image of $\mathbf{i}$. R is a line subbundle of E with $\deg(R) \geq k$. Since the bundle E/R is generically k-very ample, we conclude by induction on E. □

Definition 1.8. Let C be a smooth projective curve. Fix $k > 0$. We denote by $\tau(C)$ the minimal degree of a very ample line bundle on C and by $\tau(C,k)$ the minimal degree of a k-very ample line bundle on C.

Note that $\tau(C,k) \geq \tau(C) + k - 1$ (and usually this bound can be greatly improved).

Lemma 1.9. *Fix $k > 0$. Let C be a smooth projective curve and E a rank r k-very ample vector bundle on C. Then $\deg(E) \geq k(r-1) + \tau(C,k)$ (hence $\deg(E) \geq k(r-1) + \tau(C) + k - 1$). In particular $\deg(E) \geq rk$ and if $\deg(E) = rk$, then $C \cong \mathbf{P}^1$ and E is the direct sum of r line bundles of degree k.*

Proof. Fix $A \subset C$ with $\mathrm{card}(A) = k$; we will denote with the same symbol, A, the Cartier divisor and the degree k line bundle on C determined by the reduced scheme A. Since E is k-very ample, we have $0 \leq h^0(C, E(-A)) = h^0(C,E) - rk$ and there is an inclusion (as sheaves) $\mathbf{i}$ of A into E. Let R the subline bundle of E which is the saturation of $\mathbf{i}(A)$. By induction on r we may assume the thesis for E/R. Hence we will reduce to the case $r = 1$, which is just Definition 1.5. □

Now we can state the following result which implies Theorem 0.1.

Theorem 1.10. *Let X be an integral complete variety and E a rank r $(k, a+1)$-very ample vector bundle on X. Let U be the open subset of $\mathrm{Hilb}^k(X)$ parametrizing the 0-dimensional subschemes of embedding dimension at most a. Then the k-th morphism $\phi := \phi_{U^k}$ associated on U to $H^0(X,E)$ is an embedding.*

We omit a complete proof of 1.10 (hence of 0.1), because it is easy to follow the proof of the main theorem of [CG] and obtain 1.10 even for rank $r > 1$; alternatively, using [CG], Lemma 1.1, it is an easy exercise to modify the proof of [BS], Th. 3.1, to obtain 1.10. Here we sketch only the injectivity part in the statement of 1.10. Fix A and $A' \in U$ with $A \neq A'$ and assume $\phi(A) = \phi(A')$; using [CG], Lemma 1.1, we find a scheme Z with $A \subset Z \subseteq (A \cup A')$, $Z \neq A$, $\mathrm{length}(Z) \leq \mathrm{length}(A) + 1 \leq k+1$, $\dim(T_x Z) \leq \dim(T_x A) + 1 \leq a + 1$ for every $x \in Z_{red}$ and with $\dim(\mathrm{Im}(r_Z)) = \dim(\mathrm{Im}(r_A)) \leq r(\mathrm{length}(A))$, contradiction.

Remark 1.11. It is easy to check that both approaches to the proof of 1.10 give a criterion for the separation of two points (or of a tangent vector) by the map $\phi = \phi^k$ even when ϕ is not defined on all U.

Definition 1.12. Fix integers $r \geq 1$ and $a \geq 0$. Fix a complete scheme X, a rank r vector bundle E on X and a linear subspace $W \subseteq H^0(X,E)$. Fix a sequence $\mathbf{k} := \{k_i\}_{i \geq 0}$ of non negative integers with $k_0 = r$, $k_{i+1} \leq k_i$ for every i and $k_j = 0$ for large j. For every closed subscheme T of X, let $K(T)$ be the kernel of the restriction map from W to $H^0(T, E|T)$; set $k(T) := \dim_{\mathbf{K}}(K(T))$. W is called $(\mathbf{k}; a)$-very ample (or $(\mathbf{k}; a)$-spanned) if for every integer $t \geq 0$, for every subscheme $Z \subset X$ with $\dim(T_x X) \leq a$ for every $x \in Z_{red}$ and with $\mathrm{length}(Z) = t+1$, for every subscheme $Z' \subset Z$, $Z' \neq Z$ (including $Z' = \emptyset$ if $t = 0$) we have $k(Z) - k(Z') \geq k_t$.

In particular $(\mathbf{k}, a)$-very ampleness (or $(\mathbf{k}; a)$-spannedness) implies that E is spanned. If $k_i = r$ for $i \leq k$ and $k_j = 0$ if $j > k$, we find again the notion of (k, a)-very ampleness.

In the next result we assume $\mathrm{char}(\mathbf{K}) = 0$ to use [Hi], Th. 4.6 (which in turn is an improvement of results in [La]). We fix a smooth projective curve C of genus $g \geq 2$. Let $M(r, d)$ the moduli scheme of rank r stable vector bundle of degree d on C; $M(r, d)$ is an integral variety. To explain the numerical meaning of Proposition 1.13 below note that if $G \in M(r, d)$ and $E \in M(r', d')$, then the rank rr' vector bundle $G \otimes E$ has degree $r'd + rd'$ and by Riemann-Roch we have $\chi(G \otimes E) = r'd + rd' + rr'(1-g)$; [Hi], Th. 4.6, states that if G is general in $M(r, d)$ and E general in $M(r', d')$, then either $h^0(G \otimes E) = 0$ or $h^1(G \otimes E) = 0$ (the two cases being distinguished by the sign of $\chi(G \otimes E)$).

Proposition 1.13. *Let C be a curve of genus $g \geq 2$. Fix integers k, r, d, r', d' with $k \geq 0$, $r \geq 1, r' \geq 1$. For general $G \in M(r, d)$, $E \in M(r', d')$, the rank rr' bundle $G \otimes E$ is generically k-very ample if and only if*

$$r'd + rd' + rr'(1 - g) \geq (k + 1)rr'$$

In particular (taking $r = 1$ and $d = 0$) a general $M \in M(r', d')$ is generically k-very ample if and only if $d' + r'(1 - g) \geq (k + 1)r'$.

Proof. Note that for a general $Z \subset C$ with $\mathrm{length}(Z) = k + 1$ and a general E, the bundle $E(-Z)$ may be considered as a general element of $M(r', d' - (k + 1)r')$. Apply [Hi], Th. 4.6, to G and $E(-Z)$. □

Remark 1.14. Let C be a smooth curve of genus g and E a vector bundle on C. If $k + 1 \geq g$, the notion of k-very ampleness for E is strongly related to the stratification studied in [Hi], §3.2; in particular, if C has general moduli [Hi], Th. 3.2, gives for what r, d and k with $k + 1 \geq g$ a general stable bundle $E \in M(r, d)$ is k-very ample.

Remark 1.15. Take $\mathbf{k} = \{k_i\}$ with $k_1 > 0$. Let E be a $(\mathbf{k}, a)$-very ample bundle with $a \geq 1$ if X is smooth and $a = \max_{x \in X}\{\dim(T_x X)\}$ if X is singular. The reader can easily check that the morphism from X to the appropriate Grassmannian is an embedding. Hence (in author's opinion) E should be considered "geometrically very ample", although if $k_1 < r$ the bundle E may be not ample.

2. Examples and moduli of bundles on surfaces

In the first part of this section we will show how restrictive is the condition "E is k-very ample" if the Chern classes of E are small enough (e.g. on a surface if c_2 is low). We consider this part only as a collection of examples. We will use heavily previous works and stop before the matter begin to be interesting, i.e. where previous papers, the results proved in §1 and short computations are not enough to conclude.

In the second part of this section we will consider how the notion of k-very ampleness and order of stability interacts to give a stratification of the moduli spaces of rank 2 vector bundles on a surface.

In the last part of the section we will consider π-uniform vector bundles ([I]) on a $\mathbf{P}^1$-bundle over a curve.

Just to use freely the literature, in the first part of this section we will assume $\mathrm{char}(\mathbf{K}) = 0$. We fix an integral complete variety X and a vector bundle E on X; set $n := \dim(X)$ and $r := \mathrm{rank}(E)$; we will assume E spanned. In [B2] if $n = r$ we defined an integer $s(E)$ with $n \leq s(E) \leq nc_n(E)$ in the following way. Set $t := c_n(E)$; let M be the open subset of $\mathbf{P}(H^0(X,E)^*)$ parametrizing the sections, s, with 0-dimensional and reduced zero locus $(s)_0$; set $Y := \{(z,[s]) \in S^t(X) \times M$ with $z = (s)_0\}$ (the incidence correspondence); let $\pi_1 : Y \to S^t(X)$ be the projection; set $s(E) := \dim(\mathrm{Im}(\pi_1))$. The integer $s(E)$ seems to be very useful for classification purposes and reflects roughly the birational geometry of the morphism associated to $H^0(X,E)$; furthermore, if $s(E) = nc_n(E)$, then X is unirational (for instance by the proof of [B2], lemma 1.1).

Definition 2.1. Assume $\mathrm{char}(\mathbf{K}) = 0$. Let E be a spanned vector bundle on an integral complete variety X with $\mathrm{rank}(E) = \dim(X) > 0$ and $c_{\dim(X)}(E) \neq 0$. Set $c(E) := \max\{t \in \mathbf{N}$: for a general $A \subset X$ with $\mathrm{card}(A) = t$ there is $s \in H^0(X,E)$ with zero locus finite, reduced and containing $A\}$.

By the definition of $s(E)$, we have $s(E) \geq c(E)(\dim(X))$.

Lemma 2.2. *Assume* $\mathrm{char}(\mathbf{K}) = 0$.

(a) *If* $n = r$ *and* E *is spanned, generically* k*-very ample, then* $c(E) \geq k$ *and* $s(E) \geq kn$.

(b) *If* $n = r$, E *is spanned, generically* k*-very ample, and* $h^0(S,E) > n(k+1)$, *then* $c(E) \geq k+1$ *and* $s(E) \geq n(k+1)$.

In particular under the assumptions of (a) we have $c_n(E) \geq k$, while under the assumptions of (b) we have $c_n(E) \geq k+1$.

Proof. We will prove part (b), the proof of part (a) being very similar (but easier). Fix a dense open subset $U \subset X^{(k+1)}$ such that every $Z \in U$ is reduced and has $r_{Z,E}$ surjective. Consider the restriction J to U of the incidence correspondence $I \subset \mathbf{P}(H^0(X,E)) \times X^{(k+1)}$, $I := \{([s], z) \in \mathbf{P}(H^0(X,E)) \times X^{(k+1)} : s$ vanishes on $z\}$. By the choice of U the projection $J \to U$ is smooth and equidimensional with as fibers projective spaces of dimension $h^0(X,E) - n(k+1)$. In particular J is smooth and irreducible, while by Bertini's theorem a general section of E vanishes on a reduced finite set. We obtain the existence of an open subset J' of J corresponding to sections vanishing on a reduced finite set and such that the projection $J' \to U$ is dominant. By definition of $s(E)$ this means that $s(E) \geq n(k+1)$.

Since $c(E) \leq c_n(E)$, the last part of 2.2 is obvious. □

Remark 2.3. Assume $n = r = 2$; then there is a complete classification of pairs (X, E) with E spanned and $c_2(E) = 1$ (see [B1] for smooth X, [B4], 2.1, for normal X, while the non normal case reduces easily to the case of a normal surface X).

Using 1.3, 1.4, 1.7, 1.8, and 2.2, it is easy to translate [B1], [B2] and [B3] to the case of k-very ample (or $(k, 1)$-very ample) rank 2 vector bundles on a smooth complete surface (with either $k + 1 = c_2(E)$ or with $k + 2 = c_2(E)$ plus the assumption of "suitable conditions"). However very often one obtains in this way only a characterization of the empty set. To weaken the assumptions and arrive at interesting bundles, the interested reader should use the techniques of [B2] and [B3], not only the statements contained there, and make full use of the notion of k-very ampleness. For instance there is a strong link (using Definition 1.8) of this concept for bundles on curves and the classical notion of Clifford index.

The second part of this section points in a different direction. Fix a smooth complete surface S and an ample line bundle H on S; fix $c_1 \in \mathrm{Pic}(S)$ and an integer c_2. Consider the moduli space $M_{S,H}(c_1, c_2)$ (or $M(c_1, c_2)$ if there is no danger of misunderstanding) of H-stable rank 2 vector bundles on S with Chern classes c_i. Concerning $M(c_1, c_2)$ and the notion of k-very ampleness (and its variants) there are at least 4 natural questions: find good conditions on k and the Chern classes c_i such that:

(i) there is at least one $E \in M(c_1, c_2)$ which is k-very ample;

(ii) there is an irreducible component M of $M(c_1, c_2)_{red}$ such that a general $E \in M$ is k-very ample;

(iii) for every irreducible component M of $M(c_1, c_2)_{red}$ a general $E \in M$ is k-very ample;

(iv) every $E \in M$ is k-very ample.

Note that (i) and (ii) are equivalent for the Chern classes for which all $E \in M(c_1, c_2)$ have the same $h^0(S, E)$. And then there are the bundles which are not stable. They fit very well in families (usually parametrized by an Ext^1-group, hence unirational), although these families have no universal property. Here we will consider these questions only for families of unstable bundles in a very particular, but important case: the case $S = \mathbf{P}^2$ (see Theorem 2.4). Several results both for the unstable case and for $M_{S,H}(c_1, c_2)$ for general S are planned to be part of a joint paper (the unstable case being much easier). Of course, for $S = \mathbf{P}^2$ we can find much more precise informations.

Recall that (by definition) every rank 2 vector bundle E on $\mathbf{P}^2$ which is not stable fits in an exact sequence:

$$0 \to \mathbf{O}_S(a) \to E \to \mathbf{I}_W(b) \to 0 \tag{1}$$

with $a \geq b$ and W 0-dimensional subscheme which is locally a complete intersection. Viceversa, since $a \geq b$, by the Cayley - Bacharach property (see [C]), for all triples (a, b, W) there is a sequence (1) with a rank 2 vector bundle as middle term E. If E is given by (1) with $c :=\text{length}(W)$, then $\deg(\det(E)) = a + b$ and $c_2(E) = ab + c$; E is not stable because $a \geq b$; we will say that such a bundle is of type $(a, b; c)$). Note that, for fixed a, b, c, the set of all bundles of type $(a, b; c)$ is parametrized by an integral variety.

Theorem 2.4. *Fix integers a, b, c, k with $a \geq b$, $c \geq 0$, $k \geq 0$.*

(a) *The general bundle of type $(a, b; c)$ is k-very ample if and only if one of the following conditions is satisfied:*

$$\text{(i)}\quad c \geq 2 \quad \text{and} \quad b \geq k+2;$$
$$\text{(ii)}\quad c = 1 \quad \text{and} \quad b \geq k+1;$$
$$\text{(iii)}\quad c = 0 \quad \text{and} \quad b \geq k.$$

(b) *There is a bundle of type $(a, b; c)$ which is k-very ample if and only if $b \geq k+c$.*

Proof. For all integers $t \geq 0$, $j \geq 0$ with $j \leq 2t+1$, a 0-dimensional length j subscheme A of $\mathbf{P}^2$ has $h^1(\mathbf{I}_A(t)) \neq 0$ if and only if there is a line $D \subset \mathbf{P}^2$ such that length$(A \cap D) > t+1$([A], lemma 4.1). If $c \geq 2$ we may find a line, D, with length$(W \cap D) \geq 2$ and choose as Z $k+1$ general points of D. Thus part (a) of 2.4 is easy. Now we prove part (b). Choose W with as length$(W \cap D)$ any possible integer not greater than c. Use the fact (true because a line is a smooth curve) that length$((W \cup Z) \cap D) =$ length$(W \cap D)+$ length$(Z \cap D)$ even if $W_{red} \cap Z_{red} \neq \emptyset$. Hence we obtain part (b). □

We conclude this note with the discussion of another definition (concerning only varieties which are $\mathbf{P}^k$-bundles, say over a curve). This is relevant to the classification of pairs (X, E) with X smooth surface and E "good" bundle with small invariants, since usually the extremal cases arises on $\mathbf{P}^2$ (and here they are easily classified) and on relatively minimal ruled surfaces with as possible bundles the bundles of the following form. Let C be a smooth complete curve and G a rank s vector bundle on C; set $X := \mathbf{P}(G)$; a rank r bundle E on X is said to be π-uniform (in the sense of Ishimura ([I])) if there are integers $a_1, \ldots, a_r$ such that for every fiber T of the projection $\pi : X \to C$, $E|T$ is the direct sum of r line bundles of degree $a_1, \ldots, a_r$; if $a_i = 1$ for every i this means that $E \cong \pi^*(F) \otimes \mathbf{O}_X(1)$ for some bundle F on C. The sheaf $G \otimes F \cong \pi_*(E)$ arises again (in the particular case of splitting type $(1, \ldots, 1)$) when we consider the following definition.

Definition 2.5. With the notations just introduced, the π-uniform bundle E with splitting type $(1, \ldots, 1)$ is called π-spanned if the bundle $G \otimes F$ is spanned.

This is very strong non degeneracy condition. Indeed, classifications of such ample and π-spanned vector bundles with low invariants is much easier (for the same invariants) than the case "E ample and spanned" (usually one finds the empty list).

Example 2.6. Note that for every π-uniform rank 2 vector bundle E of splitting type (1,1) we have $c_1(E)^2 = 4c_2(E)$. Hence it is easy to check that such a bundle E is not π-spanned if $h^0(X, E) = 4$.

In the same way one can define when such a bundle is π-ample or $\pi - k$-very ample. Definition 2.5 gives another reason for the interest of statement like [Hi], th. 4.6 (see §1).

In the interpretation of Definition 2.5 with the spannedness of the bundle $G \otimes F$ (in case of splitting type $(1, \ldots, 1)$) it appears in a natural way the "multiplicative" structure of the set of vector bundles. Here several differences (even if $q = 1$) appear between the characteristic 0 case (when the tensor product of two semistable bundles is again semistable) and the positive characteristic case (in which this fails for certain bundles on certain curves). Only the case $q = 1$ is completely understood ([O], §2); in this case for a fixed $p := \mathrm{char}(\mathbf{K})$ there is a fundamental difference between elliptic curves with Hasse invariant 0 and elliptic curves with Hasse invariant $\neq 0$.

References

[A] d'Almeida, J., Courbes de l'espace projectif. Séries linéare incomplète et multisecantes, J. Reine Angew. Math. **370** (1986), 30-51.

[ABP] Andreatta, M., Palleschi, M., (with an appendix by E. Ballico), On the 2 very ampleness of the adjoint bundle, Manuscripta Math. **73** (1991), 45-62.

[B1] Ballico, E., On globally generated vector bundles on surfaces with $c_2 = 1$, Geom. Dedicata **27** (1988), 319-324.

[B2] Ballico, E., Rank-2 vector bundles on a surfaces with many sections and low c_2, Geom. Dedicata **29** (1989), 109-124.

[B3] Ballico, E., Rank-2 vector bundles on a surface with many spread sections, Indiana Univ. Math. J. **38** (1989), 471-487.

[B4] Ballico, E., Vector bundles, reflexive sheaves and algebraic surfaces, Rend. Sem. Mat. Fis. Milano (to appear).

[BFS] Beltrametti, M., Francia, P., Sommese, A.J., On Reider's method and higher order embeddings, Duke Math. J. **58** (1989), 425-439.

[BS] Beltrametti, M., Sommese, A.J., (with an appendix by L. Göttsche), Zero cycles and k-th order embeddings of smooth projective surfaces, Proc. Projective Surfaces and their Classification, Cortona (1988), Symposia Mathematica, INDAM, Academic Press (to appear).

[C] Catanese, F., Footnotes to a theorem of I. Reider, Algebraic Geometry, Proc. L'Aquila 1988, pp.67-74, Lect. Notes in Math. **1417**, Springer-Verlag, 1990.

[CG] Catanese, F., Göttsche, L., d-very ample line bundles and embeddings of Hilbert schemes of 0-cycles, Manuscripta Math. **68** (1990), 337-341.

[Gi] Gieseker, D., p-ample bundles and their Chern classes, Nagoya Math. J. **43** (1971), 91-116.

[H] Hartshorne, R., Ample vector bundles on curves, Nagoya Math. J. **43** (1971), 71-89.

[Hi] Hirschowitz, A., Problèmes de Brill-Noether en rang superieur, C.R. Acad. Sci. Paris, Sér. I, **307** (1988), 153-156.

[I] Ishimura, S., On π-uniform vector bundles, Tokyo J. Math. **2** (1979), 337-342.

[La] Lange, H., Zur Klassifikation von Regelmannigfaltigkeiten, Math. Ann. **262** (1983), 447-459.

[O] Oda, T., Vector bundles on an elliptic curve, Nagoya Math. J. **43** (1971), 41-72.

Singular curves on surfaces and linear systems

Edoardo Ballico, Luca Chiantini

Abstract. Let S be a smooth projective surface, C an irreducible curve on S, L a g_k^1 on the normalization φ of C. We study conditions on S, C, L which force L to be induced by a map $S \to P^1$.

1991 Mathematics Subject Classification: 14C21.

Introduction

The study of linear series on smooth curves is a fundamental topic in Algebraic Geometry and there is a big amount of literature on this subject. On the other hand, singular curves embedded in some suitable surface arise naturally in the theory of curves: just to give an example, it follows from Castelnuovo's theory on projective curves of high genus, that in order to map birationally a curve of genus g as a non degenerate curve of degree d in P^n, one has to allow singularities (see e.g. [Ta]).

This paper deals with linear systems on (smooth or singular) curves embedded in a smooth surface S. Given a fibration $S \to P^1$ and an integral curve $C \subset S$, one obtains, by pull back and restriction, a pencil on the normalization Y of C; in this paper we look for conditions which force a given pencil on Y to be induced by a morphism $S \to P^1$.

Through the paper, we take the curve C in an ample linear system L and (although not strictly necessary) we assume that C has only t nodes as singularities; more basically, we will assume that C ranges on a family of nodal curves, such that all their normalizatons have a g_k^1, Γ and for a general subset T of t distinct points of S, there exists a curve in the family with nodes at the points of T. We use standard Bogomolov - Reider theory to show that, under some numerical assumptions on L, Γ and t, then on the general curve of the family, Γ is induced by a fibration $S \to P^1$.

Our main result is stronger (in general) than the results we could obtain applying Reider technique to the surface S', obtained blowing up S at T. In the second section, we will apply Reider's theory to see explicitely what it gives in the motivating case of minimal ruled surface, getting stronger results in the rational case; observe that rational surface scrolls play a key role in Castelnuovo's theory for curves of high genus (see [Ha] ch.III or [ACGH] ch.III).

We want to thank very strongly M. Andreatta and A.J. Sommese. During the last part of the preparation of this paper one of us (E.B.) enjoyed a stay at Notre Dame University; he wants to thank J. Migliore and A.J. Sommese for all their help and kindness during his stay. The authors were partially supported by MURST and GNSAGA of CNR (Italy).

The main results

We give here the proof of our main theorems and we list several cases in which the assumptions are satisfied; we will give also some extension of the theorem and alternative assumptions which lead to the same result.

From now on, let S be any algebraic smooth surface, defined over an algebraically closed field of characteristic 0, with canonical class K. Let L be an ample divisor on S.

For any integral curve $C \subset S$, we call 'base point free g^r_k on C' any morphism $C \to P^r$; it corresponds to a linear series on the normalization Y of C, which factors through $Y \to C$.

The following propositions introduce Reider's fundamental method for the study of linear series on curves embedded on smooth surfaces.

Proposition 1. *Let S be a smooth surface, L an ample divisor on S. Let C be an integral, non rational curve of L and let Γ be a base point free g^1_k on C. Assume $L^2 > (k+1)^2$. Then there exists a morphism $S \to P^1$ which induces Γ on C.*

Proof. See [ABS] 1.5 or [Se].

If we assume that C has t nodes for singularities and turn our attention to linear series on the normalization Y of C, then using the same idea of prop.1.5 of [ABS], one gets:

Proposition 2. *Let C be as in the previous proposition, with t nodes as singularities, and let Y be a normalization of C. If Γ is a base point free g^1_k on Y and $L^2 > (t+k+1)^2$, then there is a morphism $u : S \to P^1$ which induces Γ.*

Proof. The proof works exactly as the proof of the previous proposition; we give it here only for the sake of completeness.

If T is the set of nodes of C, we have:

$$H^0(Y, \omega_Y) = x \in H^0(C, \omega_C) : x(P) = 0 \quad \forall P \in T.$$

Let A be the image in C of a divisor in Γ and put $Z' = T \cup A$. By Riemann-Roch, Z' cannot impose independent conditions to the sections of $H^0(C, \omega_C)$, hence there is a $Z \subset Z'$, for which $K + L$ is a Cayley-Bacharach line bundle; we have $A \subset Z$ for Γ is base point free, and $\deg Z \leq t + k$.

From the standard Reider's technique (see [BFS] or [BS1]), one gets a rank 2 bundle E on S, with an exact sequence $0 \to O_S \to E \to I_Z(L) \to 0$; the chern classes of S are $c_1 = L$ and $c_2 = \deg Z$; since in our case $L^2 > 4(t + k)$, then E is Bogolomov unstable; this implies the existence of an effective divisor D on S, containing Z, with

$$0 \leq L.D - (t + k) \leq D^2 < L.D/2 < t + k.$$

Assume $D^2 > 0$; by Hodge index theorem we have $D^2L^2 \leq (M.L)^2$ hence

$$L^2 \leq (D.L)^2/D^2 \leq (D^2 + t + k)/D^2$$

which is impossible, by ou assumptions, since $D^2 < t + k$. Thus $D^2 = 0$ and the existence of the map u follows as in [Se], claim 6.

The numerical condition $L^2 > (t+k+1)^2$ is somehow heavy; in order to weaken it, we make use of some 'uniformity' condition on the set Z formed by the nodes of C together with a divisor in Γ.

To do this, as noted in the introduction, we start with a 'family' of curves in H^0L, each having t nodes for singularity, each bearing a base point free g^1_k on its normalization; to formalize this datum, we need the following:

Definition. Let $A \subset S$ be any set of k distinct points. We put:
$W_t \subset P(H^0(L))$ =integral curves in L having exactly t nodes as singularities;
$w_t : W_t \to S^{(t)}$ = the map which sends a curve to its nodes.

Let W be any variety, let $f : W \to W_t$ be a map and let V_0 be the total space of the family induced by f by base change. Since we work in characteristic 0, we may take a desingularization V of V_0 and, replacing W with some Zariski open subset, we may assume that $V \to W$ is a flat family of non-singular curves; thus each element of the family V is the normalization of the corresponding curve in V_0.

We call V the 'desingularization' of the family V_0.

We call 'base point free g^r_k on V' any line bundle G over V which restricts to a base point free linear series g^r_k on any fiber.

When $S = P^2$, then W_t corresponds to (an open subset of) a Severi variety, which is known to be irreducible, but this may be not the case for general S.

Remark. Let C be a general curve of the family V_0 and assume that the composition $W \to W_t \to S^{(t)}$ is dominant. We can 'move' the nodes of C, so that we may interchange any two of them. It follows, by a standard argument (see [Ha]) that the position of the nodes of C is 'uniform'.

Let Γ be the linear series defined by G on the normalization of C and let A be the image in C of a general divisor in Γ. Then we may move C in the family interchanging two of its nodes and leaving Γ unchanged; it follows that the nodes of C are in uniform position even with respect to the points of A, in the following sense:

Let M be any divisor on S and let T be the set of nodes of C; if Z, Z' are subsets of T with the same cardinality, then

$$h^0 I_{Z\cup A}(M) = h^0 I_{Z'\cup A}(M).$$

Theorem 3. *Let S be a smooth surface, L an ample divisor on S. Let $f : W \to W_t$ be a map of varieties and let G be a base point free g^1_k $(k > 1)$ on the normalization V of the family V_0 induced by f. Assume:*
(1) $tL^2 > (2t+2k-1)^2$ and $L^2 > 4(t+k)$;
(2) the composition $W \to W_t \to S^{(t)}$ is dominant.
Then for $w \in W$ general, there is a g^1_k on the nodal curve $C_w = (V_0)_w$ whose pull back to the normalization $Y = V_w$ is the series induced by G on Y.

Proof. Take $w \in W$ general and call C, Y the fibers of V_0, V respectively, over w. Call T the set of nodes of C; also call ω_C, ω_Y the dualizing sheaves. Let A be the image on C of a general divisor of the g^1_k Γ induced by G on Y; set $Z = T \cup A$.

By adjunction theory and Riemann-Roch, as in the proof of prop.2, for all $P \in A$ we have $h^0(C, \omega_C \otimes I_Z) = h^0(C, \omega_C \otimes I_{Z-P})$.

Choose P in T and assume that divisors in the system $L+K$, passing through Z, separate P; by (2) and by the previous remark, it follows then that divisors of $L+K$ passing through Z separate all the points of T; in this case we must have $h^0(C, \omega_C \otimes I_A) = h^0(C, \omega_C \otimes I_{A-P})$ for all $P \in A$, thus by adjunction and by Riemann-Roch, A moves in a g^1_k over C.

Assume, on the contrary, that $L+K$ separates no points of Z, i.e., it is a Cayley-Bacharach line bundle for Z. By Reider's technique, one finds then an effective divisor D on S with $Z \subset D$ and

$$t + k \leq D.L \leq 2t + 2k - 1.$$

Since D is effective, L is ample and $D.L$ is bounded, after replacing W with some Zariski open subset, we may assume that moving C in the family and moving A in Γ, the corresponding divisors D move in an algebraic family.

Call M the movable part of D (in the sense of algebraic systems, not in the sense of linear systems) and $F = D - M$.

Since, by (2), we may move freely the nodes of C when the curve ranges in the family, then for C general, F cannot contain any of the nodes. Since Γ is base point free, we must have $A \subset M$, so that in fact $Z \subset M$. M is nef and when we move A in the linear series, then M moves in a rational family, hence in a linear family, for rational equivalence implies linear equivalence on divisors (see e.g. [ABS]). Moving A in the pencil, we see that $M^2 \geq t$: if T is not contained on a single component of M, then interchanging the nodes, we see that M has at least t algebraically equivalent components M_i, each containing at least one node of C, and the number $M_i.M_j$ does not depend on the choice of the indeces.

The Hodge index theorem tells now that $tL^2 \leq M^2L^2 \leq (M.L)^2 \leq (2t+2k-1)^2$, so we are done.

Combining proposition 1 and theorem 3, one finds:

Corollary 4. *With the previous notation, assume:*
(1) $tL^2 > (2t+2k-1)^2$ *and* $L^2 > 4(t+k)$*;*
(2) *the composition* $W \to W_t \to S^{(t)}$ *is dominant;*
(3) $L^2 > (k+1)^2$.

Then there exists a map $u : S \to P^1$ *which induces on each fiber of* V *the linear series defined by* G *(i.e.* G *is induced by the pull-back of the map* u*).*

Corollary 5. *Use the notation of the previous theorem and call* S' *the blowing up of* S *at the nodes of* C*. If condition* (1) *is replaced by:*
(1)' $(t+1)L > (2t+2k-1)^2$ *and* $L^2 > 4(t+k)$*;*
then there exists a morphism $S' \to P^1$ *which induces* Γ *on the normalization* Y *of* C*.*

Proof. The proof goes exactly as in the theorem, except for the last step, so let us keep the previous notation.

We know that, if Γ is not induced by a map $S \to P^1$, then there exists a divisor M such that $Z \subset M$ and $M^2 \geq t$. If $M^2 \geq t+1$, then $M^2.L^2 \geq (M.L)^2$ by assumption, a contradiction.

So, as above, $M^2 = t$ and T corresponds (scheme-theoretically) to the base locus of M; blowing up S at T, we get a strict transform M' of M with the properties that $M'^2 = 0$ and M' induces Γ on the strict transform (desingularization) of C.

The existence of the line bundle G, which implies that the linear series g^1_k's on the family of curves are defined globally, and the assumption (2) on the composition $W \to S(t)$ are needed only to make the 'monodromy argument' work; we list a couple of alternative sets of assumptions.

Remark. Let W be as before and call Σ the pull-back, over W, of the variety which parametrizes all base point free g^1_k's over the normalization of the curves of W_t (it exists on some suitable open subset).

Fix $\Gamma \in \Sigma$, corresponding to a linear series on the normalization Y of a nodal curve C. Call Σ_0 the g^1_k's over Y induced by Σ.

Assume:
(i) There is a component of Σ_0, of dimension d, containing Γ, such that Σ_0 has at most $t-1$ components of dimension d;
(ii) The composition $\Sigma \to \S^{(t)}$ is dominant;
(iii) $tL^2 > (2t+2k-1)^2$ and $L^2 > 4(t+k)$.

Then the conclusion of theorem 3 still works. Indeed the assumptions imply that we can still interchange the nodes of C, so that Γ is sent to a series Γ' with the same properties of Γ, absurd unless Γ' is algebraically equivalent to Γ, by (i); then we can still apply the 'monodromy' argument and everything works as in the theorem.

Remark. We need the composition $W \to S^{(t)}$ dominant only in order to be sure that we can interchange the nodes; in fact, theorem 3 and corollary 4 still work if we assume that the image of W in $S^{(t)}$ contains a subset which allows this permutation.

Now we list some cases in which some of the assumptions of theorem 3 are satisfied.

6.1. Assume $h^0(S, L) > 3t$. Then for a general $T \subset S$ with card(T)=t, there is a curve C in L with $T \subset \mathrm{Sing}(C)$. Since the symmetric products $S^{(t)}$ are irreducible, if one point of T is a node for a suitable C, then all of them are nodes.

6.2. If (S, L) is $(2t+1)$-jet ample (in the sense of [BFS] or [BS2]), then for every $T \subset S$ of length t there is a curve C in L, with exactly a node at each point of T and smooth outside T.

Let $\pi : S' \to S$ be the blowing up at t general points and let D_i be the exceptional divisors. Assume $L^2 > 9t$ and $\pi^{-1}(L) - \sum 3D_i$ spanned; then a general C in L, with $T \subset \mathrm{Sing}(C)$, has only nodes for singularities and it is smooth outside T.

6.3. Other singularities can be considered too (see also [ABS]).

If C has t' nodes and t'' cusp, for instance, with $t = t' + t''$, then the proof of theorem 3 works verbatim when $t' = 0$.

When $t't'' \neq 0$, everything works, except that we cannot interchange cusps and nodes; in this case, we may repeat theorem 3, taking Y as the partial normalization of C at the nodes (resp. at the cusps), or adding to our assumption
(4) $t'L^2 > (2t' + 2k - 1)^2$ and $t''L^2 > (2t'' + 2k - 1)^2$.

Remark. If C has singularities outside T, theorem 3 works taking Y to be the partial normalization of C only at T.

We can prove statements which are similar to the one of theorem 3, assuming for instance that C belongs to a component of W_t whose image in $S^{(t)}$ contains any union of t general points of a fixed integral curve $Q \subset S$.

Under stronger numerical condition, we may repeat the proof of theorem 3 when some of the points of T are fixed.

An example: ruled surfaces

Let us turn our attention to some application of Reider's method in the study of linear series of curves embedded on some smooth surface S.

The important case $S = P^2$ can be handled in this way (see e.g. [Ba]), but, for plane curves, stronger results were proven by a method of Coppens ([Co] and [CK]); we were unable to extend Coppens' method to other cases in which we are interested (i.e. minimal ruled surfaces) for technical reasons.

Here we show, as an example, how far one can go, without any effort, using Reider's method, in the case of minimal ruled surfaces (for the non rational ones, however, our results are weak; we put them in a remark). Reider's method works in fact for every smooth curve in a given system, not only for a general one.

Minimal rational ruled surfaces. Fix such a surface, with invariant $e \geq 0$, so that Pic(S) has a basis h, f, with h minimal section, f fibre and $h^2 = -e$, $f^2 = 0$, $hf = 1$.

Fix integers a, a' with $a \geq 2$, $a' > ea$ and $a' \geq a+2$; so $L = ah + a'f$ is ample and $L^2 = a(2a' - ea)$. Fix a smooth curve C in L and fix an integer $k \geq 2$ such that C has a base point free g^1_k, Γ.

First assume $k \leq a$. We have then $L^2 \geq 4k + 1$, hence we get, as in the proof of theorem 3, a Reider's divisor $D = M + F$ with, say, $M = ih + i'f$ and $F = uh$, with $i \geq 0$, $i' > 0$, $u \geq 0$ and $u > 0$ only if $i' = ei$.

If $i > 0$, then since $M.L \leq D.L \leq 2k - 1$ and $M^2 \geq 0$, we must have $e = i = 1$, $u = 0$, $i' = 1$; M corresponds then to a pencil of $h + f$; on the other hand $M.L = a' \geq a + 2$ but, for any 2 points of S, either $h + f$ has a unique curve through them, or the 2 points are in some curve of h, which is a fixed part of the divisors in $h + f$ through them. Since M has no fixed part, both cases are impossible and we get a contradiction.

Thus $i = 0$, then $u = 0$ and $M^2 = 0$; M induces here a g^{i+1}_a on C, which must be equivalent to our Γ. By Serre duality, since always $a' + e > ae + i'$, then $h^1(S, M - L) = 0$ so M induces a complete series on C. Hence Γ is complete, $i' = 1$ and Γ is the g^1_a induced on C by the ruling.

Even for $k > a$ we get informations on Γ. Indeed, let $s > 0$ be the integer such that $sa < k \leq (s+1)a$ and assume $L^2 = 4k + 1$; once again we get the Reider's divisor $D = M + F$, with $M = ih + i'f$. If $i = 0$ and $a' + e > ae + i'$, once again $M = i'h$ induces a complete series on C, so $i' = s + 1$ and Γ is a pencil in the g^{s+1}_k induced on C by M. When $i > 0$, $D.L = (i + u)(a' - ea) + ai' < 2k$ leads to a contradiction in several numerical cases, e.g. when $e \geq s$.

One motivation. Fix integers d, n. It is known (refined Castelnuovo's theory, see e.g. [ACGH], ch.III or [Ha], ch.III) that if the arithmetic genus of a degree d, non degenerate curve $C \subset P^n$ is near Castelnuovo's upper bound, then C is contained in a minimal degree surface, i.e. either in a minimal rational surface S, embedded as a scroll, or in the cone over a rational normal curve, or in a Veronese surface.

It follows that, outside this last case, the normalization Y of C has an interesting pencil. To know the uniqueness of this pencil is nice; for instance, from the pencil, under standard cohomological conditions, one obtains the dimension of every complete linear series which is multiple of the pencil. These data determines e, hence S.

Singular curves. When C is singular, with t nodes in general position, a similar procedure works, under stronger numerical restrictions. One gets, for instance, the non-existence of a g^1_k with $k < a$ and the uniqueness of the g^1_a when $L^2 > 4(t+k)$.

One can also handle some case of curves with nodes in special position, as, for instance, curves C with two nodes on the same line of the ruling.

Minimal irregular ruled surfaces. Fix such a surface S, with irregularity $q > 0$, invariant e and h, f as a base for the Neron Severi group of S. By [N] or [H2], ch.V, we have $e \geq -q$.

If $e \geq 0$, S is the projectivizaton of a non stable rank 2 bundle on Alb(S) and the same calculation of 6.1 works on NS(S), but we get weaker results, since a single fiber, in general does not move in a linear system and it does not induces 'in general' a g^1_a on C.

If $e < 0$, S is the projectivization of a stable bundle and the cones of effective and ample divisors become complicate (see [H1] ch.1.10). When $q = 1$, since bundles over elliptic curves are classified, everything can be computed.

Observe that however, from the point of view of Castelnuovo's theory, the most interesting ruled surface in P^n is the cone over an elliptic normal curve in P^{n-1} ([Ha] th.3.15) whose normalization has $e = n$.

References

[ABS] Andreatta M., Ballico E., Sommese A.J. (in preparation).

[ACGH] Arbarello E., Cornalba M., Griffiths P., Harris J. Geometry of Algebraic Curves I. Springer (1985).

[Ba] Ballico E. On the gonality of nodal curves. Geom.Dedic. 37 (1991), 357-360.

[BFS] Beltrametti M., Francia P., Sommese A.J. On Reider's method and high order embeddings. Duke Math. J. 58 (1989), 425-439.

[BS1] Beltrametti M., Sommese A.J. On k-spannedness for projective surfaces. Springer Lect. Notes vol.1417 (1989), 24-51.

[BS2] Beltrametti M., Sommese A.J. On k-jet ampleness. (preprint 1992).

[Co] Coppens M. Free linear system on integral Gorenstein curves. J. Algebra 145 (1992), 209-218.

[CK] Coppens M., Kato T. The gonality of smooth curves with plane models. Manuscripta Math. 70 (1990), 5-25.

[Ha] Harris J. Curves in projective space. Presses Univ. Montreal (1982).

[H1] Hartshorne R. Ample subvarieties of algebraic varieties. Springer Lect. Notes vol. 156 (1970).

[H2] Hartshorne R. Algebraic Geometry. Springer (1977).

[N] Nagata M. On self intersection number of a section on a ruled surface. Nagoya Math. J. 37 (1970), 191-156.

[Se] Serrano F. Extensions of morphisms defined on a divisor. Math.Ann. 277 (1987), 395-413.

[Ta] Tannenbaum A. Families of curves with nodes. Comp. Math. 41 (1980), 107-126.

The Strong Castelnuovo Lemma for zerodimensional schemes

M.P. Cavaliere, M.E. Rossi and G. Valla *

Abstract. In this paper we extend the proof of the Strong Castelnuovo Lemma for zero dimensional schemes in linearly general position from the reduced to the non reduced case.

1991 Mathematics Subject Classification: Primary 14M05, 14C20; Secondary 13H15, 13D40.

Dedicated to the memory of our friend Mario Raimondo

Introduction

Let k be an algebraically closed field of characteristic zero. If we cut a reduced, irreducible, non degenerate projective variety V of codimension n in $\mathbf{P}^r = \mathbf{P}^r_k$ with a general linear subspace of dimension n, we get a set X of distinct points on this $\mathbf{P}^n$ in linearly general position; this means that for every integer $h < n$, $h+2$ points of X are never contained on an h-plane.

This explains why this notion of points in linearly general position has a central role in many problems of Algebraic Geometry, especially in the so called Castelnuovo theory as developed by Castelnuovo and more recently by Eisenbud and Harris in [H] and Green in [G].

However a recent work made by Bayer and Eisenbud leads to extend Castelnuovo theory to more general finite subschemes of the projective space.

Eisenbud and Harris started in [EH1] the project of studying finite subschemes of the projective space in linearly general position. Their definition is quite natural and is suggested by the classical one.

A finite subscheme X in $\mathbf{P}^n$ is in **linearly general position** if for every proper linear subspace L in $\mathbf{P}^n$ we have:

$$\deg(X \cap L) \leq 1 + \dim(L).$$

* The authors were partially supported by MURST (Italy). This research was made in the framework of Europroj project.

In [EH1], Theorem 3.1, the authors proved that the general hyperplane section of a one dimensional ribbon is in linearly general position, thus giving a strong motivation to their work.

In two recent papers, they extend some of the standard results in the classical Castelnuovo theory to this more general setting (see [EH1] and [EH2]). But one of the most interesting results, the classical Strong Castelnuovo Lemma proved by Mark Green in [G2], Theorem 3.c.6, was not considered and in fact the authors asked explicitly the following question: Does some analog hold for schemes?

In this paper we prove the Strong Castelnuovo Lemma for zero dimensional schemes in linearly general position (see Theorem 7). Our proof follows the same ideas of Green, but we are able to treat some of the details in a more elementary way so that the extension to the general case becomes available.

Some of the results we need are well known or folklore, but we insert here a proof for the sake of completeness.

The authors thank F.O. Schreyer, D. Eisenbud and S. Gimigliano for some helpful discussion on the subject of this paper.

The main result

Let k be an algebraically closed field, R the polynomial ring $k[X_0, \dots, X_n]$, I an homogeneous ideal of R with initial degree ≥ 2 and A the graded algebra R/I. Let $\mathbf{E}$ be a minimal graded free resolution of A over R:

$$\mathbf{E} : 0 \to E_h \to \cdots \to E_1 \to R \to A \to 0$$

where $h = hd_R(A)$ and $E_i = \oplus_{j=1}^{\beta_i} R(-d_{ij})$. It is well known that $d_{ij} \geq i+1$ for every i and j. We denote by a_i the multiplicity of the shift $i+1$ in E_i.

Since the resolution is minimal we get for every $i = 1, \dots, h$,

$$a_i = \dim_k(\mathrm{Tor}_i^R(A,k)_{i+1}).$$

We can compute $\mathrm{Tor}_i^R(A,k)$ using a resolution of k which can be obtained by the Koszul complex of $X_0, \dots, X_n$. Let V be a fixed k-vector space of dimension $n+1$; then the Koszul resolution of k is given by

$$0 \to \overset{n+1}{\Lambda} V \otimes R(-n-1) \xrightarrow{\delta_{n+1}} \overset{n}{\Lambda} V \otimes R(-n) \to \cdots \to \Lambda V \otimes R(-1) \xrightarrow{\delta_1} R \to k \to 0$$

where the δ_i's are the usual Koszul maps. Since $A_1 = R_1$, by tensoring with A and taking the graded pieces, one finds that $\mathrm{Tor}_i^R(A,k)_{i+1}$ is the homology of the complex

$$\overset{i+1}{\Lambda} V \otimes R_0 \longrightarrow \overset{i}{\Lambda} V \otimes R_1 \longrightarrow \overset{i-1}{\Lambda} V \otimes A_2.$$

For any $j \geq 1$ we shall denote by K_j the Kernel of δ_j in degree $2+j$. Thus K_j is the Kernel of the usual Koszul map $\overset{j}{\Lambda} V \otimes R_2 \rightarrow \overset{j-1}{\Lambda} V \otimes R_3$ which is still denoted by δ_j.

The reading of Green and Lazarsfeld's paper [GL] suggested us the following crucial result which is proved in [CRV].

Proposition 1. *Let i be any integer, $1 \leq i \leq h$. With the above assumptions and notations, we have*

$$a_i = \dim_k[(\overset{i-1}{\Lambda} V \otimes I_2) \cap K_{i-1}].$$

Proof. The proof is the same as in Proposition 1 of [CRV]. □

We want to prove now a result for which Green in [G2] refers to [G1]. We insert here a proof which is very elementary and uses a deep analysis of the quadrics passing through a given projective variety, a theme which will be central in all this paper.

Let X be a non degenerate projective variety in $\mathbf{P}^n$. Let $I = I(X)$ be the defining ideal of X in the polynomial ring $R := k[X_0. \dots, X_n]$, and let $A := R/I$ be the homogeneous coordinate ring of X. Since we are interested in the integers $a_i(X)$ we want to study the vector spaces

$$(\overset{i-1}{\Lambda} V \otimes I_2) \cap K_{i-1}.$$

Let $e_0, \dots, e_n$ be a k-vector base of V; if α is an element in $(\overset{i-1}{\Lambda} V \otimes I_2) \cap K_{i-1}$, we may write

$$\alpha = \sum_{j=\{j_1,\dots,j_{i-1}\}} e_{j_1} \wedge \dots \wedge e_{j_{i-1}} \otimes F_j$$

with $F_j \in I_2$ and $\delta_{i-1}(\alpha) = 0$.

Since $X \subseteq \mathbf{P}^n$ is a non degenerate projective variety, after a suitable change of coordinates, we may assume that X contains the coordinate points $P_0, \dots, P_n$ where $P_0 := (1, 0, \dots, 0)$, $P_1 := (0, 1, 0, \dots, 0)$,..., $P_n := (0, 0, \dots, 1)$.

We will refer to the **extra points** of X for the points of X which are different from $P_0, \dots, P_n$.

We have some important remarks.

Remark 2.

(i) In F_j there is no term of the form X_p^2 for every $p = 0, \dots, n$.
It follows from the fact that F_j must vanish on the coordinate points.

(ii) In F_j there is no monomial of the form $X_p X_q$ with $p \in \{j_1, \dots, j_{i-1}\}$.

If for example $\alpha = e_{j_1} \wedge \cdots \wedge e_{j_{i-1}} \otimes X_{j_1} X_q + \ldots$, then $\delta_{i-1}(\alpha) = e_{j_2} \wedge \cdots \wedge e_{j_{i-1}} \otimes X_{j_1}^2 X_q + \ldots$ cannot be zero since to cancel $e_{j_2} \wedge \cdots \wedge e_{j_{i-1}} \otimes X_{j_1}^2 X_q$ we need in α a term $e_{j_2} \wedge \cdots \wedge e_{j_{i-1}} \wedge e_q \otimes X_{j_1}^2$. This is impossible by (i).

In the following if j is a k-uple $\{0 \le j_1 < \cdots < j_k \le n\}$, we denote by ϵ_j the element $e_{j_1} \wedge \cdots \wedge e_{j_k}$ of $\overset{k}{\Lambda} V$.

With this notation, if $\alpha = \sum_{|j|=i-1} \epsilon_j \otimes F_j \in (\overset{i-1}{\Lambda} V \otimes I_2) \cap K_{i-1}$, then in F_j we have only monomials of the form $X_p X_q$ with $p \neq q$ and $p, q \in \mathcal{C}j := \{0, \ldots, n\} \backslash \{j\}$. Hence it will be useful to write $F_{\mathcal{C}j}$ instead of F_j; in this way $F_{\mathcal{C}j}$ is a square free quadratic form in the variables corresponding to the indexes in $\mathcal{C}j$.

Remark 3. Let $\alpha = \sum_{|j|=i-1} \epsilon_j \otimes F_{\mathcal{C}j}$ be an element of $(\overset{i-1}{\Lambda} V \otimes I_2) \cap K_{i-1}$. We remark that as j runs among the subsets of $i-1$ elements, $\mathcal{C}j$ runs among the subsets of $n-i+2$ elements of $\{0, \ldots, n\}$.

We claim that, for every $h = \{h_1, \ldots, h_{n-i+3}\}$ such that $0 \le h_1 < \cdots < h_{n-i+3} \le n$, we have

$$\sum_{r=1}^{n-i+3} (-1)^{h_r - r + 1} X_{h_r} F_{h_1 \ldots \hat{h}_r . h_{n-i+3}} = 0$$

In fact

$$0 = \delta_{i-1}(\alpha) = \sum_{|j|=i-1} \sum_{r=1}^{i-1} (-1)^{r+1} e_{j_1} \wedge \cdots \wedge \hat{e}_{j_r} \wedge \ldots e_{j_{i-1}} \otimes F_{\mathcal{C}j} X_{j_r} =$$

$$= \sum_{|q|=i-2} \epsilon_q \otimes \left(\sum_{r=1}^{n-i+3} (-1)^{h_r - r + 1} X_{h_r} F_{h_1 \ldots \hat{h}_r . h_{n-i+3}} \right)$$

where $\{h_1 < \cdots < h_{n-i+3}\} = \{0, \ldots, n\} \backslash q$. The conclusion follows.

If we apply the above description to the case $i = n-1$, we get that if α is an element of $(\overset{n-2}{\Lambda} V \otimes I_2) \cap K_{n-2}$, then $\alpha = \sum_{|j|=n-2} \epsilon_j \otimes F_{\mathcal{C}j}$ where if $\mathcal{C}j = \{\alpha, \beta, \gamma\}$, then $F_{\mathcal{C}j} = F_{\alpha\beta\gamma} = \lambda X_\alpha X_\beta + \mu X_\alpha X_\gamma + \sigma X_\beta X_\gamma$ for some $\lambda, \mu, \sigma \in k$.

Proposition 4. *Let $X \subseteq \mathbf{P}^n$ be a non degenerate reduced irreducible variety of positive dimension. If $a_{n-1} \neq 0$ and $\alpha = \sum_{|j|=n-2} \epsilon_j \otimes F_{\mathcal{C}j}$ is a non zero element of $(\overset{n-2}{\Lambda} V \otimes I_2) \cap K_{n-2}$, we let W be the subspace of R_2 generated by the forms $F_{\mathcal{C}j}$, $|j| = n-2$.*

With this notation we have

$$\dim_k(W) \ge \binom{n}{2}.$$

Proof. Since X is a non degenerate reduced irreducible variety , every non zero element of I_2 is irreducible.

Now, by Remark 3, for any $0 \leq a < b < c < d \leq n$, we have

$$(-1)^a X_a F_{bcd} + (-1)^{b-1} X_b F_{acd} + (-1)^c X_c F_{abd} + (-1)^{d-1} X_d F_{abc} = 0$$

Here the crucial observation is that if two among $F_{bcd}, F_{acd}, F_{abd}, F_{abc}$ are zero, then they are all zero. In fact if for example $F_{abd} = F_{abc} = 0$, then $X_a F_{bcd} \pm X_b F_{acd} = 0$. This implies $F_{acd} = X_a M$ for some linear form M. As we have seen before, the quadrics are either irreducible or zero, hence we get the conclusion.

Since $\alpha \neq 0$, we may assume for example that $F_{n-2,n-1,n} \neq 0$. We order the monomials of degree two in R by the lexicographic order induced by $X_0 < X_1 < \cdots < X_n$. We will prove that for every $0 \leq a < b \leq n-1$ there is a non zero quadric in W such that its minimal monomial is $X_a X_b$. Since these monomials are $\binom{n}{2}$, the conclusion will follow.

For every $0 \leq a < b \leq n-1$, $X_a X_b$ is the minimal monomial for those which are not zero among the $n-b$ quadrics $F_{a,b,b+1}, F_{a,b,b+2}, \ldots, F_{abn}$.

First we prove that if $n - b \geq 2$, there is a form F_{abi} with $i \in \{b+1, \ldots, n\}$ which is different from zero.

If $n - b = 2$, that is $b = n-2$, then $F_{a,n-2,n-1} = F_{a,n-2,n} = 0$ cannot hold otherwise, by the above remark, also $F_{n-2,n-1,n} = 0$.

Let $n - b \geq 3$, that is $b \leq n-3$; if $F_{a,b,n-2} = F_{a,b,n-1} = F_{abn} = 0$, then again by the above remark $F_{b,n-2,n-1}$ and $F_{b,n-2,n}$ are zero and by the same reason this implies $F_{n-2,n-1,n} = 0$, a contradiction.

The last case to be considered is $b = n-1$. If $F_{a,n-1,n} = 0$, since $F_{n-2,n-1,n} \neq 0$, then $a < n-2$ and the quadratic forms $F_{a,n-2,n-1} = \lambda X_a X_{n-2} + \mu X_a X_{n-1} + \sigma X_{n-2} X_{n-1}$, $F_{a,n-2,n} = \lambda' X_a X_{n-2} + \mu' X_a X_n + \sigma' X_{n-2} X_n$ are different from zero and hence, since they are irreducible, all their coefficients are not zero. It follows that the quadric $\lambda' F_{a,n-2,n-1} - \lambda F_{a,n-2,n}$ is different from zero and it has $X_a X_{n-1}$ as minimal monomial as required. □

We remark that we can extend the above result to a set X of distinct points in $\mathbf{P}^n$ if we assume that the points are in linearly general position and that $s := |X| \geq n+3$.

For this it will be enough to prove that:
for every j, $|j| = n-2$, either $F_{\mathcal{C}j} = 0$ or $F_{\mathcal{C}j}$ is an irreducible quadric.

Let $\mathcal{C}j = \{\alpha, \beta, \gamma\}$ and assume that $F_{\alpha\beta\gamma} = \lambda X_\alpha X_\beta + \mu X_\alpha X_\gamma + \sigma X_\beta X_\gamma$ is reducible.

If for example $\lambda = 0$, then $F_{\alpha\beta\gamma} = X_\gamma(\mu X_\alpha + \sigma X_\beta)$; since the points are in linearly general position and $X_\gamma = 0$ contains n coordinate points, all the extra points, which are $s - n - 1$, are on the hyperplane $L := (\mu X_\alpha + \sigma X_\beta)$ which contains also $n-1$ coordinate points. All toghether we have $s - 2 \geq n+1$ points on L, hence $L = 0$ and $F_{\alpha,\beta,\gamma} = 0$.

If X consists of $n+2$ points the above result is not true. In fact if we consider for example 5 points in $\mathbf{P}^3$ such that 3 of them are on the plane $X_0 = 0$ and 2 out of the plane, but on the line $X_1 = X_2 = 0$, then the points are in linearly general position. Since $X_0X_1, X_0X_2 \in I$, if we let $\alpha = \delta_2(e_1 \wedge e_2 \otimes X_0)$, then $\alpha \in (\Lambda V \otimes I_2) \cap K_1$ and $\alpha \neq 0$, but $\dim_k\langle X_0X_1, X_0X_2\rangle = 2 < \binom{3}{2}$.

We need now a result which is well known, but whose proof is not easily readable.

Proposition 5. *Let X be a zero dimensional scheme in linearly general position. Then we can find a closed subscheme Y of X such that* $\deg(Y) = \deg(X) - 1$ *and Y is in linearly general position.*

Proof. Let $\wp_1, \dots, \wp_s$ be the defining prime ideals of the support of X. Then

$$I(X) = J_1 \cap J_2 \cap \cdots \cap J_s$$

where J_i is a $\wp_i$-primary ideal. If we let X_i to be the zero dimensional scheme defined by J_i, then

$$\deg(X) = \sum_{i=1}^{s} \deg(X_i).$$

Hence we may assume that $s = 1$ and $I(X) = I$ where I is a primary ideal for $\wp := (X_1, \dots, X_n)$. Also we may assume that $I \neq \wp$, otherwise we simply delete $\wp$ to get the conclusion.

Since $\wp$ is an associated prime of I, we can find an element $F \in R$ of minimal degree such that $\wp = I : (F)$. Clearly $\deg(F) \geq 1$, otherwise $I = \wp$. Further, if $F \in I + (X_0)$, then we can write $F = G + X_0T$ with $G \in I$, $\deg(T) = \deg(F) - 1$. It follows that $\wp = I : (F) = I : (X_0T) = I : (T)$, a contradiction to the minimality of F. Hence $F \notin I + (X_0)$.

We claim that X_0 is not a zero divisor modulo $I + (F)$. If $X_0L \in I + (F)$, then we can write $X_0L = G + FP$ with $G \in I$ and $\deg(P) \geq 1$, otherwise $F \in I + (X_0)$. It follows that $P = X_0^r + H$ with $r \geq 1$ and $H \in \wp$. Hence we get $X_0L = G + FX_0^r + FH$, and since $H \in \wp$ and $FX_i \in I$ for every $i = 1, \dots, n$ this implies $X_0(L - FX_0^{r-1}) \in I$ so that $L - FX_0^{r-1} \in I$ and $L \in I + (F)$, as claimed.

This implies that $I + (F)$ is the defining ideal of a zero dimensional scheme, say Y, for which we get: $\deg(Y) = \dim_k(R/(I + (F, X_0))$. We have $\deg(X) = \dim_k(R/(I + (X_0)))$ and since $F \notin (I + (X_0))$ but $FX_i \in (I + (X_0))$ for every $i = 0, \dots, n$ we get also $\dim_k((I + (X_0, F))/(I + (X_0))) = 1$. Now we can use the exact sequence of vector spaces

$$0 \to (I + (X_0, F))/(I + (X_0)) \to R/(I + (X_0)) \to R/(I + (X_0, F)) \to 0$$

to get $\deg(Y) = \deg(X) - 1$. Since it is clear that every subscheme of a zero dimensional scheme in linearly general position is still in linearly general position, the conclusion follows. □

The following result is possibly well known. We insert here a proof for the sake of completeness.

Theorem 6. *Let X be a zero dimensional scheme in $\mathbf{P}^n$, $\deg(X) = n+3$. If X is contained on a rational normal curve of $\mathbf{P}^n$ and A is the homogeneous coordinate ring of X, the minimal free resolution of A is linear at least for the first $n-2$ steps.*

Proof. Let C be the rational normal curve of $\mathbf{P}^n$ containing X. After a linear change of coordinates we may assume that C is defined by the 2×2 minors of the matrix

$$\begin{pmatrix} X_0 & X_1 & \cdots & X_{n-1} \\ X_1 & X_2 & \cdots & X_n \end{pmatrix}.$$

By deleting one of the internal columns we get the defining ideal of a surface S of $\mathbf{P}^n$, which is a rational normal scroll containing C as a directrix curve (see [H], pg. 90). Then S is a surface of degree $n-1$ whose Picard group is generated by the *hyperplane class* H and the *ruling* R:

$$Pic(S) = \mathbf{Z}H \oplus \mathbf{Z}R.$$

The intersection product is given by

$$H^2 = n-1, \quad H \cdot R = 1, \quad R^2 = 0.$$

The dualizing sheaf of S is $\omega := \omega_S = -2H + (n-3)R$. Since it is clear that as a divisor on S we have $C = H + R$, we get

$$(-\omega)C = [2H - (n-3)R][H+R] = \\ = 2H^2 - (n-5)HR = 2(n-1) - (n-5) = n+3.$$

This means that $-\omega$ cuts on C the scheme X so that $\mathcal{O}_C(-X) = \mathcal{O}_C(\omega)$. We have a short exact sequence:

$$0 \to \mathcal{O}_C(-X) = \mathcal{O}_C(\omega) \to \mathcal{O}_C \to \mathcal{O}_X \to 0.$$

Further, by tensoring with ω the short exact sequence

$$0 \to \mathcal{O}_S(-C) \to \mathcal{O}_S \to \mathcal{O}_C \to 0,$$

we get a short exact sequence

$$0 \to \mathcal{O}_S(\omega - C) \to \mathcal{O}_S(\omega) \to \mathcal{O}_C(\omega) \to 0.$$

Now

$$\omega - C = -2H + (n-3)R - H - R = -3H + (n-4)R.$$

By using a classical result of Buchsbaum and Eisenbud (see [S], pg.112, Corollary), we get resolutions of $\mathcal{O}_S(\omega)$ and $\mathcal{O}_S(\omega - C)$ as $\mathcal{O}_{\mathbf{P}^n}$-modules. The canonical embedding $\mathcal{O}_S(\omega - C) \to \mathcal{O}_S(\omega)$ gives a morphism of complexes

$$
\begin{array}{ccc}
0 & \longrightarrow & 0 \\
\uparrow & & \uparrow \\
\mathcal{O}_S(\omega - C) & \longrightarrow & \mathcal{O}_S(\omega) \\
\uparrow & & \uparrow \\
R(-3)^{n-3} & \longrightarrow & R(-2)^{n-2} \\
\uparrow & & \uparrow \\
R(-4)^{(n-1)(n-4)} & \longrightarrow & R(-3)^{(n-1)(n-3)} \\
\uparrow & & \uparrow \\
\vdots & & \vdots \\
R(-n+2)^{2\binom{n-1}{n-5}} & \longrightarrow & \cdots \\
\uparrow & & \uparrow \\
R(-n+1)^{\binom{n-1}{n-4}} & \longrightarrow & R(-n+2)^{2\binom{n-1}{n-4}} \\
\uparrow & & \uparrow \\
\cdots & \longrightarrow & R(-n+1)^{\binom{n-1}{n-3}} \\
\uparrow & & \uparrow \\
\vdots & & \vdots
\end{array}
$$

By using the mapping cone of this morphism we get a resolution of $\mathcal{O}_C(\omega)$; further, a resolution of $\mathcal{O}_C$ is the well known Eagon-Northcott complex. The canonical embedding $\mathcal{O}_C(\omega) \to \mathcal{O}_C$ gives a morphism of complexes

$$\begin{array}{ccc}
0 & \longrightarrow & 0 \\
\uparrow & & \uparrow \\
\mathcal{O}_C(\omega) & \longrightarrow & \mathcal{O}_C \\
\uparrow & & \uparrow \\
R(-2)^{n-2} & \longrightarrow & R \\
\uparrow & & \uparrow \\
R(-3)^{n(n-3)} & \longrightarrow & R(-2)^{\binom{n}{2}} \\
\uparrow & & \uparrow \\
\vdots & & \vdots \\
\uparrow & & \uparrow \\
R(-n+2)^{2\binom{n}{4}} & \longrightarrow & \cdots \\
\uparrow & & \uparrow \\
R(-n+1)^{\binom{n}{3}} & \longrightarrow & R(-n+2)^{(n-3)\binom{n}{n-2}} \\
\uparrow & & \uparrow \\
\cdots & \longrightarrow & R(-n+1)^{n(n-2)} \\
\uparrow & & \uparrow \\
\vdots & & \vdots
\end{array}$$

By using again the mapping cone of this morphism we get a resolution of $\mathcal{O}_X$:

$$\cdots \to R(-n+1)^{(n-2)\binom{n+2}{3}-(n+3)\binom{n}{3}} \to \cdots \to R(-3)^{2\binom{n+2}{3}-n(n+3)} \to$$
$$\to R(-2)^{\binom{n+2}{2}-(n+3)} \to R \to \mathcal{O}_X \to 0$$

Even if this resolution is not necessarily minimal, the conclusion clearly follows. □

We can prove now the main result of the paper. It is the extension to the non reduced case of the Strong Castelnuovo Lemma proved by Green in [G].

Theorem 7. *Let $X \subseteq \mathbf{P}^n$ be a non degenerate zero dimensional scheme of degree at least $n+3$.*

If X is in linearly general position and $a_{n-1} \neq 0$, then X lies on a rational normal curve.

Proof. By Proposition 5 we can find a subscheme Y of X of degree $n+3$ which is in linearly general position. By Theorem 2.1 in [EH1] Y lies on a rational normal curve C, so that $I(C) \subseteq I(Y)$. Further by Theorem 3.2 in [EH1], Y imposes independent conditions on forms of degree two, hence the Hilbert Function of Y is $H(0) = 1, H(1) = n+1, H(i) = n+3$ for every $i \geq 2$. This implies that the homogeneous coordinate ring of Y is a short graded algebra, according to the definition given in [CRV]. Since, by the above theorem, the resolution of Y is linear for the first $n-2$ steps, by Proposition 1.6 in [CRV], we get $a_{n-1}(Y) = n-1$. But also it is well known and easy to prove that $a_{n-1}(C) = n-1$, hence the inclusion of vector spaces $(\overset{n-2}{\Lambda} V \otimes I(C)_2) \cap K_{n-2} \subseteq (\overset{n-2}{\Lambda} V \otimes I(Y)_2) \cap K_{n-2}$ is indeed an equality.

If $a_{n-1}(X) \neq 0$, and $\alpha = \sum \epsilon_j \otimes F_{Cj}$ is a non zero element of $(\overset{n-2}{\Lambda} V \otimes I(X)_2) \cap K_{n-2}$, then $F_{Cj} \in I(X)_2$ for every j and $\alpha \in (\overset{n-2}{\Lambda} V \otimes I(Y)_2) \cap K_{n-2} = (\overset{n-2}{\Lambda} V \otimes I(C)_2) \cap K_{n-2}$. By Proposition 4, if we let W to be the subspace of R_2 generated by the forms F_{Cj}, we get

$$\dim(I(C)_2) \geq \dim(W) \geq \binom{n}{2}.$$

Since the number of independent quadrics passing through the rational normal curve is $\binom{n}{2}$, this implies $W = I(C)_2$, hence every quadric which contains C has to contain X. The conclusion follows since the defining ideal of the rational normal curve is generated by quadrics. □

Remark 8. We can compare the above result with the Classical Castelnuovo Lemma which in the non reduced version was proved in [EH2],Theorem 2.2. It says:

Let X be a finite subscheme of $\mathbf{P}^n$ in linearly general position. If $\deg(X) = s \geq 2n+3$, and $H_X(2) \leq 2n+1$ then X is contained on a rational normal curve.

We would like to prove this result as a consequence of the Strong Castelnuovo Lemma. This is easy if $n = 3$.

In this case we have a finite subscheme X of $\mathbf{P}^3$ with $\deg(X) = s \geq 9$ and $H_X(2) \leq 7$. Since X is in linearly general position, by Theorem 3.2 in [EH1], we know that $H_X(3) \geq \min(s, 10) \geq 9$, which implies $\dim(I_3) = \binom{6}{3} - H_X(3) \leq 11$. On the other hand $H_X(2) \leq 7$ implies $\dim(I_2) = \binom{5}{2} - H_X(2) \geq 3$. If F_1, F_2, F_3 are linearly independent forms in I_2, then $\{X_i F_j\}$, $0 \leq i \leq 3$, $1 \leq j \leq 3$, is a set of 12 vectors in I_3. These vectors have to be linearly dependent: this means exactly that $a_2 > 0$.

Some of the results here were discovered or confirmed with the help of the computer algebra program CoCoA written by A.Giovini and G.Niesi (see [GN]).

References

[CRV] Cavaliere, M.P., Rossi, M.E., Valla, G., On the resolution of certain graded algebras, to appear in Trans. Amer. Math. Soc.

[EH1] Eisenbud, D., Harris, J., Finite projective schemes in linearly general position, J. Algebraic Geometry 1 (1992), 15-30.

[EH2] Eisenbud, D., Harris, J., An intersection bound for rank 1 loci, with application to Castelnuovo and Clifford theory, J. Algebraic Geometry 1 (1992), 31-60.

[G1] Green, M., The canonical ring of a variety of general type, Duke Math. J. 49 (1982), 1087-1113.

[G2] Green, M., Koszul cohomology and the geometry of projective varieties, J. Differential Geometry 19 (1984), 125-171.

[GL] Green, M., Lazarsfeld, R., Some results on the syzygies of finite sets and algebraic curves, Compositio Math. 67 (1988), 310-314.

[GN] Giovini, A., Niesi, G., CoCoA: A User-Friendly System for Commutative Algebra. In: Design and Implementation of Symbolic Computation Systems, Lecture Notes in Comput. Sci. 429, 20-30. Springer-Verlag, 1990.

[H] Harris, J., Curves in projective space, Sém. Math. Sup., Université de Montreal 1982.

[S] Schreyer, F.O., Syzygies of canonical curves and special linear series, Math. Ann. 275 (1986), 105-137.

Hilbert Functions of Dots in Linear General Position

Karen A. Chandler

Abstract. An **α-dot** is the first infinitesimal neighbourhood of a point with respect to an $(\alpha - 1)$-dimensional affine space. We estimate the Hilbert function of a collection of dots over a set of points in linear general position, with the result that

Theorem 4.2. *Let Γ be a collection of d α-dots in linear general position in $\mathbb{P}^n$, $\alpha \geq 2$. Then the Hilbert function h_Γ of Γ satisfies*

$$h_\Gamma(r) \geq \min(rn + 1, 2d) + (\alpha - 2)\min((r - 1)n - 1, d)$$

for $r \geq 3$. Equality occurs for some r with $rn + 2 \leq 2d$ if and only if Γ_{red} is contained in a rational normal curve C, and the tangent directions to this curve at these points are all contained in Γ. Equality occurs for some r with $(r - 1)n \leq d$ if and only if Γ is contained in the first infinitesimal neighbourhood of C with respect to a subbundle, of rank $\alpha - 1$ and of maximal degree, of the normal bundle of C in $\mathbb{P}^n$.

By a Castelnuovo argument (see [C]), this implies an upper bound on the degree of a subbundle of rank $\alpha - 1$ of the normal bundle of a curve of degree d in $\mathbb{P}^n$.

1991 Mathematics Subject Classification: 13D40, 14H99, 14A15

1. Introduction

Definition 1.1. We define an **α-rope** to be a projective scheme X such that
(a) $C = X_{red}$ is an irreducible nondegenerate smooth curve,
(b) the ideal sheaf $\mathcal{I} = \mathcal{I}_{X,C}$ has $\mathcal{I}^2 = 0$ and hence is an $\mathcal{O}_C$-module,
(c) $\mathcal{I}$ is locally free of rank $\alpha - 1$ over C.

An α-rope $X \subset \mathbb{P}^n$ with $C = X_{red}$ is the first infinitesimal neighbourhood of a rank $\alpha - 1$ subbundle of the normal bundle of C in $\mathbb{P}^n$. For example, a 1-rope is just a curve $C \subset \mathbb{P}^n$, and a 2-rope is a ribbon in the sense of [BE], whereas an n-rope in $\mathbb{P}^n$ is the first infinitesimal neighbourhood of the curve in its entire normal bundle. We would like to find a lower bound for the Hilbert function h_X of any rope X, and thereby obtain an upper bound for its genus. This in turn gives us an estimate for the degree of any subbundle E of the normal bundle of a curve $C \subset \mathbb{P}^n$ in terms of the degree and genus of C and the rank $\alpha - 1$ of E.

If E corresponds to a rope X, then the degree of E is $G+\alpha(1-g)-1$, where G is the genus of X and g is the genus of C. As a first step, we look at the Hilbert function of a hyperplane section of X which is a collection of dots.

The inspiration for this problem is Castelnuovo's classical result that the genus of a curve of degree d in $\mathbb{P}^n$ is bounded. Castelnuovo obtains this upper bound on the genus by bounding from below the number of independent linear conditions imposed by the curve on hypersurfaces of each fixed degree, viz. the Hilbert function of the curve. His idea is to estimate the Hilbert function h_X of $X \subset \mathbb{P}^n$ by considering a general hyperplane section $\Gamma = H \cap X$ of X, using the following two observations:

Lemma 1.2. *Let $X \subset \mathbb{P}^n$ be a subscheme, and let $Y = X \cap H$ be a hyperplane section, where H is a hyperplane not containing any component of X. Then*

$$h_X(r) - h_X(r-1) \geq h_{Y,H}(r).$$

(If H is any hyperplane, then let X' be the union of the components of X not lying in H and $Y = X' \cap H$ to obtain the same result.)

Proof. We may find $h_X(r-1)$ homogeneous polynomials of degree $r-1$ no linear combination of which vanishes on X. Now if we multiply each of these by the linear form L that defines the hyperplane H, we obtain $h_X(r-1)$ homogeneous polynomials of degree r that vanish on H and such that no linear combination of them vanishes on X. Likewise, we may find $h_{Y,H}(r)$ homogeneous polynomials of degree r, no linear combination of which vanishes on either Y or H. □

Definition 1.3. A zero-dimensional subscheme $\Gamma \subset \mathbb{P}^n$ is in **linear general position** if the linear span of any subscheme Σ of Γ has dimension at least $\min(\deg \Sigma - 1, n)$.

Lemma 1.4 (Castelnuovo). *Let C be a curve in $\mathbb{P}^{n+1}$ and let $\Gamma = H \cap C$ be a general hyperplane section of C. Then $\Gamma \subset H = \mathbb{P}^n$ is a collection of d (reduced) points in linear general position.*

For such a Γ, then, we have $h_\Gamma(1) = n+1$. Assume by induction that every subset of Γ consisting of $(r-1)n+1$ points imposes independent conditions on $(r-1)$-ics. Then given subsets of Γ of degrees $(r-1)n+1$ and $n+1$, respectively, that meet at a point p, we may find an $(r-1)$-ic F, and a linear form L so that FL vanishes on the rn points other than p and fails to vanish at p.

Thus, we get:

Lemma 1.5 (Castelnuovo). *Let $\Gamma \subset \mathbb{P}^n$ be a collection of $d \geq 2n+1$ points in linear general position. Then $h_\Gamma(r) \geq \min(rn+1, d)$. Moreover, we have equality for some r with $rn+2 \leq d$ if and only if Γ lies on a rational normal curve.*

In [EH1], Eisenbud and Harris extend Castelnuovo's result to ribbons. They show that if $\Gamma \subset \mathbb{P}^n$ is a general hyperplane section of a ribbon in $\mathbb{P}^{n+1}$, then Γ

is in linear general position. Eisenbud and Harris then prove that any collection Γ of d arrows (2-dots) in linear general position in $\mathbb{P}^n$ has Hilbert function at least

$$h_\Gamma(r) \geq \min(rn + 1, 2d),$$

and that equality occurs for some r with $rn + 2 \leq 2d$ if and only if Γ is contained in a rational normal curve.

In this paper, we consider Hilbert functions of collections of dots having the following property:

Definition 1.6. We say that a zero-dimensional subscheme $\Gamma \subset \mathbb{P}^n$ is in **weak linear general position** if Γ_{red} is in linear general position.

We start in section 2 by examining criteria for determining whether two subschemes of Γ impose the same number of conditions on a linear system $\mathcal{D}$. In general, we expect that given two subschemes of the same degree, the one containing "fatter" dots will impose more conditions than the other. Nonetheless, general subschemes of sufficiently similar shape will impose the same number of conditions.

Next in section 3, we examine the Hilbert functions of collections of dots in weak linear general position. From the linear general position property, we find a formula for $h_\Gamma(r+1)$ in terms of $h_\Gamma(r)$ as before; again if we join at a point two subschemes, one of which is independent in degree r and the other in degree 1, they form a subscheme that is independent in degree $r+1$. But, further, we find that if two such subschemes are joined at several points then these points may be "promoted" to dots of higher degree. Hence, it follows that $h_\Gamma(r)$ grows as at least $(\alpha - 1)n$, for small r.

In section 4, we look at ropes over a rational normal curve and find the α-rope of minimal Hilbert function. We conclude that a collection of α-dots achieves the lower bound obtained in section 3 exactly when it lies on that α-rope.

Acknowledgements: This problem was suggested to me by Joe Harris. The ideas in this paper arose from discussions with Joe Harris and David Eisenbud. During part of this work I was supported by a Sloan Doctoral Dissertation Fellowship.

2. Collections of Dots

Definition 2.1. An **α-dot** is a scheme isomorphic to the first infinitesimal neighbourhood of a point in $\mathbb{A}^{\alpha-1}_{\mathbb{C}}$, i.e., $\operatorname{Spec}(\mathbb{C}[z_1, \ldots, z_{\alpha-1}]/(z_1, \ldots, z_{\alpha-1})^2)$.

Thus such a scheme is given by a reduced point p and its $(\alpha-1)$-dimensional Zariski tangent space V; we denote this α-dot by $p(V)$. Where appropriate, we denote by $p(\alpha)$ any α-dot at the point p. If $\alpha \leq 0$ we define an α-dot $p(\alpha) = \emptyset$.

So, an α-dot has degree α, e.g., a 1-dot is a (reduced) point, and a 2-dot or arrow is a point together with a one-dimensional tangent space. We consider collections of dots in $\mathbb{P}^n$ whose underlying reduced set of points is in linear general position. Such a scheme arises naturally as the general plane section of a projective scheme X where X_{red} is an irreducible nondegenerate variety.

Collections of Dots

Definition 2.2. We say that a zero-dimensional projective scheme **imposes independent conditions** on a linear system $\mathcal{D}$ (or briefly, that the subscheme is **$\mathcal{D}$-independent**) if it imposes as many conditions on $\mathcal{D}$ as its degree.

Lemma 2.3. *Let Γ be a collection of dots and $\mathcal{D}$ a linear system. Let*

$$\{p(V), q(W)\} \cup \Sigma$$

be a subscheme of Γ that is $\mathcal{D}$-independent. Suppose that $W' \supset W$, codim$(W, W') = 1$ such that $q(W') \subset \Gamma$ and $\{p, q(W')\} \cup \Sigma$ is $\mathcal{D}$-independent. Then there exists $V' \subset V$ with codim$(V', V) = 1$ so that

$$\{p(V'), q(W')\} \cup \Sigma$$

is also $\mathcal{D}$-independent.

Proof. Note that a subscheme of the form $p(k) \cup \Sigma$ of Γ is $\mathcal{D}$-independent if and only if $p \cup \Sigma$ is $\mathcal{D}$-independent and $p(\alpha) \cup \Sigma$ imposes at least k more conditions on $\mathcal{D}$ than does Σ.

Let $k = \dim V + 1$ and $l = \dim W + 1$. Since $\{p(V), q(W)\} \cup \Sigma$ is $\mathcal{D}$-independent, we have that this subscheme imposes $k + l$ more conditions on $\mathcal{D}$ than Σ alone. Then of course $\{p(V), q(W')\} \cup \Sigma$ imposes at least $k + l$ more conditions on $\mathcal{D}$ than Σ does. Now if $\{p, q(W')\} \cup \Sigma$ is $\mathcal{D}$-independent then $\{p(V), q(W')\} \cup \Sigma$ imposes at least $k - 1$ more conditions on $\mathcal{D}$ than $\{q(W')\} \cup \Sigma$. Hence, there exists $V' \subset V$, codim$(V', V) = 1$ so that $\{p(V'), q(W')\} \cup \Sigma$ is $\mathcal{D}$-independent. □

Corollary 2.4 ("Demotion"). *Let Γ be a collection of dots and let $\mathcal{D}$ be a linear system. Suppose that there exist integers $k_i \geq 1, l \geq 0$ such that for every*

$$\{q, p_1, \ldots, p_m\} \subset \Gamma,$$

there is a subscheme

$$\{q(l), p_1(k_1), \ldots, p_m(k_m)\}$$

of Γ which imposes independent conditions on $\mathcal{D}$. Suppose that there exist

$$\{q(l+1), p_1, \ldots, p_m\} \subset \Gamma,$$

so that every subscheme of the form

$$\{q(l+1), p_1(k_1), \ldots, p_m(k_m)\}$$

fails to impose independent conditions on $\mathcal{D}$. Let

$$k_i' = \begin{cases} 2, & \text{if } k_i \geq l \\ 1, & \text{otherwise.} \end{cases}$$

Then there exist $p_i(k_i') \subseteq \Gamma$ so that for every $q(l+1) \subseteq \Gamma$ we have

$$\{q(l+1), p_1(k_1'), \ldots, p_m(k_m')\}$$

fails to impose independent conditions on $\mathcal{D}$.

Proof. Let us call $\Sigma = \{p_2(k_2), \ldots, p_m(k_m)\}$.

Suppose $k_1 \leq l-1$. By lemma 2.3, given that

$$\{q(l+1), p_1(k_1)\} \cup \Sigma$$

fails to be $\mathcal{D}$-independent for all $(l+1)$-dots $q(l+1) \subset \Gamma$ at q), we have that

$$\{q(l+1), p_1\} \cup \Sigma$$

fails to be $\mathcal{D}$-independent.

On the other hand, suppose $k_1 \geq l$. Then

$$\{q(l+1), p_1\} \cup \Sigma$$

is $\mathcal{D}$-independent (since $\{q(k_1), p_1\} \cup \Sigma$ is $\mathcal{D}$-independent by hypothesis). But

$$\{q(l+1), p_1(k_1)\} \cup \Sigma$$

is $\mathcal{D}$-dependent, so there is an arrow $p_1(2) \subseteq p_1(k_1)$ for which

$$\{q(l+1), p_1(2)\} \cup \Sigma$$

is $\mathcal{D}$-dependent.

Now, applying the same argument to each p_i, $i \geq 1$ the result follows. □

Promotion

Now let $\Gamma \subset \mathbb{P}^n$ be a collection of α-dots in weak linear general position. We now develop tools to estimate the growth of the Hilbert function of Γ.

Lemma 2.5 ("Promotion"). *Suppose that for $1 \leq i \leq n+1$*

$$\Sigma \cup \{p_1, \ldots, \hat{p}_i, \ldots p_{n+1}\} \cup \{p_i(k)\} \subset \Gamma$$

imposes independent conditions on hypersurfaces of degree r. Then if $\Sigma' \subset \Gamma$ is a general subscheme of the same shape as Σ then

$$\Sigma' \cup \{p_1(k), \ldots, p_{n+1}(k)\}$$

imposes independent conditions in degree $r+1$.

Proof. We can find $F_1, F_2, \ldots, F_k$ of degree r, each vanishing on the subscheme $\{p_1, \ldots, \hat{p}_i, \ldots, p_{n+1}\} \cup \Sigma$ and so that no linear combination of them vanishes on $p_i(k)$. We can find a linear form L that vanishes on $p_j, j \neq i$ and fails to vanish on p_i. Thus F_jL vanishes on $\{p_1(\alpha), p_2(\alpha), \ldots, \hat{p}_i, \ldots, p_{n+1}(\alpha)\} \cup \Sigma$ for each j, and no linear combination of the F_jL vanishes at $p_i(k)$. □

Our strategy for estimating the Hilbert function of a collection Γ of α-dots in weak linear general position will be to find, for each r, a subscheme Σ of Γ that is r-independent and is maximal with respect to this property i.e.

$$h_\Gamma(r) = h_\Sigma(r) = \deg \Sigma.$$

If such a Σ contains an α-dot and at least n reduced points then the promotion lemma implies that

$$h_\Gamma(r+1) \geq h_\Gamma(r) + (\alpha - 1)n.$$

3. Estimate of the Hilbert function of dots in linear general position

Throughout this section, Γ will be a collection of $d \geq n+1$ α-dots over points in linear general position in $\mathbb{P}^n$. In this section we establish that the Hilbert function h_Γ of Γ satisfies the lower bound, for $r \geq 3$:

$$h_\Gamma(r) \geq \min(rn+1, 2d) + (\alpha - 2)\min((r-1)n - 1, d).$$

First, we verify this for $r = 2, 3$ using a combination of the "promotion" lemma with special tricks. Then we use induction to deduce the bound in higher degrees.

Computations in low degrees

Let $\Gamma \subset \mathbb{P}^n$ be a collection of α-dots whose reduced points are in linear general position. We will show that

$$h_\Gamma(3) \geq \min(3n+1, 2d) + (\alpha - 2)\min(2n - 1, d).$$

We start by considering the case $\alpha = n+1$. Let $p_1, \ldots, p_{n+1} \in \Gamma$. The subscheme $\{p_{n+1}(n+1)\}$ is linearly independent, hence it is quadrically independent as well. Also, a

$$\{p_n(n), p_{n+1}\} \subset \Gamma$$

is linearly independent, according to the linear general position of Γ_{red} so by the promotion lemma

$$\{p_n(n), p_{n+1}(n+1)\} \subset \Gamma$$

is also quadratically independent. Indeed, for each $i \leq n$, some

$$\{p_i(i), p_{i+1}, \dots, p_{n+1}\} \subset \Gamma$$

is linearly independent so inductively applying the lemma we arrive at a

$$\{p_{n+1}(n+1), p_n(n), \dots, p_2(2), p_1(1)\} \subset \Gamma$$

that is 2-independent. Hence,

$$h_\Gamma(2) \geq \binom{n+2}{2}.$$

In particular, a subscheme $\{p_{n+1}(V_{n+1}), \dots, p_1(V_1)\} \subset \Gamma$ is 2-independent provided $\dim V_i \leq i-1$ and for each $i \leq n$, V_i does not contain $p_{i+1}, \dots, p_{n+1}$.

Immediately, by the promotion lemma, we get that $\{p_{n+1}(n+1), \dots, p_1(n+1)\}$ imposes independent conditions on cubics, and hence $h_\Gamma(3) \geq (n+1)^2$. (Indeed, one can use the promotion lemma to prove that a subscheme of Γ of the form

$$\{p_{n-1}(n-1), \dots, p_1(1), q_0(n+1), q_1(n+1), \dots, q_n(n+1)\}$$

is 3-independent. However, we will derive a stronger result than this in the following.)

Lemma 3.1 ($\mathbb{P}^3$). *There is a unique quadric Q that is singular at a point p and vanishes at an additional 5 points in linear general position.*

Remark. As we'll see in section 4, if C is the twisted cubic determined by these 6 points, then Q is the quadric that vanishes on the "canonical" ribbon determined by C and p.

Proof. $\{p(4), p_1, \dots, p_5\}$ impose 9 conditions on quadrics in $\mathbb{P}^3$. □

Corollary 3.2. *There is a unique quadric Q that is singular on $n-2$ points and and vanishes at an additional 5 points, if all $n+3$ points are in linear general position.*

Remark. Likewise, Q is the quadric that vanishes on the "canonical" rope determined by the rational normal curve C through the $n+3$ points, and with basepoints the given $n-2$ points.

Proof. If (X_0, X_1, X_2, X_3) is the ideal of the $n-2$ points then the ideal of the $n-2$ dots is exactly $(X_0, X_1, X_2, X_3)^2$. By the lemma, exactly one of these vanishes on an additional 5 points in linear general position. □

We now look at the number of conditions that Γ imposes on cubics. Let us consider first the case $\alpha = n+1$. We will find a subscheme of Γ having $2n-1$ reduced points and imposing at least $2n^2+2$ conditions on cubics. To warm up, let us first find a subscheme having $2n$ reduced points and imposing $2n^2+2$

conditions on cubics. Take $2n$ $(n+1)$-dots, say Γ_0, of Γ. Now, fix $n+1$ of these, say $q_1, \ldots, q_{n+1}$. For a cubic or a quadric to be singular at all of these points it must contain every line between any pair of them, by Bezout's theorem, since it meets each such line to degree at least 4. Moreover, this is an equivalent condition by linear general position. So, let Y be the union of these lines. Let $\Sigma = p_1, \ldots, p_{n-1}$ be the remaining points, and let K be the $(n-2)$-plane spanned by the p_i's, and let $Z \subset K$ be the $(n-1)$-dots at the p_i's. Let H be a general hyperplane containing K.

Claim: We have that

(a) Z imposes independent conditions on cubics in $K = \mathbb{P}^{n-2}$.

(b) No quadric in $H = \mathbb{P}^{n-1}$ contains $Y \cap H$.

(c) No quadric in $\mathbb{P}^n$ contains Y.

The claim implies that

$$\begin{aligned} h_{\Gamma_0,\mathbb{P}^n}(3) &\geq h_{Z\cup(Y\cap H),H}(3) + h_{Y,\mathbb{P}^n}(2) \\ &\geq h_{Z,K}(3) + h_{Y\cap H,H}(2) + h_{Y,\mathbb{P}^n}(2) \\ &= (n-1)^2 + \binom{n+1}{2} + \binom{n+2}{2} \\ &= 2n^2 + 2 \end{aligned}$$

and hence we need now only verify the conditions of the claim to get our bound on h_{Γ_0}.

We have already seen that the first and third conditions hold, so let us prove the middle one. We can write Y as $Y = \cup l_{ij}$ where l_{ij} is the line between q_i and q_j. Then $Y \cap H = \{r_{ij}\}$ where $r_{ij} = l_{ij} \cap H$. Our goal is to prove that the points r_{ij} impose independent conditions on quadrics, and hence we seek for each i, j a quadric that vanishes on each $r_{k,l}$ other than r_{ij} and fails to vanish at r_{ij}. For each i, $1 \leq i \leq n+1$, let H_i be the hyperplane of $\mathbb{P}^n$ spanned by $\{q_k : k \neq i\}$. Then H_i does not contain r_{ij} by linear general position. That is, if H_i did contain $r_{i,j}$, then it would contain the whole line between $r_{i,j}$ and q_j ($r_{i,j} \neq q_j$ since H does not contain any q_j). But this is ruled out by the fact that H_i does not contain q_i, which is on that line. Hence the union of the two hyperplanes of H given by $H \cap H_i$ and $H \cap H_j$ gives a (reducible) quadric in H as desired.

Let us now prove that actually $2n-1$ $(n+1)$-dots impose $2n^2+2$ conditions on cubics. To see this, let H be the hyperplane spanned by some $p_1, \ldots, p_n \in \Gamma$. Then

Claim: $\Gamma_0 = \{q_1(n+1), \ldots, q_{n-1}(n+1), p_1(H), \ldots, p_n(H)\}$ imposes independent conditions on cubics.

Proof. Set $q_n = p_n$ and $q_{n+1} = p_{n-1}$ in the above, so that at least

$$\{q_1(n+1), \ldots, q_{n-1}(n+1), p_1(K), \ldots, p_{n-2}(K), p_{n-1}(H), p_n(H)\}$$

imposes independent conditions on cubics, where K is the $(n-2)$-plane spanned by $p_1, \dots, p_{n-1}$. In particular the subscheme

$$\{r_{i,j}\}_{i,j \le n-1}, \cup \{p_1(K), \dots, p_{n-2}(K), p_{n-1}(H), p_n(H)\}$$

imposes independent conditions on cubics **in** H, with $r_{i,j}$'s as before, $i, j < n$. For each k with $1 \le k \le n-2$ we should find a cubic in H that vanishes on $\{r_{i,j}\}_{i,j \le n-1}$, is singular at p_i's for $i \ne k$, vanishes on $p_k(K)$, but fails to be singular at p_k. Let H_1 be the span of $q_1, \dots, q_{n-1}, p_n$, so $H_1 \cap H$ contains the $r_{i,j}$'s for $i, j \le n-1$. Let $H_2 = K$, so H_2 contains $p_i(K)$ for each $i < n$. Let H_3 contain p_i, $i \ne k$. Then the cubic on H given by $H_1 \cap H$, H_2, H_3 is singular on p_i's for $i \ne k$, vanishes on $p_k(K)$ but fails to be singular at p_k, vanishes on $r_{i,j}$ for $i, j \le n-1$.

Lemma 3.3. *A collection of* $2n-1$ $(n+1)$*-dots in linear general position in* $\mathbb{P}^n$ *imposes at least* $2n^2+2$ *conditions on cubics.*

Proof. We have

$$\{p_1(n+1), \dots, p_{n-1}(n+1), p_n(H), \dots, p_{2n-1}(H)\}$$

is 3-independent, where H is the hyperplane spanned by

$$\{p_n, \dots, p_{2n-1}\}.$$

Now we may find a quadric Q that vanishes on $\{p_1(n+1), \dots, p_{n-1}(n+1), p_n, p_{n+1}\}$ but fails to vanish on p_{n+2}.

(Indeed, we may take Q given by the two hyperplanes span$(p_1, \dots, p_n)$ and span$(p_1, \dots, p_{n-1}, p_{n+1})$.)

Then if L is a linear form vanishing on $p_n, \dots, p_{2n-1}$, we have that QL vanishes on

$$\{p_1(n+1), \dots, p_{n+1}(n+1), p_{n+2}(H), \dots, p_{2n-1}(H)\}$$

and yet fails to vanish on $p_{n+2}(n+1)$. Similarly for $j = n, n+1$ we may find cubics vanishing on $p_i(n+1), 1 \le i \le n+2, i \ne j$ and on $p_i(H), n \le i \le 2n-1$, and failing to vanish on $p_j(n+1)$. Hence

$$\{p_1(n+1), \dots, p_{n+2}(n+1), p_{n+3}(H), \dots, p_{2n-1}(H)\}$$

is 3-independent. □

We would like to show now that this bound can be attained only by a collection of points lying on a rational normal curve.

Lemma 3.4. *Suppose that* Γ *is a collection of arrows and points such that* Γ_{red} *is a collection of points in linear general position, and* $\deg \Gamma \ge 3n+2$. *If* Γ *imposes only* $3n+1$ *conditions on cubics then it is contained in a rational normal curve.*

Proof. By the result of Eisenbud-Harris in [EH1] it suffices to prove that Γ is in linear general position. Let $\Sigma \subset \Gamma$ be a subscheme of minimal degree with respect to the property of not imposing independent conditions on cubics. So Σ contains a total of d points with arrows at some m of them, where $m+d = \deg \Sigma \leq 3n+2$. According to lemma 3.3, we have $d > n+2$. Suppose that there is a hyperplane H that contains a subscheme of Σ of degree $n+1$, say

$$\{p_1(2), \ldots, p_t(2), p_{t+1}, \ldots, p_{n+1-t}\}$$

(where $t \geq 1$ by the linear general position of the points). Then choose such a subscheme for which t is minimal. If Σ has only t arrows, then we may put another $2n$ points of Σ on two hyperplanes H_1,H_2 so that $H \cup H_1 \cup H_2$ contains a subscheme of Σ of degree $\deg \Sigma - 1$, hence contains an additional point of Γ, contrary to linear general position of the points and the minimality of t.

Hence we may assume that there is an additional arrow, say $p_d(2) \subset \Sigma$. Find hyperplanes H_1, H_2 each containing exactly n reduced points of Σ and such that $H_1 \cup H_2 \cup H$ contains a subscheme of Σ of degree $\deg \Sigma - 1$. Choose H_2 to contain $p_{d-n-1}, \ldots, \hat{p_i}, \ldots, p_d$ and $p_d \notin H_1$. Since $H_1 \cup H_2 \cup H$ contains a 3-independent subscheme of Σ of degree one less than that of Σ this implies that $\Sigma \subset H_1 \cup H_2 \cup H$. In particular $p_d(2) \subset H_2$. But varying i with $d-n-1 \leq i \leq d-1$, and likewise varying H_2 we get that $p_d(2)$ is contained in every hyperplane containing p_d and $n-1$ of the points $p_{d-n-1}, \ldots, p_{d-1}$. By linear general position, however, the intersection of such hyperplanes is exactly the point p_d, contradiction. □

Corollary 3.5. *If $\{p_1, \ldots, p_d\}$ is a collection of points in linear general position in $\mathbb{P}^n$ and $m+d \leq 3n+1$ for some m then*

$$\{p_1(n+1), \ldots, p_m(n+1), p_{m+1}, \ldots, p_d\}$$

is 3-independent. Moreover, if p_1 is not in the rational normal curve determined by $p_2, \ldots, p_{n+4}$ then

$$\{p_1(n+1), \ldots, p_{m+1}(n+1), p_{m+2}, \ldots, p_d\}$$

is 3-independent.

Proof. Suppose that

$$\{p_1(n+1), \ldots, p_m(n+1), p_{m+1}, \ldots, p_d\}$$

fails to be 3-independent, for some minimal m. Then by corollary 2.4 there is a subcollection

$$\{p_1(l_1), \ldots, p_m(l_m), \ldots, p_d\}$$

of arrows and points that imposes only $d+m-1$ conditions on cubics. Thus according to lemma 3.4, either $d+m = 3n+2$ and the points and arrows are on a rational normal curve, or else $d+m \geq 3n+3$. □

Lemma 3.6. *A collection of* $d \leq 2n-1$ $(n+1)$*-dots in linear general position imposes at least* $\min(3n+1, 2d) + (n-1)d$ *conditions on cubics with equality* ($2d \geq 3n+2$) *if and only if its reduced points lie on a rational normal curve.*

Proof. Let $\Gamma = \{p_1(n+1), \ldots, p_d(n+1)\}$ where $p_1, \ldots, p_d$ are in linear general position and $d \leq 2n-1$. We may find Γ', such that

$$\Gamma \supset \Gamma' = \{p_1(n+1), \ldots, p_m(n+1), p_{m+1}(n), \ldots, p_d(n)\}$$

imposes independent conditions on cubics, so that any cubic containing Γ' contains Γ (indeed $m \geq n+2$). By corollary 2.4, this implies that

$$\{p_1(n+1), \ldots, p_{m+1}(n+1), p_{m+1}, \ldots, p_i(n+1), \ldots, p_d\}$$

fails to be cubically independent, i.e. that there is some arrow $p_{m+1}(l_{m+1}) \subset p_{m+1}(n+1)$ for which any cubic vanishing on

$$\{p_1(n+1), \ldots, p_m(n+1), p_{m+1}, \ldots, p_d\}$$

must vanish on $p_{m+1}(l_{m+1})$. By lemma 3.5 this implies that $m + d \geq 3n+1$, with equality if and only if all the points are on a rational normal curve. □

We are now in a position to establish the

Corollary 3.7. *A collection* Γ *of* d α*-dots in weak linear general position in* $\mathbb{P}^n$ *imposes at least* $\min(3n+1, 2d) + (\alpha - 2)\min(2n-1, d)$ *conditions on cubics. We have equality for* $2d \geq 3n+2$ *if and only if the reduced points of* Γ *lie on a rational normal curve, and at each reduced point* p *of* Γ *the tangent direction to that curve at* p *is contained in the Zariski tangent space to* Γ *at* p.

Proof. We have just established the first part of the lemma in case $\alpha = n+1$. For $\alpha \leq n+1$, we may include Γ in the scheme Γ_+ consisting of the $(n+1)$-dots over the reduced points of Γ. Then, as we have just seen, any subscheme of Γ_+ consisting of the $(n+1)$-dots at some $q_1, \ldots, q_{n+2}$ together with general n-dots at $p_1, \ldots, p_{n-3}$ imposes at least $2n^2 + 2$ conditions on cubics. So, if we take those α-dots at $q_1, \ldots, q_{n+2}$ contained in Γ and general $(\alpha-1)$-dots at $p_1, \ldots, p_{n-3}$ contained in Γ then this subscheme is cubically independent. Indeed, by considering Γ_+, we get that either Γ_{red} lies on a rational normal curve and Γ contains the tangent arrow to the curve at each point or else a subscheme of Γ consisting of $n+3$ α-dots and $n-4$ $(\alpha-1)$-dots is cubically independent. If $d \leq 2n-1$ then apply the same argument to the result of lemma 3.6. □

Induction Step

Assume that a subscheme of Γ consisting of $n+2$ dots of degree α and $(r-2)n-3$ dots of degree $(\alpha-1)$ imposes independent conditions in degree $r-1$. By the demotion lemma, if $d \geq (r-1)n-1$ then a subscheme consisting of two α-dots, n

reduced points, and $(r-1)n-3$ $(\alpha-1)$-dots also imposes independent conditions in degree $r-1$. (To see this, for $t=0$ to $n-1$ apply corollary 2.4 — with $l=0$ — to a subscheme of Γ consisting of t reduced points, 2 α-dots and $(r-1)n-3$ $(\alpha-1)$-dots. By hypothesis, such a subscheme is $(r-1)$-independent for $t=0$.)

So, by promotion, $n+2$ α-dots and $(r-1)n-3$ $(\alpha-1)$-dots imposes independent conditions in degree r. Thus, if $d \geq (r-1)n-1$ we have

$$h_\Gamma(r) \geq rn+1+(\alpha-2)((r-1)n-1).$$

If $(r-2)n-1 \leq d < (r-1)n-1$, let $s=d-(r-2)n+1$. Then at least $n+2-s$ α-dots, s reduced points, and $d-n-2$ $(\alpha-1)$-dots is $(r-1)$-independent, according to lemma 3.8. Now in degree r let us promote the s reduced points and $n-s$ of the $(\alpha-1)$-dots to get an r-independent subscheme consisting of $2n+2-s$ α-dots and $d-(2n+2-s)$ $(\alpha-1)$-dots for a total of

$$\begin{aligned} 2n-s+2+(\alpha-1)d &= 2n-d+(r-2)n+1+(\alpha-1)d \\ &= rn+1+(\alpha-2)d \end{aligned}$$

conditions. For each subsequent r, we can promote n of the $(\alpha-1)$-dots, hence for every r with $d \leq (r-1)n-1$ we get

$$h_\Gamma(r) \geq \min(rn+1+(\alpha-2)d, \alpha d).$$

Using a similar argument together with corollary 3.7 we see that if $rn+1 \leq 2d$ the latter bound can be attained only if Γ_{red} lies on a rational normal curve as before.

So, we get the

Theorem 3.8. *Let Γ be a collection of $d \geq n+1$ α-dots in weak linear general position in $\mathbb{P}^n$, $\alpha \geq 2$. Then the Hilbert function h_Γ of Γ satisfies*

$$h_\Gamma \geq \min(rn+1, 2d) + (\alpha-2)\min((r-1)n-1, d),$$

for $r \geq 3$. Equality occurs for some r with $rn+2 \leq 2d$ only if Γ_{red} is contained in a rational normal curve C, and the tangent directions to this curve at these points are all contained in Γ.

4. Ropes over Rational Normal Curves

In the previous section we obtained a lower bound for the Hilbert function of a collection Γ of dots in weak linear general position and showed that if that bound is attained then the reduced points of Γ lie on a rational normal curve. Let us now see that equality can actually occur, by looking at collections of dots whose reduced points lie on a rational normal curve.

Let C be a rational normal curve in $\mathbb{P}^n$, and let $\mathcal{N}$ be the normal bundle of C in $\mathbb{P}^n$. For any $p \in C$ we may define a line bundle $\mathcal{L}_p$ so that a section of $\mathcal{L}_p$ over a point $q \in C, q \neq p$ is the line between p and q in $\mathbb{P}^n$. Then, if $p_1, \ldots, p_{n-1} \in C$

are distinct points then $\mathcal{N} = \mathcal{L}_{p_1} \oplus \ldots \oplus \mathcal{L}_{p_{n-1}}$. We have $\deg \mathcal{L}_p = n+2$ and $\mathcal{L}_p$ has the maximal degree $n+2$ among all line subbundles of $\mathcal{N}$.

Consider the ribbon X associated to $\mathcal{L}_p$. Then $\deg X = 2n$ and $g(X) = n+1$ so the Hilbert polynomial of X is $p_X(k) = (2k-1)n$. On the other hand, by taking $n+1$ general 3-dots on X we see that $h_X(2) = 3n$, and hence $h_X(r) = p_X(r)$ for $r \geq 2$.

We refer to the ribbon X as a "canonical" ribbon over C. Indeed, X is a canonical curve, except for the fact that it is nonreduced. That is, X is a one-dimensional nondegenerate subscheme of $\mathbb{P}^n$ of degree $2n$ and genus $n+1$.

Now, take the "canonical" ribbons $X_1, \ldots, X_{n-1}$ corresponding to direct summands $\mathcal{L}_{p_1}, \ldots, \mathcal{L}_{p_{n-1}}$. Let $W_1, \ldots, W_{n-1}$ be the spaces of quadrics vanishing on $X_1, \ldots, X_{n-1}$ respectively. If W is the vector space of quadrics vanishing on C then we have

$$\begin{aligned}
\operatorname{codim}(W_i) &= n-1 \\
\operatorname{codim}(W_i \cap W_j) &\geq 2n-3 \\
\operatorname{codim}(W_{i_1} \cap \ldots \cap W_{i_j}) &\geq \sum_{k=n-j}^{n-1} k \\
&\vdots \\
\operatorname{codim}(W_{i_1} \cap \ldots \cap W_{i_{n-3}}) &\geq \dim W - 3 \\
\operatorname{codim}(W_{i_1} \cap \ldots \cap W_{i_{n-2}}) &= \dim W - 1, \text{ and} \\
\operatorname{codim}(W_1 \cap \ldots \cap W_{n-1}) &= \dim W.
\end{aligned}$$

(That we really have equality in the last two inequalities is a consequence of the $\mathbb{P}^3$-lemma and its corollary in section 3.) It follows (from elementary linear algebra) that *each* of these inequalities is an equality. So an α-rope X_α corresponding to $\alpha - 1$ direct summands $\mathcal{L}_{p_i}$ has Hilbert function

$$h_{X_\alpha}(2) = p_{X_\alpha}(2) - \binom{\alpha-2}{2}$$

which is exactly the lower bound that we found for the Hilbert function of $(\alpha+1)$-dots. A similar calculation shows that

$$h_{X_\alpha}(r) = p_{X_\alpha}(r)$$

for $r \geq 3$.

Thus, a collection of α-dots lying on $X_{\alpha-1}$ will have the Hilbert function given in theorem 3.9. Conversely, suppose that a collection Γ of α-dots has Hilbert function that agrees with the minimal Hilbert function for some r. If $h_\Gamma(r) = h_{X_{\alpha-1}}(r)$ for some r with $(r-1)n - 1 \leq d$ then by our estimate of the growth of h_Γ we have that the same holds for $r = 3$. We have seen that a collection of dots in weak linear general position having this Hilbert function must have reduced points that lie on a rational normal curve C. We may then find an $(\alpha - 1)$-rope X containing some $\min(2n, d)$ α-dots of Γ.

Suppose $C \subset \mathbb{P}^n$ is a rational normal curve and that $p_1, \ldots, p_{2n} \in C$ are points on C. Let $\Gamma = \{p_1(V_1), \ldots, p_{2n}(V_{2n})\}$ be a collection of 3-dots at the points p_i.

Lemma 4.1. *There is a ribbon X supported on C containing Γ, corresponding to a line bundle $\mathcal{L} \cong \mathcal{O}_{\mathbb{P}^1}(m)$, where $2 \le m \le n+2$.*

Proof. To find such a ribbon X of genus $m-1$ over C is to find a line bundle $\mathcal{L} \cong \mathcal{O}_{\mathbb{P}^1}(m) \subset \mathcal{N}$ for which $\mathcal{L}|_{p_i} = V_i$.

Since $\mathcal{N} \cong \mathcal{O}_{\mathbb{P}^1}(n+2) \otimes V$, where $\dim V = n-1$, the data of such a line bundle are equivalent to that of its Gauss map $\phi : \mathbb{P}^1 \to \mathbb{P}V = \mathbb{P}^{n-2}$ where $\phi(p_i) = [V_i]$, $i = 1, \ldots, 2n$ which is then of degree $d = \deg \phi = n + 2 - m$.

We seek a $(n-1)$-tuple $\phi = (\phi_1, \ldots, \phi_{n-1})$ of homogeneous polynomials on $\mathbb{P}^1$, each of degree d (or less), having prescribed values at $2n$ points.

Taking each pair of coordinates separately, then we look for some $(n-2)$ maps from $\mathbb{P}^1$ to $\mathbb{P}^1$ with prescribed values at $2n$ points. Each time this is possible if the degree d has $2(d+1) > 2n$ or $d \ge n$. Hence we can find an appropriate Gauss map of degree $d \le n$, i.e. $m = n + 2 - d \ge 2$.

We have

$$\begin{array}{ccccc}
I_C(3) & \longrightarrow & H^0(\mathcal{N}^* \otimes \mathcal{O}_{\mathbb{P}^n}(3)) & \xrightarrow{\text{rest}} & H^0(\mathcal{L}^* \otimes \mathcal{O}_{\mathbb{P}^n}(3)) \\
 & & \| & & \| \\
 & & H^0(\mathcal{N}^*(3n)) & \longrightarrow & H^0(\mathcal{L}^*(3n)) \\
 & & \| & & \| \\
 & & H^0(\mathcal{O}_{\mathbb{P}^1}(n+2) \otimes V) & \longrightarrow & H^0(\mathcal{O}_{\mathbb{P}^1}(3n-m))
\end{array}$$

We have seen that the map $I_C(3) \xrightarrow{\text{rest}} H^0(\mathcal{N}^* \otimes \mathcal{O}_{\mathbb{P}^n}(3))$ is surjective (since the Hilbert function of the n-rope over C agrees with its Hilbert polynomial in degree 3), whereas the map $H^0(\mathcal{N}^*(3n)) \to H^0(\mathcal{L}^*(3n))$ (multiplication) is surjective for $m \ge 2$.

Now since the map $I_C(3) \to H^0(\mathcal{O}_{\mathbb{P}^1}(3n-m))$ is surjective, then the points $p_1, \ldots, p_{2n}$ fail to impose independent conditions on the linear series $|\mathcal{O}_{\mathbb{P}^1}(3n-m)|$ if and only if $3n - m \le 2n - 2$ i.e. $m \ge n+2$.

We may put any $2n$ dots of Γ on an $(\alpha - 1)$-rope X corresponding to some subbundle $\mathcal{E} \cong \mathcal{O}_{\mathbb{P}^1}(m_1) \oplus \ldots \mathcal{O}_{\mathbb{P}^1}(m_{\alpha-1})$ of the normal bundle of C in $\mathbb{P}^n$. Then

$$2\alpha n - h_\Sigma(3) = \#\{i : m_i = n+2\}$$

and in particular $h_\Gamma = 3n + 1 + (\alpha - 2)(2n-1)$ if and only if Γ is contained in a ribbon corresponding to a direct sum of $\mathcal{O}_{\mathbb{P}^1}(n+2)$'s.

Likewise, if any one of these direct summands fails to be $\mathcal{O}_{\mathbb{P}^1}(n+2)$, then we have

$$\begin{aligned} h_\Gamma(2) &= h_X(2) \\ &\geq 2n+1+(\alpha-2)(n-1)-\binom{\alpha-2}{2}+1. \end{aligned}$$

□

In summary, we get the

Theorem 4.2. *Let Γ be a collection of d α-dots in weak linear general position in $\mathbb{P}^n$, $\alpha \geq 2$. Then the Hilbert function h_Γ of Γ satisfies*

$$h_\Gamma \geq \min(h_{X_{\alpha-1}}(r), rn+1+(\alpha-2)d, \alpha d)$$

where $X_{\alpha-1}$ is the $(\alpha-1)$-rope described above. We have equality for some r with $(r-1)n \leq d$ if and only if Γ is contained in an $X_{\alpha-1}$. Equality occurs for some r with $rn+2 \leq 2d$ if and only if Γ_{red} is contained in a rational normal curve, and the tangent directions to the curve at these points are all contained in Γ.

References

[AH] J. Alexander, A. Hirschowitz, La méthode d'Horace éclatée: application à l'interpolation en degré quatre. Invent. Math. **107** (1992), 585-602.

[ACGH] E. Arbarello, M. Cornalba, P.A. Griffiths, J. Harris, Geometry of Algebraic Curves, Vol. I. Springer-Verlag, New York, 1985.

[BE] D. Bayer, D. Eisenbud, Rational ribbons. Preprint.

[C] K. Chandler, Hilbert functions of zero-dimensional schemes in uniform position. Ph.D. Thesis, Harvard University, June, 1992.

[EH1] D. Eisenbud, J. Harris, Finite projective schemes in linearly general position. J. Algebraic Geom. **1** (1992), 15-30.

[EH2] D. Eisenbud, J. Harris, Schemes: The Language of Algebraic Geometry, Wadsworth and Brooks, Belmont, California, 1992.

[H1] J. Harris with D. Eisenbud, Curves in Projective Space, Les Presses de l'Université de Montréal, Montréal (Québec), Canada, 1982.

[H2] J. Harris, Algebraic Geometry: A first course, Graduate Texts in Math. 33, Springer-Verlag 1992.

Curves of maximal genus in P^4

Luca Chiantini, Ciro Ciliberto

Abstract. If C is an irreducible curve in P^r of degree d, not lying on surfaces of degree $< s$, then for $d \gg s \gg 0$ we know (see [ccd]) that its genus is bounded by a function G (r.d.s.). We classify here all such curves in P^4 whose genus is maximal.

1991 Mathematics Subject Classification: 14H45.

Introduction

Let $C \subset P^r$ be a reduced, irreducible curve, not contained on surfaces of degree $< s$; when $d = \deg C$ is 'big' with respect to s, the genus g of C is bounded by a function $G(d, r, s)$ which is of the type $(d^2/2s) + O(d)$.

The existence of this bound for $C \subset P^3$ was announced by Halphen in 1870 and proved by Gruson and Peskine in [GP]; for curves in $P^r, r \geq 4$ such a bound is stated and proved in [CCD]. The existence of curves whose genus attains the bound is known in P^3 for $d > s^2 - s$ and in $P^r, r \geq 4$, at least for $d \gg s$.

[GP] contains a precise description of those space curves which do not lie on surfaces of degree $< s$ and whose genus is maximal; they are arithmetically Cohen-Macaulay, lie on a surface of degree s and directly linked to plane curves.

The aim of this note is to give a similar description for curves $C \subset P^4$, not contained on surfaces of degree $< s$, whose genus is $G(d, 4, s)$. We know yet ([CCD] section 6) that such a C must be arithmetically Cohen-Macaulay and it must lie on a surface S of degree s, whose general hyperplane section Γ is a 'Castelnuovo curve' in P^3, i.e. a space curve of maximal genus. As in [GP] for the case of space curves, we describe our curves C of genus $G(d, 4, s)$ in terms of the curve C' obtained by linking C with S and a hypersurface F of minimal degree passing through C and not containing S.

Theorem. *Let $C \subset P^4$ be an integral,non degenerate curve, not contained on surfaces of degree $< s$; put $d = \deg C$ and let $g = G(d, 4, s)$ be its genus. Assume $d > 12s^2$.*

Then C is arithmetically Cohen-Macaulay and lies on a surface S of degree s.

Put $d - 1 = sm + \epsilon, 0 \leq \epsilon < s$ and $s - 1 = 2w + v, v = 0, 1$; also divide $\epsilon = kw + \delta, \delta = 0, \ldots, w - 1$ (note that $k \leq 3$; if $k \geq 2$ then $\delta \leq v$ and $k = 3$ only if $s = 4$, $\epsilon = 3$, $v = \delta$).

Then C is contained on a hypersurface F of degree $m+1$, not passing through S. If C' is the curve linked to C by F and S, we have:

a) *when $k=3$ or $k=2$ and $v=\delta$ then $C'=\emptyset$, i.e. C is complete intersection on S;*

b) *when $k=1$, or $k=2$ and $\delta<v$ (i.e. $v=1, \delta=0$) then C' is a plane curve of degree $(m+1)s-d=s-1-\epsilon$;*

c1) *when $k=0$ and s is even, then C' is a space curve, which lies on a quadric surface;*

c2) *when $k=0$ and s is odd, then the curve C' is contained on a cubic surface in P^4.*

Finally, for all s,d with $d>12s^2$, one can find a curve of degree d, genus $G(d,4,s)$ which does not lie on any surface of degree $<s$.

The proof is based on the analysis of the Hilbert function of a general hyperplane section Z of C.

It should be observed that, while in P^3 all curves of maximal genus are directly linked to degenerate curves ([GP]), this is no longer true in P^4 (case c.2 of the previous classification) as we pointed out in the following:

Example (see [CCD] ex. 6.6). Let $S \subset P^4$ be the rational quintic surface obtained blowing up a plane at 9 points $O, A, P_1, \dots, P_7$ embedded with the linear system $5L-3O-2A-\sum P_i$; S contains a rational normal quartic C', image of a conic passing through O, A, P_1.

For $m \gg 0$, let C be the curve residual to C' in the intersection of S with a general hypersurface of degree $m+1$. We have $d=\deg(C)=5m+1$ and one computes that the genus of C is $G(5m+1,4,5)$; by Bezout C is not contained on surfaces of degree <5; on the other hand, no curve directly linked to C is degenerate.

The paper has been written while the authors were members of CNR - GNSAGA and supported by MURST. This work has been performed in the framework of Europroj's project 'Hyperplane sections'.

Preliminaries

We start recalling a few results of [CCD] which will be used frequently in this paper.

From now on, let C be an integral, nondegenerate curve of degree d and genus g in P^4, with $d>12s^2$. Assume C not contained on surfaces of degree $<s$ and define $m, \epsilon, w, v, k, \delta$ as in the statement of the theorem.

Then the genus g is bounded by the function:

$$G(d,4,s)=1+(1/2)(d(w+m-2)-(m+1)(w-\epsilon)+(w+1)vm-\delta(w-\delta))$$

([CCD] section 5).

If Z is a general hyperplane section of C and h_Z is the Hilbert function of Z, then h_Z must be bigger than the function h whose difference Δh is defined by:

$$\Delta h(n) = \begin{cases} 0 & \text{if } n < 0 \\ 2n+1 & \text{if } 0 \leq n \leq w \\ s & \text{if } w < n \leq m \\ s+k-2(n-m) & \text{if } m < n \leq m+\delta \\ s+k-2(n-m)-1 & \text{if } m+\delta < n \leq w+m+e \\ 0 & \text{if } w+m < n \end{cases}$$

and $e = 0$ if $\epsilon < w(3-v)$ while $e = 1$ otherwise.

Proposition 1. *If $g = G(d,4,s)$, then $\Delta h_Z(n) = \Delta h(n)$ for all n and C is arithmetically Cohen-Macaulay.*

Moreover Z is contained on a space curve Γ of degree s and (maximal) genus $w^2-(1-v)w$. Since $d > s^2$, Γ is unique and, when we move the hyperplane, all this curves Γ's patch together giving a surface $S \subset P^4$ of degree s through C.

Proof. See [CCD] 0.1, 6.1, 6.2, 6.3.

S is a 'Castelnuovo surface' in P^4, i.e. a surface whose general hyperplane section is a curve of maximal genus in P^3.

Proposition 2. *S is irreducible.*
When s is even, then S is complete intersection of type $2, w+1$.
When s is odd, S is linked to a plane by a complete intersection of type $2, w+1$.

Proof. Everything follows from the assumption $d \gg s^2$ and the fact that C is arithmetically Cohen-Macaulay.

S is irreducible for reasons of degree, since C is irreducible.

A general hyperplane section Γ of S lies on a quadric surface Y which contains Z; it must lift to a quadric hypersurface Q containing C, hence also S.

Finally, Γ is a Castelnuovo curve, hence it lies on a surface of degree $w+1$ which does not contain Y; such a surface must lift to a hypersurface of degree $w+1$ in P^4, containing C, hence containing S, for reasons of degree.

Observe that, for s odd, the quadric through S cannot be smooth.

Proposition 3. *There exists a hypersurface F of degree $m+1$, passing through C and not containing S.*

Proof. For a general hyperplane section Γ of S, the Hilbert function h_Γ is known (see e.g. [EH]) so one computes:

$$h_Z(m+1) = h(m+1) < h_\Gamma(m+1).$$

The classification

Definition 4. Call C' the curve linked to C by S, F; $\deg C' = s - \epsilon - 1$.

Call C'' the curve linked to C by hypersurfaces of degree $2, w+1$ passing through S and by F; $\deg C'' = s - \epsilon + m$ and C'' is arithmetically Cohen-Macaulay.

Of course $C' \subset C''$. In fact:

$$\begin{array}{ll} \text{for } s \text{ even} & C'' = C' \\ \text{for } s \text{ odd} & C'' = C' \cup \text{ plane curve of degree } m+1. \end{array}$$

Call Z', Z'' general hyperplane sections of C', C'' respectively. Z'' is linked to Z by surfaces of degree $2, m+1, w+1$ while Z' is linked to Z by a surface of degree $m+1$ and a Castelnuovo curve Γ in P^3. We have:

$$\begin{array}{ll} \text{for } s \text{ even} & Z'' = Z' \\ \text{for } s \text{ odd} & Z'' = Z' \cup m+1 \text{ points on a line.} \end{array}$$

Lemma 5. *Let Y be a set of points in P^3 contained in the smooth part of an irreducible quadric surface Q. Let Y be linked, on Q, by surfaces of degree b, c, to a 0-dimensional scheme Y' which is locally Cohen-Macaulay.*

Then for $i \leq b, c$ one has:

$$h^1\mathcal{I}_{Y,Q}(b+c-2-i) = h^0\mathcal{I}_{Y',Q}(i).$$

Proof. Take a resolution of $\mathcal{I}_{Y',Q}$:

$$0 \to E \to F \to \mathcal{I}_{Y',Q} \to 0$$

where $F = \oplus\mathcal{O}_Q(-\alpha_i)$ and E is a vector bundle such that $h^1E(n) = 0\ \forall n$. By the mapping cone, one gets a resolution

$$0 \to F^*(-b-c) \to E^*(-b-c) \oplus \mathcal{O}_Q(-b) \oplus \mathcal{O}_Q(-c) \to \mathcal{I}_{Y,Q} \to 0.$$

For $i \leq b, c$, $h^1\mathcal{I}_{Y,Q}(b+c-2-i) = r$ implies $\dim \operatorname{Ker}(H^2F^*(-2-i) \to H^2E^*(-2-i)) = r$ hence also, by duality, $\dim \operatorname{Coker}(H^0E(i) \to H^0F(i)) = r$.

Remark 6. Since C is irreducible, taking general hyperplane sections we find that Z is contained in the smooth part of an irreducible quadric surface in P^3 and Z'' is disjoint from Z, hence it is locally complete intersection.

Remark 7. If Y is any set of points in P^n, we have

$$h^1\mathcal{I}_Y(n) = \sum_{i=n+1}^{\infty} \Delta h_Y(i).$$

Indeed

$$\begin{aligned} h^1\mathcal{I}_Y(n) &= d - h_Y(n) = d + \Delta h_Y(n+1) - h_Y(n+1) = \\ &= d + \textstyle\sum_{i=n+1}^{t} \Delta h_Y(i) - h_Y(t) = \sum_{i=n+1}^{\infty} \Delta h_Y(i) \end{aligned}$$

for large t.

Observe that $h^1\mathcal{I}_Z(n) = h^1\mathcal{I}_{Z,Q}(n)$ for all n, by $0 \to \mathcal{I}_Q \to \mathcal{I}_Z \to \mathcal{I}_{Z,Q} \to 0$.

PROOF OF THE THEOREM.

a) $k = 2$ and $v = \delta$.
We have $d = (m+1)s$, hence $\epsilon = s - 1$. It follows $\deg C' = s - \epsilon - 1 = 0$ and we are done.

b1) $k = 1$, s even or $k = 2, v = 1, \delta = 0$.
Here $s = 2w + 2$. If $k = 2$ and $v = 1$ then one computes $\deg C' = 1$ and we are done. Assume $w \leq \epsilon < 2w$. By remark 6, remark 7 and lemma 5:

$$h^1\mathcal{I}_Z(w+m-1) = \sum_{j=w+m}^{\infty} \Delta h_Z(j) = \Delta h(w+m) = 2$$

thus $h^0\mathcal{I}_{Z''}(1) = 2$ and Z'' is contained on a line. Since C'' is arithmetically Cohen-Macaulay, it is a plane curve; here $C' = C''$.

b2) $k = 1$, s odd.
Here $s = 2w+1$, $w \leq \epsilon < 2w$; further $h^1\mathcal{I}_Z(w+m-1) = 1$ implies $h^0\mathcal{I}_{Z''}(1) = 1$; we also have $h^1\mathcal{I}_Z(w+m-2) \geq 4$ so that $h^0\mathcal{I}_{Z'',Q}(2)) \geq 4$ and Z'' lies on some conic Λ. Remind that $Z'' = Z' \cup m+1$ points on a line and $m > 2$ otherwise we would have $d \leq 2s$; it follows that Λ splits in two lines.
C'' is thus contained on a reducible quadric surface; on the other hand, C'' is contained on a surface of degree $m+1$ and $C'' = C' \cup$ a plane curve of degree $m+1$. It follows that C' lies on a plane.

c1) $k = 0$, s even.
We have $\epsilon < w$, $h^1\mathcal{I}_Z(w+m-1) = 1$ and $h^1\mathcal{I}_Z(w+m-2) \geq 4$, so Z'' lies on some conic. This means that $C' = C''$ lies on some quadric surface and it is degenerate. Something more can be said about the structure of C' in this case (see prop.8)

c2) $k = 0$, $s = 2w+1$ odd.
As we saw in ex.6.6 of [CCD], here C' need not to lie on a hyperplane.
We have $h^1\mathcal{I}_Z(m+w-2) \geq 2$, so that Z'' lies on 2 independent quadrics; since Z'' contains $m+1$ aligned points and $m > 1$, these 2 quadrics meet along a line l; it follows that C'', which is arithmetically Cohen-Macaulay, lies on 2 independent quadric hypersurfaces which have a common plane π. Hence C' lives on a cubic surface directly linked to π, for $C'' - C'$ is a curve of degree $m+1$ on π and C'' is contained on a hypersurface of degree $m+1$, not containing some quadric through C''.

We give a closer description for cases c1) and c2) above.

Proposition 8. *When $k = 0$ and $s = 2w+2$ is even, either:*

i) *$\epsilon = 0$ and C' is directly linked to a line;*

ii) *$\epsilon = 1$ and C' is directly linked to a conic;*

iii) *$\epsilon > 1$ and C' splits in the union of 2 plane curves of degree $w - \epsilon$ and $w+1$, meeting at $w - \epsilon$ points.*

Proof. Let S, Γ be as usual. We know from the theorem that the surfaces of degree $m+1$ through Z cut on Γ a linear series γ of degree $s-\epsilon-1$, whose divisors are contained on planes. Since Γ is arithmetically Cohen-Macaulay, it follows that γ is cut out on Γ by the linear system of planes through $\epsilon+1$ points which have to lie on a line l if $\epsilon > 1$. When $\epsilon > 1$, l must lie on the quadric surface Q containing Γ, so l hits Γ in a set of degree $w+1$, hence it contains $w-\epsilon$ points of Z'; by construction, the remaining $w+1$ points of Z' lie on another line.

Since $C' = C''$ is arithmetically Cohen-Macaulay, the claim follows.

For the case c2), one has:

Proposition 9. *Let $k=0$, $s=2w+1$.*

If either $w+\epsilon+1>4$ or $\epsilon=0$ and $w>2$, then all cubic surfaces containing C' are reducible.

If $\epsilon>1$, then C' is the union of a plane curve of degree w or $w+1$ with a curve of degree $w-\epsilon$ or $w-\epsilon-1$, lying on some surface of degree 2.

Proof. Call Q an irreducible quadric surface which contains a general hyperplane section Γ of S. Γ is arithmetically Cohen-Macaulay and it is of type $(w+1, w)$ if Q is smooth. Look at the linear series γ cut on Γ by surfaces of degree $m+1$ through Z; by the description of h_Γ, h_Z, it follows that γ is at least a pencil and $\dim \gamma \geq 2$ when $\epsilon=0$, for in this case also $\delta=0$. By the theorem, all the divisors of γ lie on a cubic curve contained in Q; these curves are of type $(2,1)$ when Q is smooth.

The linear series cut on Γ by such cubic curves is thus complete; so these curves meet Γ in another fixed set Z_0 of degree $(3w+1)-(s-\epsilon-1)=w+\epsilon+1$. When $w+\epsilon+1>4$ or $w+\epsilon+1>3$ and $\epsilon=0$, they must have a fixed part X.

If $\deg X = 1$, then the moving part is formed by conics passing through (at most) 2 fixed points; it follows $\epsilon \leq 1$.

Otherwise, the cubics split in Y plus a moving line l which meets Γ in w or $w+1$ points (accordingly with the type of l on Q) which must lie in Z'; hence Z' has w or $w+1$ points on a line.

The surface of degree 2 of the previous statement may be reducible in 2 planes which span P^4.

Remark. It follows easily, from the previous classification, that in case c2), the example 6.6 of [CCD], mentioned in the introduction, is (numerically) the only one in which C' is non-degenerate and can be realized on an irreducible cubic surface.

Using the classification of Castelnuovo surfaces given in [H], one can prove that when S is smooth, also for s odd one has C' arithmetically Cohen-Macaulay.

Lastly, we need an effective construction for curves of degree d, genus $G(d,4,s)$ in P^4, not lying on surfaces of degree $< s$.

Example 10. For all $d \geq 12s^2$, there is an irreducible curve $C \subset P^4$ of degree d and genus $G(d, 4, s)$ contained on an irreducible surface S of degree s.

By Bezout, such a curve cannot lie on surfaces of degree $< s$.

a) $k = 2$ and $\delta = v$.
In this case, take S to be complete intersection of type $2, w+1$ (when s is even) or linked to a plane by hypersurfaces of degree $2, w+1$. In both cases the complete intersection of S with a general hypersurface of degree $m+1$ gives the required curve.

b1) $k \geq 1$, s even.
Take a plane curve C' of degree $s - \epsilon - 1$ and let C be a curve linked to C' by general hypersurfaces of degree $2, w+1, m+1$. We may get C irreducible, for C' is cut in degree $w+1$. C lies on a surface of degree $s = 2w+2$ and it has the required numerical characters.

b2) $k = 1$, s odd.
Let S_0 be the union of two planes in P^4, meeting along a line. Let $C'' \subset S_0$ be the union of two plane curves, C' and C_0' of degree $s - \epsilon - 1, m+1$, meeting at $s - \epsilon - 1$ points. By construction C'' is arithmetically Cohen-Macaulay, since $m+1 > s - \epsilon - 1$.
Take general surfaces F_0, F_1, F of degree $2, w+1, m+1$ passing through C'' and let C be the residual curve. $F_0 \cap F_1$ must contain the plane of C_0', since $m+1 > w+1$, thus C lies on an irreducible surface S of degree $s = 2w+1$ and since C' is cut in degree $m+1$, we may take C irreducible. Then C is the required curve.

c1) $k = 0$, s even.
Let S_0 be the union of two planes, meeting at a line. Let $C' \subset S_0$ be the union of two plane curves of degree $w - \epsilon, w+1$, meeting at $w - \epsilon$ points. By construction, C' is arithmetically Cohen-Macaulay and it is cut in P^4 by hypersurfaces of degree $w+1$. If C is linked to C' by a general complete intersection of type $2, w+1, m+1$, then C lies on an irreducible surface of degree s and it has the required properties.

c2) $k = 0$, s odd.
In this case, one may take $S_0 =$ union of a smooth quadric surface q and a plane π which span P^4 and meet along a line l. Take further $C' =$ union of a curve of degree w on π and an arithmetically normal curve C_0 of degree $w - \epsilon$ on q, which meet at $w - \epsilon$ points. Let C'' be the union of C' with a plane curve of degree $m+1$, meeting C' at $s - \epsilon - 1$ points, $w - \epsilon$ of which in C_0. C'' is arithmetically Cohen-Macaulay.
Link C'' with general hypersurfaces of degree $2, w+1, m+1$; as in case b2), one gets a curve $C$with the required properties.

We built C' as a reducible curve in case c1) and c2); as we saw in prop.8 or in ex.6.6 of [CCD], C' need not to be reducible for some value of s, even in the case $k = 0$.

Example 11. In case c2) there are alternative ways for constructing curves of maximal genus.

For instance, take 4 planes $\alpha_1, \alpha_2, \beta_1, \beta_2$ in P^4, such that $\alpha_i \cap \beta_j$ =line, while $\alpha_1 \cup \alpha_2$ and $\beta_1 \cup \beta_2$ span P^4. Take a number z with $0 < z < w - \epsilon$ and let E_1, E_2 be curves of degree $w - \epsilon - z, z$ on α_1, α_2 respectively. Let $E_3 \subset \beta_1$ be a curve of degree w which meets E_1, E_2 in $w - \epsilon - z, z$ points respectively. Finally let $E_4 \subset \beta_2$ be a curve of degree $m+1$ meeting E_1, E_2 in $w - \epsilon - z, z$ points respectively.

Put $C' = E_1 \cup E_2 \cup E_3$ and $C'' = C' \cup E_4$; then C' is cut by hypersurfaces of degree $w+1$ and C'' is cut by hypersurfaces of degree $m+1$. C'' is arithmetically Cohen-Macaulay. Linking C'', by means of hypersurfaces of degree $2, w+1, m+1$, to an irreducible curve of degree C one gets a curve with the required properties.

More could be said about curves of maximal genus in case c2), with a closer analisys of the possible construction of the residual curve C', which would require a long list of cases, which we do not intend to go through.

References

[CCD] L. Chiantini, C. Ciliberto, V. Di Gennaro. The genus of projective curves; to appear in Duke Math. J.

[EH] J. Harris, D. Eisenbud. Curves in projective spaces. Presses de l'Université de Montreal (1982).

[H] J. Harris. A bound on the geometric genus of projective varieties. Ann. Scuola Norm. Sup. Pisa (1981), vol.(4), 35–68.

[GP] L. Gruson, C. Peskine. Genre des courbes dans l'espace projectif. Springer Lecture Notes, vol. 687 (1978).

Graph Curves, Colorings, and Matroids

Ciro Ciliberto and Rick Miranda

Abstract. Coloring properties of planar maps determined by planar graphs are examined in terms of algebro-geometric properties of a degenerate algebraic curve associated to the graph. In particular, q-colorings of the planar map are shown to be equivalent.

1991 Mathematics Subject Classification: 05C14, 14H99, 14G15.

1. Introduction

In this article we would like to present some relationships between the theory of graphs and the theory of algebraic curves. To every simple trivalent graph G one can associate a (degenerate) algebraic curve C_G, which is a union of lines (one for each vertex of G) joined as the edges of G indicate.

These so-called "graph curves" have been useful in proving various theorems about Riemann surfaces, since they are natural degenerations of smooth Riemann surfaces, and any semi-continuous property of curves which is maximal for any graph curve must be maximal for the general Riemann surface also. This was the basis of the proof in [5] that the general curve of genus $g \geq 10$, $g \neq 11$ had a surjective Gaussian map: for these genera, there exists a graph curve for which the map is surjective. See also [11] for a discussion of graph curves and Gaussian maps in general, [3], [4], and [10] for computations involving gaussian maps and graph curves, and [8] for applications to Prym-canonical curves. The reader may also wish to consult [6], [7], and [9] for a discussion of Gaussian maps for various families of curves. A preliminary version of parts of this article appeared in [12]. The basic reference for Gaussian maps is Wahl's paper [14].

This article focuses on the colorability problems for planar graphs, and we show that colorings are related to the existence of certain regular 1-forms on the graph curve. We obtain no new results, but we do put into one package a set of coloring criteria which a priori are quite different. The results always reduce the coloring of the planar map to a certain vertex-edge assignment problem in a finite field. We also develop a relationship between certain matroid representations and the Picard Group of the graph curve. In particular the component of the Picard Group containing the canonical bundle is naturally associated to certain representations

Research supported in part by the National Science Foundation under grant DMS-9104058.

of matroids defined on the edges of the graph, for which the stars of the vertices are dependent sets.

The authors would like to thank David Eisenbud and Joe Harris for several useful conversations.

2. Graphs and Graph Curves

Let G be a connected graph. Formally, this would be a pair (V, E), where V is a set (of vertices) and E is the set of edges. Each edge e can be viewed as a map from an ordered two-point set to V, representing the vertices at the ends of e. The two vertices at the ends of e will be denoted by e_1 and e_2. The edge e is a *loop* if both ends are at the same vertex. We allow multiple edges, so that there may be many edges joining two vertices. The graph G will be called *simple* if it has no loops or multiple edges; this is an important special case.

The *valency* of a vertex $v \in V$ is the number of edges containing v. A loop at a vertex v would count for 2 in the valency of v. A vertex is *trivalent* if there are exactly 3 edges containing it, and it is *sub-trivalent* if there are at most 3 edges containing it. A graph G is *trivalent*, respectively *sub-trivalent*, if every one of its vertices are.

Denote by $\text{val}(v)$ the valency of a vertex v. Thus a connected subtrivalent graph satisfies $1 \leq \text{val}(v) \leq 3$ for all vertices v.

The *arithmetic genus* (or simply *genus*) of a connected sub-trivalent graph is the integer

$$g = |E| - |V| + 1.$$

Note that this is always a non-negative integer. Also note that this definition differs from the graph-theoretic use of the word genus. If G is actually trivalent, then we have

$$3|V| = 2|E|$$

and so $|V| = 2g - 2$ and $|E| = 3g - 3$. Note in this case that there are exactly $6g - 6$ incident vertes-edge pairs (v, e).

Fix a field $\mathbf{K}$. To every connected sub-trivalent graph G we can associate a connected degenerate algebraic curve C_G defined over $\mathbf{K}$ as follows. To every vertex v of G associate an abstract projective line $C_v \cong \mathbf{P}^1_{\mathbf{K}}$. Mark $\text{val}(v)$ points on each C_v, corresponding to the $\text{val}(v)$ edges of G which meet at the vertex v. (Since the automorphism group of $\mathbf{P}^1$ acts triply transitively on the points, there is a unique way to do this, up to isomorphism of the line C_v. Indeed, we may choose a coordinate on C_v so that these $\text{val}(v)$ points are, in some order, $\{0\}$ or $\{0, 1\}$ or $\{0, 1, \infty\}$.) Let e be an edge of G joining vertices v and w. Therefore on the line C_v there is a marked point corresponding to e, as there is on the line C_w. (If e is a loop, then there are two marked points on C_v corresponding to e.) For each such edge, glue the line C_v to the line C_w transversally at these marked points associated to the edge e. This union of lines, glued via the edge data, is the curve C_G.

The curve C_G is an algebraic curve, reducible of course, and is called the *graph curve* associated to the connected sub-trivalent graph G. It is connected, since G is. It is an easy exercise to check that the graph curve C_G associated to a connected sub-trivalent graph G with genus g has arithmetic genus g; this is the source of the notation. Moreover if G is trivalent, it is a stable curve in the sense of Deligne and Mumford.

Note also that the construction of C_G does not depend on the orientation of the edges of G. The graph curve is actually associated to the unoriented graph.

The "smallest" simple trivalent graph curve constructible in this way is the only graph curve of arithmetic genus 3, consisting of 4 vertices and all six possible edges joining any 2 of the 4. This graph is the 1-skeleton of the tetrahedron. Its associated graph curve can be realized in the plane as 4 general lines (each meeting the other three).

Note that if a vertex v has a loop at it, then in the construction of the graph curve C_G, one would glue C_v to itself at the two points $x_1(e)$ and $x_2(e)$ corresponding to the loop, creating a rational nodal curve $\overline{C}_v$. This would then be glued to another line C_w corresponding to the third edge (unless it is the only vertex.)

Secondly, suppose that G has multiple edges joining vertices v and w. Then, in the construction of the graph curve C_G, the lines C_v and C_w would be glued together at more than one point.

In any case we see that each component of a graph curve is either a line or a rational nodal curve. If a component is a line, it meets at most three others; if a component is a rational nodal curve, it meets at most one other.

The smallest trivalent graph curves have arithmetic genus 2, with two vertices v and w. The two possibilities are that there are three edges each joining v to w, or that there are loops at both v and w with one edge joining them. The first has associated graph curve equal to two lines meeting transversally at 3 points; the second is two rational nodal curves meeting transversally at one point.

We will use the following notation in general.

G	the graph
C_G	the graph curve associated to G
$\overline{C}_v$	the component of C_G corresponding to the vertex v
C_v	the normalization of $\overline{C}_v$, isomorphic to $\mathbf{P}^1$
n_e	the node of C_G corresponding to the edge e
e_1, e_2	the first and second vertex lying on e
$x_1(e), x_2(e)$	the points of C_{e_1} and C_{e_2} corresponding to the edge e
$x_{v,e}$	the point of C_v corresponding to the non-loop edge e

3. Line Bundles on Graph Curves

Let L be a line bundle on a graph curve C_G. By restriction and pull-back, we obtain line bundles L_v on the normalization C_v of each component $\overline{C}_v$ of C_G. (I.e., we restrict, and in case there are loops in G, so that the component $\overline{C}_v$ is a nodal rational curve, we also pull back L to the normalization C_v.) The bundle L_v is a line bundle on a $\mathbf{P}^1$, and therefore has a degree, which we will denote by $d_v(L)$. The *total degree* of the line bundle L on C_G is the $|V|$-tuple of integers $(d_v(L))_{v\in V}$, or, equivalently, the function $d(L) : V \to \mathbf{Z}$, sending v to $d_v(L)$.

We therefore see that the Picard group of C_G decomposes into components, each component corresponding to a possible total degree. The connected component of the identity, $\mathrm{Pic}^0(C_G)$, forms a connected algebraic group over $\mathbf{K}$, which we describe next.

The total degree does not in general determine the line bundle L. The individual bundles L_v on the components C_v are of course determined by their degrees d_v; however they must be glued together, and this glueing data must be specified.

To specify this glueing data for elements of $\mathrm{Pic}^0(C_G)$, note that any line bundle $L \in \mathrm{Pic}^0(C_G)$ has degree 0 on each component. Therefore we may choose an isomorphism

$$\phi_v : L_v \to C_v \times \mathbf{K}$$

for each vertex v. This in particular induces an isomorphism of each fiber of L_v with $\mathbf{K}$, which we will also denote by ϕ_v.

Now recall that each edge is oriented. This is one place in the theory where an orientation helps! (Its affect will disappear in the end.)

At each node n_e of C_G, corresponding to an edge e of G, the 1-dimensional vector spaces $L_{e_1}|_{x_1(e)}$ and $L_{e_2}|_{x_2(e)}$ are glued together, via a linear isomorphism, to form L. Let us denote this fiber $L_{e_i}|_{x_i(e)}$ of the bundle L_v by simply $L_i(e)$. The glueing data for these two 1-dimensional spaces is then determined by a nonzero number λ_e in the field $\mathbf{K}$; the glueing isomorphism is the map

$$t_e = {\phi_{e_2}}^{-1} \circ \lambda_e \circ \phi_{e_1} : L_1(e) \to L_2(e)$$

where λ_e in the above expression of course means multiplication by the constant λ_e in $\mathbf{K}$.

Therefore we see that glueing data is determined by specifying, for each edge e, a non-zero element λ_e in the field $\mathbf{K}$. This is essentially a 1-cochain for the graph G, with values in the multiplicative group $\mathbf{K}^\times$.

The only choices made in the above assignments was the choice of isomorphism ϕ_v. Suppose the isomorphism ϕ_v changes; it can only change by multiplying by a non-zero scalar μ_v. That is, the isomorphism ϕ_v may be replaced by an isomorphism $\phi'_v = \mu_v\phi_v$. In this case, the glueing isomorphism becomes

$$\begin{aligned} t_e &= {\phi_{e_2}}^{-1} \circ \lambda_e \circ \phi_{e_1} \\ &= \mu_{e_2}{\phi'_{e_2}}^{-1} \circ \lambda_e \circ {\mu_{e_1}}^{-1}\phi'_{e_1} \\ &= {\phi'_{e_2}}^{-1} \circ \lambda'_e \circ \phi'_{e_1} \end{aligned}$$

where $\lambda'_e = \mu_{e_2}\mu_{e_1}{}^{-1}\lambda_e$. Thus under this change, the set of glueing constants $\{\lambda_e\}$ is replaced by the set $\{\lambda'_e\}$. We see that what we have is exactly that the 1-cochain $\{\lambda_e\}$ changes by a coboundary, namely the coboundary of the 0-chain $\{\mu_v\}$. Thus we have the following.

Proposition 3.1. *Let G be a connected sub-trivalent graph with associated graph curve C_G. Then*

$$\mathrm{Pic}^0(C_G) \cong H^1(G, \mathbf{K}^\times).$$

It follows from the general theory that, given another possible total degree $\underline{d}$, the line bundles on C_G with total degree $\underline{d}$ form a subset $\mathrm{Pic}^{\underline{d}}(C_G)$ which is a torsor for $\mathrm{Pic}^0(C_G)$. The torsor structure is easy to see. Suppose that L is any line bundle in $\mathrm{Pic}^{\underline{d}}(C_G)$. Then we can vary L, using a cocycle $\{\lambda_e\}$ in $H^1(G, \mathbf{K}^\times)$, by changing the isomorphisms describing the glueing data for L by the λ's.

There is one line bundle, the canonical bundle, which of course has more importance than others. Note that since C_G is a singular curve, we mean the dualizing sheaf ω_{C_G} when we speak of the canonical bundle.

Lemma 3.2. *Let G be a connected sub-trivalent graph. Then the total degree of the dualizing sheaf ω of the associated graph curve C_G is $d_v(\omega) = \mathrm{val}(v) - 2$ for all vertices v. In particular, if G is trivalent, $d_v(\omega) = 1$ for all v.*

Proof. Sections of the dualizing sheaf on C_G correspond to rational 1-forms on each C_v, with only simple poles allowed at the $\mathrm{val}(v)$ identified points $x_i(e)$; furthermore, the residues at these points must cancel in the two components identified there.

Hence we see that the dualizing sheaf ω, when restricted to each C_v, is the sheaf $\omega_{C_v}(D)$, where D is the divisor of the $\mathrm{val}(v)$ points identified on C_v. Since C_v has genus 0, the degree of ω_{C_v} is -2, so that the degree of the restriction of the dualizing sheaf is $\mathrm{val}(v) - 2$. □

4. 1-Forms, Twisted 1-Forms, and Residues on Graph Curves

The sections of the dualizing sheaf ω_{C_G} on a graph curve C_G will be called *regular 1-forms* on C_G, by abuse of language. As noted above, a regular 1-form on C_G is a rational 1-form on each component C_v, with at worst simple poles at the nodes; at the nodes, the two residues must cancel. This residue cancellation is the compatibility condition for patching together the "local" 1-form data. For more details, the reader may consult [2], [5], or [10]. We will denote the space of global sections of ω_{C_G} by Ω_G.

We will concentrate on the case when G is simple and trivalent; the reader can make the necessary modifications in the case where G is sub-trivalent, with possible loops or multiple edges. Recall that in this case C_G is a stable curve of genus g.

A convenient way to represent a 1-form on C_G is by using residues. Let $\omega \in \Omega_G$ be a regular 1-form on C_G. For any component C_v of C_G (corresponding to a vertex v of G), denote by ω_v the restriction of ω to the component C_v. As noted above, ω_v is a rational 1-form on C_v with at worst simple poles at the three points of C_v identified together in C_G. Let e be an edge of G emanating from v, so that n_e is one of the nodes of C_G lying on the component $\overline{C}_v$, and $x_{v,e}$ is one of the three points of C_v identified in C_G. Denote by $\mathrm{Res}_\omega(v,e)$ the residue of ω_v at the point $x_{v,e}$. Thus to every regular 1-form ω we have attached $6g-6$ residue numbers, namely the elements $\mathrm{Res}_\omega(v,e)$ of the field k as (v,e) ranges over the $6g-6$ incident vertex-edge pairs.

Since the sum of the residues of any rational 1-form is zero, we have the equation

$$\mathrm{Res}_\omega(v,e_1)+\mathrm{Res}_\omega(v,e_2)+\mathrm{Res}_\omega(v,e_3)=0 \tag{4.1}$$

if $\{e_1,e_2,e_3\}$ are the 3 edges of G emanating from the vertex v.

The compatibility condition passing from one component to another is expressed by the formula

$$\mathrm{Res}_\omega(v_1,e)+\mathrm{Res}_\omega(v_2,e)=0 \tag{4.2}$$

if $\{v_1,v_2\}$ are the 2 vertices lying on the edge e.

This gives a total of $5g-5$ conditions on the possible residue numbers, which are not quite independent: the sum of all the residue numbers must be zero, and this is seen using either of the above sets of conditions. In fact this is the only dependence of these conditions, and so there are exactly $5g-6$ independent linear conditions on the $6g-6$ residue numbers. Thus the space of residue numbers satisfying the above conditions has dimension g. This is also the dimension of the space Ω_G of regular 1-forms on any stable curve of genus g, so the map from the vector space Ω_G of regular 1-forms to the space of residue numbers satisfying the above conditions is an isomorphism, since it is clearly injective. (A 1-form with zero residues must be zero.) Thus we have:

Proposition 4.3. *Let G be a connected simple trivalent graph. Then the space of global sections Ω_G of the dualizing sheaf ω_{C_G} for the associated graph curve C_G has dimension g. Moreover it is isomorphic, via the above assignment of residue numbers, to the subspace of $\mathbf{K}^{6g-6}$ (indexed by incident vertex-edge pairs (v,e)) consisting of those vectors $(R(v,e))$ satisfying the residue conditions* (4.1) *and* (4.2).

Thus we can "encode" the 1-forms in Ω_G by their residue numbers.

We want to include in this section a discussion of line bundles L on C_G whose total degree is the same as that of the canonical bundle. By Lemma 3.2, if G is trivalent, these line bundles are exactly those which restrict to a bundle of degree one on each component. For this reason we denote the set of isomorphism classes of such bundles by $\mathrm{Pic}^1(C_G)$. This is the connected component of the Picard group containing the canonical bundle. We will call a line bundle $L \in \mathrm{Pic}^1(C_G)$ a *twisted canonical bundle*.

Suppose that $L \in \mathrm{Pic}^1(C_G)$. Therefore L differs from the canonical bundle by a bundle M in $\mathrm{Pic}^0(C_G)$. Assume that the edges of G are oriented, and that with this orientation the bundle M is given by a 1-cochain $\{\lambda_e\}$, as described in Section 3. The difference between the values λ_e and 1 is the "twisting" of the bundle from the canonical bundle. The space of global sections of a twisted canonical bundle can be interpreted as a space of "twisted" homology on the graph itself; see [2] and [8].

A section of a twisted canonical bundle will be called a *twisted 1-form* on C_G. A twisted 1-form is, locally, a 1-form; the patching data moving from component to component across the nodes is varied, however. In particular, every twisted 1-form is represented by a 1-form on each component, with only simple poles allowed at the nodes. Hence to each twisted 1-form there are also residue numbers, as is the case for regular 1-forms. I.e., if σ is a twisted 1-form, the residues $\mathrm{Res}_\sigma(v,e)$ are defined for all incident vertex-edge pairs (v,e). The sum of the residues on each component must still be zero, and so the condition (4.1) must be satisfied.

It is in the compatibility condition (4.2) that the twisting shows up. This condition must be replaced by the condition

$$\mathrm{Res}_\sigma(v_1,e) + \lambda_e \mathrm{Res}_\sigma(v_2,e) = 0 \tag{4.4}$$

if e is the edge directed from v_1 to v_2.

Clearly these are the only conditions on the residue numbers, and this gives a concrete description of all the twisted 1-forms for a given twisted canonical bundle L.

Using Riemann-Roch, we have the following elementary remark:

Lemma 4.5. *Let G be a simple sub-trivalent connected graph of genus g. Let L be a twisted canonical bundle on C_G. Then:*
1) *If L is the canonical bundle, $h^0(L) = g$.*
2) *If L is not the canonical bundle, $h^0(L) = g - 1$.*

Thus for a twisted canonical bundle which is not the canonical bundle, the $5g-5$ conditions (4.1) and (4.4) on the $6g-6$ residue numbers are independent.

There are many questions one can ask concerning twisted canonical bundles in general: when are they base-point-free, when are they very ample, etc. For our purposes we want to address the following problem: when is the linear series cut out on each component of C_G by a twisted canonical bundle complete? For this, we require a definition.

Definition 4.6. The *support* of a 1-cochain $\{\lambda_e\}$ defining a twisted canonical bundle L on C_G (by comparison with the canonical bundle) is the set of edges e with $\lambda_e \neq 1$. Let L be a twisted canonical bundle on C_G. The *twisted norm* of L is the minimum number k of edges e such that there exists a 1-cochain $\{\lambda_e\}$ defining L with exactly k λ's not equal to 1.

I.e., the twisted norm is the minimum cardinality of support among all 1-cochains defining a twisted canonical bundle. Thus the only twisted canonical bundle on C_G with twisted norm 0 is the canonical bundle itself.

Proposition 4.7. *Suppose that G is a simple trivalent 2-edge-connected graph. Suppose that $L \in \mathrm{Pic}^1(C_G)$, and that either:*
a) *L is the canonical bundle, i.e., L has twisted norm 0; or*
b) *L has twisted norm 2, but no 1-cochain defining L has support on exactly two incident edges; or*
c) *L has twisted norm at least 3.*
Then the map from $H^0(C_G, L)$ to $H^0(C_v, L_v)$ is surjective for all vertices v.

Proof. Assume first that L has twisted norm at least 3, and that C_G has genus g.

Fix a vertex v, and let e_1, e_2, and e_3 be the edges incident to v. Note first that the graph G is not disconnected by the removal of any 2 of these edges. Indeed, if it were, then it is in fact disconnected by the removal of the third, and this cannot happen by the assumption of 2-edge-connectedness.

Now to show that the restriction map from $H^0(C_G, L)$ to $H^0(C_v, L_v)$ is onto, it suffices to show that two of these edges, say e_1 and e_2, impose independent conditions on $H^0(C_G, L)$. Let G' be the graph obtained by removing e_1 and e_2; G' is a connected sub-trivalent graph, and its associated graph curve C' maps naturally to C_G, identifying the points as indicated by the two missing edges. The genus of C' is $g-2$.

Now the subspace V of $H^0(C_G, L)$ consisting of those sections vanishing at the nodes corresponding to e_1 and e_2 is isomorphic to the space of global sections of the pull-back L' of L to C'. A 1-cochain defining L' is obtained by taking a 1-cochain for L and deleting the elements λ_{e_1} and λ_{e_2}. Therefore, since L has twisted norm at least 3, L' has twisted norm at least one. In particular, L' is not the canonical bundle of C', and so by Lemma 4.5, $H^0(C', L')$ has dimension $g-3$. This is two less than the dimension of $H^0(C_G, L)$ by Lemma 4.5, so the two edges do impose independent conditions.

This proves case c). the proof of case b) is identical; the hypothesis of this case is exactly to ensure that the pull-back bundle L' has twisted norm at least one. The proof of case a) is also similar, noting that both L and L' are the canonical bundles, so the dimension drops by the genus drop, which is two. In the case of the canonical bundle, this is a result of Bayer and Eisenbud (see Proposition 2.4 of [2]). □

Note that if L has twisted norm equal to 1, Then the sections of L have a base point, namely that node corresponding to the edge e where $\lambda_e \neq 0$, but all other λ's are zero.

The set of line bundles in $\mathrm{Pic}^1(C_G)$ with twisted norm at most k forms a closed set. The general line bundle has twisted norm at least g (since $\mathrm{Pic}^1(C_G)$ is a g-dimensional space). Hence, if $g \geq 3$, the set of line bundles with twisted norm at least 3 forms a Zariski open set in $\mathrm{Pic}^1(C_G)$. In particular, the general twisted canonical bundle has twisted norm at least 3 and those that do form an irreducible set.

We call a twisted line bundle *component-complete* if the restriction map from $H^0(C_G, L)$ to $H^0(C_v, L_v)$ is surjective for all vertices v. It is easy to see that the above conditions characterize the component-complete twisted canonical bundles on C_G. Moreover, we see that the component-complete twisted canonical bundles form an irreducible subset of $\mathrm{Pic}^1(C_G)$: the union of the open set of bundles with twisted norm at least 3, with the single point representing the canonical bundle, and the union of the closed sets representing bundles with twisted norm 2 having a defining 1-cochain with support on exactly two incident edges. We will denote the component-complete twisted canonical bundles by $\mathrm{Pic}^1_{cc}(C_G)$.

5. Matroid Representations Associated to Line Bundles

In this section we want to mention a relationship between line bundles on graph curves and representations of a certain matroid associated to the graph.

Recall that a *matroid* M is a finite set E, together with a non-empty collection $\mathcal{I}$ of subsets of E (called the *independent sets*), such that any subset of $\mathcal{I}$ is also in $\mathcal{I}$, and which satisfies the following:

> If I and J are independent sets and $|J| > |I|$,
> then there exists an element $x \in J - I$
> such that $I \cup \{x\}$ is independent.

The sets not in $\mathcal{I}$ are the *dependent* sets. For an introduction to the theory of matroids, one may consult [1], [15], or [17].

An important example of a matroid is the *cutset* matroid of a graph G. This matroid is defined on the set of edges E; the dependent sets are those sets of edges which, when removed from G, will disconnect G. (Some authors call this the *bond* matroid of G.) An example of such a cutset is the set of edges E_v incident to a single vertex v of G: the vertex v is split off when these edges are removed.

We will call such a set of edges, incident to a single vertex v, the *star* of the vertex. Any matroid defined on the edges E of a graph G which has all the stars as dependent sets will be called a *star matroid*. The cutset matroid of the graph is the first example of a star matroid.

An *n-dimensional projective representation* of a matroid $M = (E, \mathcal{I})$ over a field $\mathbf{K}$ is a function $\phi : E \to \mathbf{P}^n_{\mathbf{K}}$ such that dependent sets in E go to dependent sets in $\mathbf{P}^n_{\mathbf{K}}$. The representation ϕ is *nondegenerate* if its image does not lie in a hyperplane of $\mathbf{P}^n$. The representation ϕ is *faithful* if independent sets in E go to independent sets in $\mathbf{P}^n_{\mathbf{K}}$. (The usual use of the word representation in matroid theory corresponds to a faithful representation; we want to allow non-faithful representations, where independent sets may become dependent, so we will abuse the standard language in this way.) Two representations are said to be *isomorphic* if they differ by a linear automorphism of the projective space.

Let $\phi : E \to \mathbf{P}^n$ be an n-dimensional projective representation of M over $\mathbf{K}$, and suppose that $\pi : \mathbf{P}^n \to \mathbf{P}^{n-k}$ is a linear projection, with center not intersecting $\phi(E)$. Then $\pi \circ \phi$ is an $(n-k)$-dimensional projective representation of M, since dependency is preserved under π. Such a representation will be called *projected*, and a representation which is non-degenerate and which is not projected will be called *linearly normal.*

We can now describe how representations of star matroids associated to a graph correspond to certain line bundles on the graph curve C_G. We assume that G is a connected simple trivalent graph of genus g.

Recall that $\mathrm{Pic}^1(C_G)$ is the set of isomorphism classes of line bundles L on C_G which have total degree identically one, i.e., $d_v(L) = 1$ for each vertex $v \in V$. From the results of the last section, this set of line bundles forms a component of the Picard group of C_G, and is in fact the component containing the canonical bundle ω_{C_G}.

Suppose first that $\phi : E \to \mathbf{P}^n$ is a non-degenerate projective representation of a star matroid of the graph G. Suppose further that it is 1-1 on each star, that is, the three edges incident to any vertex go to distinct points in $\mathbf{P}^n$. Note that for each vertex v of G, the three edges incident to v are dependent in this star matroid, and hence must be sent to collinear points in $\mathbf{P}^n$ by ϕ. Hence we can "extend" the map ϕ to a map $\psi : C_G \to \mathbf{P}^n$ by sending each component C_v onto the corresponding line in $\mathbf{P}^n$, such that the three nodes of C_v go to the three images of the corresponding edges. Moreover this extension is unique, since a linear map on $\mathbf{P}^1$ is determined by where it sends 3 points. Let L_ϕ be the hyperplane bundle $\mathcal{O}_{C_G}(1)$ for this extension, i.e., $L_\phi = \psi^*(\mathcal{O}_{\mathbf{P}^n}(1))$. Since each component C_v is mapped by ψ to a line, we have $d_v(L_\phi) = 1$ for all vertices v; moreover, L_ϕ is obviously component-complete. I.e., $L_\phi \in \mathrm{Pic}^1_{cc}(C_G)$.

Call a representation of a star matroid *star-separating* if it is 1-1 on each star.

If ϕ is linearly normal, then the map ψ cannot arise by mapping C_G to a higher-dimensional space (such that every component maps to a line) and then projecting; this would imply that the representation ϕ of this star matroid was projected. Hence if ϕ is linearly normal, we have that the map ψ arises from the full space of global sections of the hyperplane bundle L_ϕ, and that the dimension of the representation $n = h^0(L_\phi) - 1$.

By Lemma 4.5, if $L \in \mathrm{Pic}^1(C_G)$, then either $L = \omega_{C_G}$ (in which case $h^0(L) = g$) or $h^0(L) = g - 1$. Therefore:

Proposition 5.1. *Let G be a connected simple trivalent graph of genus g. Then the only star-separating linearly normal projective representations of star matroids* *of G are $(g-1)$- or $(g-2)$-dimensional. Moreover, if ϕ is a star-separating $(g-1)$-dimensional linearly normal projective representation of a star matroid, then the hyperplane bundle L_ϕ is the canonical bundle ω_{C_G}.*

Notice that if $\mathcal{M}_1$ and $\mathcal{M}_2$ are two star matroids, such that the dependent sets of $\mathcal{M}_1$ are a subset of the dependent sets of $\mathcal{M}_2$, then any representation of $\mathcal{M}_2$ is also a representation of $\mathcal{M}_1$. We will say that a representation of a star matroid $\mathcal{M}$ is *maximal* if it cannot be extended to a "larger" star matroid.

Clearly two isomorphic representations give the same line bundle. Hence we have a constructed a map

$$\left\{\begin{array}{c}\text{isomorphism classes of}\\ \text{maximal star}-\text{separating linearly normal}\\ \text{projective representations}\\ \text{of star matroids of } G\end{array}\right\} \to \mathrm{Pic}^1_{cc}(C_G)$$

sending a representation ϕ to the line bundle L_ϕ.

Proposition 5.2. *Suppose that G is a simple trivalent 2-edge-connected graph. Then the above assignment of isomorphism classes of maximal star-separating linearly normal projective representations of star matroids of G to component-complete twisted canonical bundles on C_G is bijective.*

Proof. The inverse map is easily described. Let L be a component-complete twisted canonical bundle on C_G. Consider the map $\psi : C_G \to \mathbf{P}^n$ given by choosing a basis for the $n+1$-dimensional space of global sections of L. This map exists because L has no base points, and it maps each component C_v of C_G to a line, since L is component-complete. If E is the edge-set of G, then composing ψ with the identification of E with the nodes of C_G gives a map $\phi(L) : E \to \mathbf{P}^n$, which is 1-1 on each star. This induces a star matroid $\mathcal{M}(L)$ on E, by declaring the dependent sets to be those which are dependent in $\mathbf{P}^n$ after applying the map $\phi(L)$. The map $\phi(L)$ clearly becomes a maximal star-separating linearly normal projective representation of $\mathcal{M}(L)$. The isomorphism class of $\phi(L)$ depends only on L, and obviously by the construction we have both $L_{\phi(L)} \cong L$ and $\phi(L_\phi) \cong \phi$. □

The canonical bundle ω_{C_G} is component-complete, and hence corresponds to a representation of a star matroid. That matroid is the cutset matroid, by Proposition 2.3 of [2], which states that the conditions imposed on the canonical series by a set of nodes are independent if and only if that set of nodes does not disconnect the graph. Hence:

Corollary 5.3. *Let G be a simple trivalent 2-edge-connected graph of genus g. Then there is a unique $(g-1)$-dimensional maximal star-separating linearly normal projective representation of the cutset matroid of G, up to isomorphism.*

6. The Generic Star Matroid

This whole story is clarified in our view by introducing the scheme X_G of $(g-1)$-dimensional representations of star matroids of G. Consider the ambient space $(\mathbf{P}^{g-1})^{3g-3}$, representing (ordered) sets of $3g-3 = |E|$ points in a projective space of dimension $g-1$, which is the maximum possible dimension for representations as we have seen in Proposition 5.1. Think of homogeneous coordinates for such a set of points as a rectangular matrix of size $(g) \times (3g-3)$, whose columns index the points, and whose rows index the ordinary homogeneous coordinates of the points.

Consider the subscheme X of $(\mathbf{P}^{g-1})^{3g-3}$ defined by the vanishing of all 3×3 minors coming from 3 columns representing stars of points in G. That is, X is exactly the subscheme of those sets of points for which the stars are collinear. We call X the *scheme of representations* of star matroids of G. Let Y be the subset of X representing those sets of points which are not star-separating and linearly normal; the complement $X - Y$ maps to $\mathrm{Pic}^1_{cc}(C_G)$ via the associated representation. The fibers of this map are all orbits under the projective group, and therefore are irreducible.

Hence we see that, since $\mathrm{Pic}^1_{cc}(C_G)$ is irreducible, there is a "generic" component X' of $X - Y$, (namely that one dominating $\mathrm{Pic}^1_{cc}(C_G)$). For the general point in this component X', there is an induced matroid, which is constant on an open set of the component.

Definition 6.1. The matroid $\mathcal{S}(G)$ on the edges of the graph G of which the general element in X' is a representation is called the *generic star matroid* of G.

The scheme $X-Y$ itself is not irreducible; the cutset matroid and its representations provide us with a component different from X', since it has independent sets which do not appear in the other component (namely sets of g independent points).

Indeed, a naive parameter count gives that both the cutset matroid representations and the component X' dominating $\mathrm{Pic}^1_{cc}(C_G)$ have dimension g^2-1. For the cutset matroid, this is given by considering the unique canonical embedding of the graph curve C_G (if the curve is 3-edge-connected) and the g^2-1 projective automorphisms of the ambient space. For the generic star matroid component X', one has $g-1$ parameters for the choice of hyperplane in which the curve lives, $(g-1)^2 - 1 = g^2 - 2g$ projective automorphisms in that hyperplane, a final g parameters coming from the choice of twisted canonical bundle; this gives a total of g^2-1 also.

We do not know whethere there are other components of the scheme of representations X; however any others must lie in the set Y.

In our view the generic star matroid $\mathcal{S}(G)$ is a very natural object associated to the graph G. There does not seem to be any discussion of this matroid in the standard texts. It is not the cutset matroid, as we have seen above; for example, the cutset matroid has sets of g independent points, which the generic star matroid does not.

Let us give several small examples to fill out our understanding. Consider first the graph G_3 of genus 3 given by the 4 vertices and 6 edges of a tetrahedron. The generic star matroid for G_3 is the 2-uniform matroid; that is, the dependent sets are those sets of cardinality 3 or more. Indeed, the representations of the twisted canonical bundles are all in dimension $g - 2 = 1$, so any three points will be collinear, hence dependent. Notice that any cutset of this matroid has at least 3 edges, so that the cutset matroid is strictly "smaller" (in the sense of dependent sets) than the generic star matroid.

Now consider the graph G_4 of genus 4 which is a prism over a triangle. This graph has 6 vertices $\{a, b, c, d, e, f\}$ and 9 edges $\{ab, ac, bc, de, df, ef, ad, be, cf\}$. The cutset matroid has, as a dependent set, the cutset $\{ad, be, cf\}$. The generic star matroid however is realized by the map of the graph curve C to $\mathbf{P}^2$ sending each of the 6 components to a line. The reader can easily check that for a generic choice of 6 lines $\{a, b, c, d, e, f\}$ in $\mathbf{P}^2$, the intersection points $\{a \cap d, b \cap e, c \cap f\}$ are *not* collinear. Hence the cutset matroid is not smaller than the generic star matroid in this case.

In general, there are always sets of cardinality g which span canonical space, and hence are independent for the cutset matroid; any such set however must be dependent in the generic star matroid, since it is faithfully represented in dimension $g-1$. Therefore, the generic star matroid is never smaller than the cutset matroid.

Thus for this example of the graph G_4, we see that the generic star matroid and the cutset matroid are incomparable.

Finally take the other graph $K_{3,3}$ of genus 4, which is the complete bipartite graph with 6 vertices, 3 in each part. The canonical image of the associated graph curve consists of 6 lines on a smooth quadric, 3 in each of the two rulings. Notice that the only cutsets of cardinality 3 are the stars themselves. Since any subset of cardinality 4 or more is dependent in the generic star matroid, we see that every cutset must be dependent in the generic star matroid. Thus in this example, the cutset matroid is strictly smaller than the generic star matroid.

We consider the determination of the generic star matroid for a general connected trivalent graph G to be a fundamental problem in the projective geometry of point sets. Indeed, any nontrivial statement about the generic star matroid of any particular graph would yield theorems similar in spirit to those of Pappus and Desargues.

7. Planar Graphs

Let G be a connected simple trivalent graph of genus g. Suppose that this trivalent graph G is embedded in the 2-sphere S^2 to form a map M, with a set of faces (or countries) F, bordered by the set of edges E of G and the set of vertices V of G.

Lemma 7.1. *There are $g+1$ faces in the map M.*

Proof. Since G is trivalent, we have $|V| = 2g-2$ and $|E| = 3g-3$. Euler's equation for the decomposition of the sphere gives

$$|V| - |E| + |F| = 2,$$

forcing $|F| = g+1$. □

Index the faces of F from 0 to g. Let k be a field. Consider the projective space $\mathbf{P}^g = \mathbf{P}^g_k$ over k with coordinates also indexed from 0 to g. For the i-th face of F, denote the corresponding coordinate point of $\mathbf{P}^g$ by f_i. Thus f_i is the point with homogeneous coordinates all 0, except for a 1 in the i-th position.

Note that each vertex of V lies a priori on three faces of F. If these faces are different, the vertex v defines a plane in $\mathbf{P}^g$, namely the span of the three points f_i corresponding to the three faces on which v lies.

There is a convenient criterion on the graph G which ensures that, for each vertex v, the three faces of the map on which v lies are different. Recall that a graph is *n-edge-connected* or simply *n-connected*, if the removal of less than n edges never disconnects the graph. A graph is 1-connected if and only if it is connected. A graph is 2-connected if and only if there is no disconnecting edge.

Lemma 7.2. *Let G be a connected simple trivalent planar graph, forming the map M. Then G is 2-connected if and only if for every vertex v of G, the three faces of the map M on which v lies are different.*

Proof. Let v be a vertex of G such that two of the faces on which v lies are the same. Then removing the edge emanating from v which separates (locally) those two regions will disconnect G. Conversely, if removing an edge e separates G, then the two faces on each side of e are actually the same, so the two vertices of e fail to lie on 3 distinct faces of the map M. □

So if G is trivalent and 2-connected, each vertex v of G does define a plane P_v. Note that in general, if G is trivalent and 2-connected, then G is 3-connected if and only if any two distinct faces meet along at most one edge.

Now assume that G is trivalent and 2-connected, so that by the above lemma each vertex v of G lies on 3 distinct faces of the map M. Then v determines the plane P_v, the span of the 3 coordinate points associated to the 3 faces. Define the surface S_M to be the union of the planes P_v over all vertices v in V:

$$S_M = \bigcup_{v\in V} P_v.$$

This "face surface" S_M is a union of planes, one plane for each vertex. It is a geometric realization of the combinatorial object M, and we would like it to reflect as much as possible the combinatorics of the planar map. In particular, if v and w are distinct vertices, we would like the planes P_v and P_w to be distinct, and if so, to intersect along a line if and only if v and w are joined by an edge in G. Indeed, we would like the edges of G to encode the line intersections of the planes of S_M. If these conditions are true, we will say that S_M *faithfully realizes* the map M.

Of course, if v and w are joined by the edge e, then the planes P_v and P_w both contain the line spanned by the coordinate points corresponding to the two faces on either side of e. If two planes P_v and P_w intersect in a line, they must intersect in a coordinate line, and so the vertices v and w must both lie at the boundary of two faces f and g of the map M. If v and w are not joined by an edge, then f and g share two edges in common: one emanating from v and one from w. Therefore, G is not 3-connected. In particular, if the two planes P_v and P_w are the same, they certainly share lines, and so G cannot be 3-connected. Conversely, if G is not 3-connected, then there are two faces f and g which meet along more than one edge. This implies that there are two vertices which are not joined, but which both have the line joining the coordinate points of f and g in their associated planes. This proves the following.

Proposition 7.3. *Let G be a 2-connected simple trivalent planar graph, inducing the map M. Then the surface S_M faithfully realizes the map M if and only if G is 3-connected.*

For the rest of this article we will assume that G is a 3-connected simple trivalent planar graph, so that the surface S_M faithfully realizes the map M.

Let H_∞ be the hyperplane of $\mathbf{P}^g$ defined by setting the sum of the coordinates to zero:

$$H_\infty : \sum_{i=0}^{g} x_i = 0.$$

Note that none of the face points f_i lie on H_∞, so H_∞ cuts each plane P_v in a line C_v. Define the curve C_G to be the union of the lines C_v over all vertices v in V (or, equivalently, to be the intersection of S_M with H_∞):

$$C_G = S_M \cap H_\infty = \bigcup_{v \in V} C_v.$$

The curve C_G is a union of $2g-2$ lines in a projective space of dimension $g-1$ (namely H_∞). Two lines C_v and C_w meet in C_G exactly when the corresponding vertices v and w are connected by an edge in G. Indeed, we see that this embedded curve is exactly the associated graph curve C_G to G, realized in a projective space.

Note that if e is an edge bounding faces i and j, then the node n_e of the graph curve C_G corresponding to e is the point in H_∞ with homogeneous coordinates equal to 1 in position i, -1 in position j, and 0 in all other positions.

The curve C_G is embedded in the projective space $H_\infty \cong \mathbf{P}^{g-1}$ with sections of a bundle which is degree 1 on each $\mathbf{P}^1$ component of C_G: the image of each component is a line. This bundle is actually therefore the canonical bundle of the stable curve C_G, since as in the general theory the canonical bundle is the only bundle with this total degree, having g linearly independent global sections.

Therefore the embedding of C_G into $\mathbf{P}^{g-1}$ is given by the complete linear system of regular 1-forms on C_G, i.e., it is the canonical embedding of C_G. (See [2] for a discussion in general of the canonical embedding for graph curves of trivalent 3-connected graphs.)

The upshot of these remarks is that any canonical divisor on C_G, that is, a divisor of zeroes of a regular 1-form on C_G, is exactly the intersection of C_G with a hyperplane in H_∞. Moreover, since two 1-forms have the same zero locus if and only if they are multiples of each other, we see that the hyperplanes in H_∞ are in $1-1$ correspondence with the 1-dimensional subpaces $\mathbf{P}\Omega_G$ of the space Ω_G of regular 1-forms on C_G.

If x is a point of a component C_v of C_G identified to a node, and ω is a regular 1-form on C_G, then the restriction of ω to C_v near x can be written as $f(z)dz/z$, where z is a local coordinate on C_v at x and f is a regular function. In other words, at the identified point x, the form dz/z is a local basis for the space of regular 1-forms. Hence we see that the 1-form ω has a "zero" at x if and only if the coefficient function f in the above expression is zero at x, i.e., if $f(0) = 0$. This is equivalent to having the residue of ω_v being zero at the point x. Since the residue of ω_v at x is opposite to the residue of the corresponding 1-form at the point identified to x by (4.2), the regular 1-form ω vanishes at the node of C_G to which these points are identified if and only if these residues are zero.

Rephrased in terms of hyperplanes, this is the following.

Proposition 7.4. *Let e be an edge of G, incident to vertices v and w, corresponding to the node n_e of C_G. Let the hyperplane $H \subset H_\infty$ correspond the the regular 1-form $\omega \in \Omega_G$. Then H passes through n_e if and only if the residue numbers* $\mathrm{Res}_\omega(v,e)$ *and* $\mathrm{Res}_\omega(w,e)$, *which are opposite in any case by* (4.2), *are both zero. The set of hyperplances in H_∞ which do not pass through any node of C_G are in $1-1$ correspondence with the 1-dimensional subspaces of Ω_G which are spanned by regular 1-forms all of whose residue numbers are non-zero.*

Because of this proposition, we will denote the set of hyperplanes in H_∞ by $\mathcal{H}_M$, and those which do not pass through any node of C_G by $\mathcal{H}'_M$. Let us also denote the subset of Ω_G consisting of 1-forms all of whose residue numbers are non-zero, by Ω'_G. The above proposition shows that $\mathcal{H}'_M$ is in $1-1$ correspondence with $\mathbf{P}(\Omega'_G)$.

When the graph G is planar, there is another convenient description of the space Ω_G of regular 1-forms on C_G, which we now develop. Fix once and for all an orientation of the 2-sphere S^2, giving therefore an orientation to each face of the map M.

Given a face f, we can then orient (clockwise, say) the edges at its boundary, which puts directions on two of the three edges emanating from each vertex on the face f. (The edge not on the boundary of the face f is not directed by this procedure.) Define a regular 1-form $\omega_f \in \Omega_G$ by assigning its residue numbers as follows.

If v is not on the boundary of f, set $\mathrm{Res}_{\omega_f}(v,e) = 0$ for all 3 edges e emanating from v.

If v is on the boundary of f, and e is the edge emanating from v which is not on f, set $\mathrm{Res}_{\omega_f}(v,e) = 0$.

If v is on the boundary of f, and e is an edge emanating from v on the boundary of f, set $\mathrm{Res}_{\omega_f}(v,e) = 1$ if e is directed away from v, and set $\mathrm{Res}_{\omega_f}(v,e) = -1$ if e is directed towards v.

It is easily checked that these assignments satisfy the residue number conditions (4.1) and (4.2), and define a regular 1-form $\omega_f \in \Omega_G$ on C_G. The following lemma is Lemma 2.5 of [10].

Lemma 7.5. *The* 1*-forms* $\{\omega_f : f \text{is a face of } M\}$ *span the space of regular* 1*-forms* Ω_G *on* C_G*. Moreover, the only linear relation among the* ω_f*'s is that their sum is zero.*

8. Colorings of the Planar Map

Suppose now that the field k over which we have been working is the finite field $\mathbf{F}_q$ with q elements. Let Q be a q-coloring of the map M, that is, an assignment of q symbols to the faces of M which has the property that if two faces share an edge, then they are assigned different symbols. We take the symbols to come from the field $\mathbf{F}_q$, and therefore we will consider a q-coloring of M to be a function Q from the set F of faces of M to the field $\mathbf{F}_q$, satisfying the coloring condition that if two faces share an edge, then Q has different values on them. Denote by $\mathrm{Col}_q(M)$ the set of q-colorings of M.

To every function $Q : F \to \mathbf{F}_q$ we may assign the regular 1-form

$$\omega(Q) = \sum_{f \in F} Q(f)\omega_f$$

which is a 1-form on the graph curve C_G.

Let us analyze the coloring condition in terms of this regular 1-form.

Proposition 8.1. *Let* $Q : F \to \mathbf{F}_q$ *be a function. Then* Q *is a* q*-coloring of* M *if and only if the regular* 1*-form* $\omega(Q)$ *has all non-zero residue numbers.*

Proof. Suppose f and g are two faces of M sharing the edge e and the two vertices v and w at the ends of e. The residue numbers $\mathrm{Res}_{\omega(Q)}(v,e)$ and $\mathrm{Res}_{\omega(Q)}(w,e)$ are determined only by the coefficients of ω_f and ω_g, since if h is a face not having e on its boundary, then $\mathrm{Res}_{\omega_h}(v,e) = \mathrm{Res}_{\omega_h}(w,e) = 0$. Suppose that we have oriented the sphere such that for the face f, the edge e is directed from v to w; then necessarily for the face g, e is directed from w to v. Then by the definition of ω_f and ω_g, we have

$$\mathrm{Res}_{\omega_f}(v,e) = \mathrm{Res}_{\omega_g}(w,e) = 1 \text{ and } \mathrm{Res}_{\omega_f}(w,e) = \mathrm{Res}_{\omega_g}(v,e) = -1.$$

Therefore the residue numbers for $\omega(Q)$ are

$$\mathrm{Res}_{\omega(Q)}(v,e) = Q(f) - Q(g) \text{ and } \mathrm{Res}_{\omega(Q)}(w,e) = Q(g) - Q(f).$$

The coloring condition on Q is that $Q(f) \neq Q(g)$ whenever f and g share an edge. Therefore the coloring condition is equivalent to both of these residue numbers given above being non-zero. Since this is true for any edge e, the coloring condition is equivalent to the regular 1-form ω_Q having all non-zero the residue numbers. □

Using our notation from above, we see that

$$Q \in \mathrm{Col}_q(M) \text{ if and only if } \omega(Q) \in \Omega'_G.$$

Algebraically, let $\mathbf{F}_q^F$ denote the vector space of functions from the set of faces F of the map to the field $\mathbf{F}_q$. The assignment

$$Q \longmapsto \omega(Q)$$

gives an $\mathbf{F}_q$-linear map from $\mathbf{F}_q^F$ to Ω_G, which is onto, with kernel the 1-dimensional space of constant functions on F. The inverse image of the set Ω'_G of 1-forms all of whose residue numbers are nonzero is exactly the set $\mathrm{Col}_q(M)$ of q-colorings of the map M.

This brings us to our main result, which is a consequence of the preceding discussion.

Theorem 8.2. *A q-coloring of the planar map M exists if and only if there is a regular 1-form on the graph curve C_G over $\mathbf{F}_q$ all of whose residue numbers are non-zero (i.e., if and only if $\Omega'_G \neq \emptyset$).*

Indeed, we see that

$$\text{the number of } q\text{-colorings of } M = |\mathrm{Col}_q(M)| = q|\Omega'_G| \tag{8.3}$$

since the fibers of the map from q-colorings of M onto Ω'_G have cardinality q. Expressed geometrically in terms of the hyperplanes in H_∞, Proposition 7.4 gives that

$$|\Omega'_G| = (q-1)|\mathcal{H}'_M| \tag{8.4}$$

since every 1-dimensional subspace has $q-1$ spanning vectors. This gives

$$|\mathrm{Col}_q(M)| = q(q-1)|\mathcal{H}'_M|. \tag{8.5}$$

Let us apply these results in the cases of low q, where the interest is highest. All of these results are well-known and rather simple; however the unified treatment is the point here. For graph-theoretic discussions of the statements any number of books may be consulted, for example [16] or [1].

We begin with $q = 2$; no map can be 2-colored, so $\mathrm{Col}_2(M) = \emptyset$. On the other hand it is easy to see that Ω'_G is also empty if $q = 2$. The only nonzero number in $\mathbf{F}_2$ is 1; therefore any 1-form in Ω'_G must have every residue number equal to

1. But $1+1=0$, so that if we fix a vertex v, the sum of the 3 residue numbers at v is 1; hence the assignment of all residue numbers equal to 1 does not give a regular 1-form on C_G.

It is more interesting in the case $q=3$. Write $F_3=\{0,\pm 1\}$; we see that any 1-form in Ω'_G has only ± 1 as residue numbers. Suppose $\omega \in \Omega'_G$. The only way for the three residue numbers of ω at any vertex v to sum to 0 is if all three numbers are the same: either all are 1 or all are -1. Therefore the residue numbers are actually associated to simple vertices, not vertex-edge pairs. Moreover, since the two residue numbers on any edge must sum to zero, if two vertices are connected by an edge they must have opposite residue numbers. Thus an element of Ω'_G is determined exactly by a 2-coloring of the vertices V of G, namely a function from V to $\{\pm 1\}$ such that if vertices v and w are joined by an edge of G, the function has different values on them. It is elementary to see that a 2-coloring of the vertices of a planar graph G is possible if and only if every face f of the map M has an even number of edges. If this is true, then the 2-coloring is determined by the color of any one vertex v_0: the other vertices are colored according to the parity of the distance from v_0, which is well-defined with the even-face assumption. Thus we recover the well-known 3-coloring criterion for a planar map:

Corollary 8.6. *A planar map M is 3-colorable if and only if every face of M has an even number of edges. If this is so, there are exactly* 6 *3-colorings of M.*

Finally let us treat the most interesting case of $q=4$. Write $\mathbf{F}_4=\{0,a,b,c\}$. Any 1-form in Ω'_G has only the nonzero numbers $\{a,b,c\}$ as residues. Since the field $\mathbf{F}_4$ has characteristic 2, and the sum of the two residue numbers at each edge must be zero, we see that the two residue numbers at each edge must be the same. Therefore the residue numbers for any ω are determined by assigning numbers to each edge. Suppose that $\omega \in \Omega'_G$. Then the three (non-zero) residue numbers at each vertex must sum to zero, and it is easy to see that this means that at each vertex, all three non-zero numbers a, b, and c must each be used (once). In other words, one obtains a 1-form in Ω^1_G exactly with a 3-coloring of the edges E of G, namely, a function from E to $\{a,b,c\}$ such that if two edges share a vertex, the function has different values on them. Therefore we obtain the (slightly less well-known but equally easy) 4-coloring criterion for a planar map:

Corollary 8.7. *The faces of a planar map M are* 4*-colorable if and only if the edges of the map are* 3*-colorable. The number of* 4*-colorings of the faces is* 4 *times the number of* 3*-colorings of the edges.*

Since we have related colorings of G to regular 1-forms on C_G, and we have seen that regular 1-forms on C_G correspond to hyperplanes in the space H_∞, we can express the coloring results in more geometric terms. The next corollary follows directly from Proposition 7.4 and Theorem 8.2.

Corollary 8.8. *The faces of a planar map M bounded by a 3-connected simple trivalent graph G can be colored by q colors if and only if there is a hyperplane H in the space H_∞ (which contains the canonically embedded graph curve C_G) which is defined over F_q and which misses every node of the graph curve C_G.*

From this point of view, the 4-color theorem says that a sufficiently general hyperplane exists over $\mathbf{F}_4$ for any planar graph. Therefore one may view the 4-color theorem as a "Bertini"-type result!

9. The Chromatic Function

As noted above, Proposition 7.4 and Theorem 8.2 show us that the existence of q-colorings of the map M is related to the existence of hyperplanes in H_∞ missing the nodes of the graph curve C_G. It is possible to count these hyperplanes (those of $\mathcal{H}'_M$), therefore counting possible colorings, by studying the configuration of the set of nodes.

For any set of nodes N of C_G, denote by $\mathcal{L}_N$ the (projective) linear span of N.

It is convenient to introduce the *dual* graph G' of G, with vertices equal to the faces in F and edges corresponding to the edges in E, now viewed as joining faces instead of joining vertices. A set of nodes N of C_G can then be viewed also as a set of edges E'_N of the dual graph G', which form (with their adjacent vertices) a subgraph G'_N of the dual graph G'.

Form the matrix A_N whose columns are the homogeneous coordinates of the nodes in N. The dimension of the linear span $\mathcal{L}_N$ of the nodes is then one less than the rank of this matrix (since we are speaking of projective dimension).

This matrix A_N has, in the column for a node n corresponding to an edge e, one 1 and one -1, in the rows corresponding to the faces of G adjacent to the given edge e. In terms of the dual graph G', the 1 and -1 occur in the rows corresponding to the two vertices of G' which lie on the edge e. Thus this matrix A_N is exactly the matrix for the boundary operator in homology, from the 1-chains of the subgraph G'_N to the 0-chains of the subgraph G'_N (given a suitable orientation of the dual graph). Therefore the kernel of A_N is identified with $H_1(G'_N)$, and the cokernel of A_N is identified with $H_0(G'_N)$. Hence the rank of A_N is equal to $|N| - \mathrm{rank} H_1(G'_N)$, and also to $|V(G'_N)| - \mathrm{rank} H_0(G'_N)$, where $V(G'_N)$ is the set of vertices of G'_N.

Thus we have the following formulae for the dimension of the linear span $\mathcal{L}_N$ of a set of nodes N:

$$\dim(\mathcal{L}_N) = |N| - \mathrm{rank} H_1(G'_N) - 1$$

or

$$\dim(\mathcal{L}_N) = |V(G'_N)| - \mathrm{rank} H_0(G'_N) - 1,$$

i.e.,

$$\dim(\mathcal{L}_N) = \#\text{ of vertices of } G'_N - \#\text{ of connected components of } G'_N - 1.$$

If N is empty, then we will set $\dim(\mathcal{L}_N) = -1$; this implies that the empty graph has trivial H_1.

Thus the dimension of the space of hyperplanes in H_∞ which pass through the nodes of N is

$$g - 1 - |N| + \text{rank} H_1(G'_N).$$

Fix a prime power q. Working over the field $\mathbf{F}_q$ with q elements, a projective space of dimension n has

$$\mathcal{P}_q(n) = \frac{q^{n+1} - 1}{q - 1}$$

elements.

It is now trivial to find a formula for the number $\mathcal{H}'_M$ of hyperplanes of H_∞ which contain no node of the graph curve C_G. For this we simply use the inclusion-exclusion principle: one then obtains

$$|\mathcal{H}'_M| = \sum_{k=0}^{3g-3} \sum_{|N|=k} (-1)^k \mathcal{P}_q(g - 1 - |N| + \text{rank} H_1(G'_N)).$$

One realizes immediately that this sum is simply over all sets of nodes N, and so we can re-write this as

$$|\mathcal{H}'_M| = \sum_{N \subset E} (-1)^{|N|} \mathcal{P}_q(g - 1 - |N| + \text{rank} H_1(G'_N)).$$

Given the results of the previous section, this is actually a version of the chromatic function for the map M. In graph theory, the chromatic function of a map M is the function $c_M(n)$ whose value at an integer n is the number of ways of coloring the map M with n colors. Using our results concerning the hyperplanes and the residues of 1-forms, we see that

$$c_M(q) = q(q-1)|\mathcal{H}_M| = q(q-1) \sum_{N \subset E} (-1)^{|N|} \mathcal{P}_q(g - 1 - |N| + \text{rank} H_1(G'_N))$$

for any prime power q.

Using the definition of $\mathcal{P}_q$, the factor of $q - 1$ can be brought into the sum to obtain

$$c_M(q) = q \sum_{N \subset E} (-1)^{|N|} (q^{g-|N|+\text{rank} H_1(G'_N)} - 1).$$

Since

$$\sum_{N \subset E} (-1)^{|N|} = 0$$

by the binomial theorem, this can be re-written as

$$c_M(q) = q \sum_{N \subset E} (-1)^{|N|} q^{g-|N|+\mathrm{rank} H_1(G'_N)}.$$

It is interesting to compare this with the standard formula for the chromatic function of a graph. Coloring the planar map M is equivalent to giving a coloring of the vertices of the dual graph G'. The chromatic function of any graph G' is the function $\pi(G', \lambda)$ which is the number of colorings of G' with λ colors. A formula for the chromatic function of a graph is developed in Section 14.4 of [O]; it is

$$\pi(G', \lambda) = \sum_{s,c} (-1)^s p(s,c) \lambda^c,$$

where $p(s,c)$ is the number of subgraphs H of G' with s edges and c connected components. This reduces to the above formula in our case of $\lambda = q$, because of the planarity (giving the Euler formula) and the obvious relationship between the arithmetic genus g and the ranks of the homology groups.

References

[1] Aigner, M., Combinatorial Theory. Springer-Verlag, New York, 1979.

[2] Bayer, D. and D. Eisenbud, Graph Curves. preprint.

[3] Ciliberto, C. and A. Franchetta, Curve poligonali, grafi e applicazione di Gauss. Rendiconti di Matematica 12 (1992), 165-196.

[4] Ciliberto, C. and A. Franchetta, L'applicazione di Gauss per grafi trivalenti. preprint.

[5] Ciliberto, C., J. Harris, and R. Miranda, On the surjectivity of the Wahl map. Duke Math. J. 57 (1988), 829-858.

[6] Ciliberto, C. and R. Miranda, On the Gaussian map for canonical curves of low genus. Duke Math. J. 61 (1990), 417-443.

[7] Ciliberto, C. and R. Miranda, Gaussian maps for certain families of canonical curves. *Complex Projective Geometry*, edited by G. Ellingsrud, C. Peskine, G. Sacchiero, and S.A. Stromme, London Math. Soc. Lecture Note Series No. 179 (1992), 106-127.

[8] Ciliberto,C. and A. Verra, On the surjectivity of the Gaussian map for Prym-canonical line bundles on a general curve. preprint.

[9] Duflot, J. and R. Miranda, The Gaussian map for rational ruled surfaces. Trans. Amer. Math. Soc. 330 (1992), 447-459.

[10] Miranda, R., The Gaussian map for certain planar graph curves. in: Algebraic Geometry: Sundance 1988. Contemporary Math. 116 (1991), 115-124.

[11] Miranda, R., Six lectures on graph curves and curves on K3 surfaces. *Lectures on Riemann surfaces*, Proc. of the College on Riemann surfaces, ICTP, Trieste, Italy, 9 Nov. - 18 Dec. 1987; pp. 119 - 176. World Scientific (1989).

[12] Miranda, R., Colorings of Planar Maps and Residues of 1-forms. In: *Methods in Module Theory*, G. Abrams, J. Haefner, and K.M. Rangaswamy, eds. Lecture Notes in Pure and Applies Mathematics, Marcel Dekker, Vol. 140, (1993), 237-248.

[13] Ore, O., Theory of Graphs. AMS Colloquium Publications, vol. XXXVIII (1962).

[14] Wahl, J., The Jacobian Algebra of a graded Gorenstein singularity. Duke Math. J. 55 (1987), 843-871.

[15] Welsh, D.J.A., Matroid Theory. London Mathematical Society Monograph No. 8, Academic Press, London, New York, San Franscisco, (1976).
[16] Wilson, R., Introduction to Graph Theory. Academic Press (1972).
[17] Wilson, R., An Introduction to Matroid Theory. American Math. Monthly 80 (1973), 500-525.

Algorithms for constructing minimal generators of ideals of projective points in polynomial time

Francesca Cioffi

Abstract. Two polynomial algorithms are presented: the first one constructs a set F of generators of the ideal I of a set of s distinct points in the projective space $\mathbf{P}^n$; the related computational cost is $O(s^4 n)$; the second one minimizes F with the computational cost $O(s^4 n^3)$.

1991 Mathematics Subject Classification: 13P10.

1. Introduction

The aim of this paper is to reduce the bound for the computational cost to compute a minimal set of generators of the ideal of s distinct points of the projective space of dimension n over a field K. A method is proposed following results of Orecchia [4] and of Ramella [5]. Other approaches have been proposed by Mora and Robbiano [3].

We propose now a polynomial algorithm to construct a set F of generators which is also a Gröbner bases. We know that the upper bound of the cardinality of this set is polynomial. Then, it is possible to obtain a minimal set of generators in polynomial time making use of the result of [5, Cap. I Corollary 4.11]. In Section 2 and in Section 3 we describe the method adopted to reach our aim and prove its correctness. Finally, we show why we obtain an $O(s^4 n^3)$ algorithm.

2. Construction of a set of generators

Let K be a field and $\mathbf{P}^n_K$ the projective space over K of dimension n. We want to construct a set of homogeneous generators of an ideal $I \subset K[X_0, X_1, \dots, X_n]$ of s points $P_1, \dots, P_s$ with a polynomial algorithm of order $O(s^4 n)$.

Let T be the monoid of all terms ordered by a term-ordering $<$, i.e. a good ordering compatible with the multiplication of the monoid. We say that a term H is a "multiple" of a term K if there is a term K' such that $H = KK'$.

Orecchia [4] yields the conductor of the coordinate ring of a finite set of projective points in polynomial time. We use here an analogous method to reach our aim. We describe the adopted method and recall its theoretical supports.

The algorithm constructs the generators of degree d, starting from $d = 1$, in the following way. Let $T_0 = 1$ be the term of degree 0 and let $T_1, T_2, \ldots, T_{n+1}$ be the terms of degree 1 ordered with respect to (w.r.t.) $<$. We evaluate every term T_j, $j = 1, \ldots, n+1$, at $P_1, \ldots, P_s$ and construct the matrix G_1^s whose j^{th} column is the vector of the evaluations of T_j. If the jth column c_j^1 of G_1^s is linearly dependent on some previous independent columns $\mathrm{c}_{j_1}, \ldots, \mathrm{c}_{j_r}$, then there is an r-ple $(b_1^{1,j}, \ldots, b_r^{1,j})$ such that $\mathrm{c}_j^1 = \sum_{k=1}^{r} \mathrm{b}_k^{1,j} \mathrm{c}_{j_k}^1$. The term which corresponds to the jth column is the leading term, w.r.t. $<$, of a generator

$$T_{j+1} - \sum_{k=1}^{r} b_k^{1,j} T_k.$$

The r-ple $(b_1^{1,j}, \ldots, b_r^{1,j})$ can be determined by reducing the matrix G_1^1 to reduced row-echelon form $\tilde{G}_1^1$ with elementary operations (i.e. with the Gauss algorithm). In fact, the first r elements of the jth column of $\tilde{G}_1^1$ are the coefficients $b_1^{1,j}, \ldots, b_r^{1,j}$ and all other elements are null [4]. Hence, for every term of degree 1, corresponding to a dependent column of G_1^s, we construct a polynomial.

To yield polynomials of degree d higher then 1 we do not consider all terms of degree d. Let $\bar{G}_1^s = G_1^s$. Now we assume to have constructed the matrix $\bar{G}_{d-1}^s$ and we want to construct the matrix $\bar{G}_d^s$. We consider only the terms of degree d which are not multiples of the terms of degree $d-1$ corresponding to the dependent columns of the matrix $\bar{G}_{d-1}^s$. As in the case $d = 1$, we determine the row-echelon form $\tilde{\bar{G}}_d^s$ of $\bar{G}_d^s$ and construct the generator polynomials of degree d in the same way.

The algorithm for each following degree d repeats the proposed construction and it ends when the matrix $\bar{G}_d^s$ has the same rank as $\bar{G}_{d-1}^s$. The criterion by which the algorithm ends is the same one used by Ramella to compute a minimal set of generators [5]. In fact, there is a degree σ which is the minimum degree such that $\mathrm{rk}(\bar{G}_\sigma^s) = \mathrm{rk}(\bar{G}_{\sigma-1}^s)$ [5, Cap.I Lemma 4.6] and every polynomial of a minimal set of generators of I has degree $d \leq \sigma$ [5, Cap. I Th. 4.10, Cap. II Th. 1.1].

Let F be the set of all polynomials determined by the method described. We shall call *independent terms of degree d* (*dependent terms of degree d*) all the terms corresponding (not corresponding) to the independent columns of the matrix $\bar{G}_d^s$.

Note that the proposed algorithm consists of three fundamental stages: first the algorithm determines the terms of degree d which will be considered for constructing the matrix $\bar{G}_d^s$; then it constructs the matrix $\bar{G}_d^s$; finally it constructs the row-echelon form $\tilde{\bar{G}}_d^s$ and the generators of degree d.

ALGORITHM I

Input: s distinct projective points and a term-ordering $<$.

Output: a set of generators of the ideal I of the assigned points.

STEP1. Set $d = 0$.

STEP2. $d = d + 1$. Construction and order of the terms of degree d which are not multiples of the dependent terms of degree $d - 1$.

STEP3. Construction of the matrix $\bar{G}^s_d$.

STEP4. Computation of the row-echelon form $\tilde{\bar{G}}^s_d$ of $\bar{G}^s_d$, determination of the independent terms of degree d and computation of $rk(\bar{G}^s_d)$.

STEP5. Construction of the generators of degree d.

STEP6. If $rk(\bar{G}^s_d) \neq rk(\bar{G}^s_{d-1})$, then GOTO STEP2, else STOP.

We want to prove now that F is a set of generators for I.

Lemma 2.1. *If a set S of non-null polynomials in I is such that the monoideal generated in T by the leading terms (w.r.t. a term-oredering $<$) of the elements of S is equal to the monoideal of the leading terms of all polynomials in I, then S is a set of generators for I.*

Proof. [7, Corollary 13.7]. □

Proposition 2.2. *The algorithm procedure described above constructs a set of generators F for the ideal I.*

Proof. To prove the result we need the construction of a set of generators proposed in [5]. The ideal I is graded, i.e. $I = \oplus_{d \geq 0} I_d$ where I_d is the vector space of homogeneous polynomials of degree d. Let T^d_j be the jth term w.r.t.$<$ among all terms of degree d. For each d we can consider the matrix G^s_d whose j^{th} column is the vector of the evaluations of T^d_j at the s assigned points $P_1, \ldots, P_s$. Let Σ_d be the linear system with G^s_d as matrix of coefficients. Let R_d a basis of the solutions of Σ_d. Clearly, a vector of R_d gives the coefficients of a polynomial of I_d, hence R_d gives a basis B_d of the K-vector space I_d. Now we show how we want to construct a basis R_d. We reduce G^s_d to the reduced echelon-form $\tilde{G}^s_d$. Let $\mathbf{c}_{j_1}, \ldots, \mathbf{c}_{j_r}$ be the columns of $\tilde{G}^s_d$ containing the leading non null entries of $\tilde{G}^s_d$. We consider a dependent columns $\mathbf{c}_j = \begin{pmatrix} a_{1_j} \\ \vdots \\ a_{s_j} \end{pmatrix}$, $j \notin \{j_1, \ldots, j_r\}$. The vector v_j which has the components a_{i_j} in the $j_1^{th}, \ldots, j_r^{th}$ places, 1 in the jth place and zero elsewhere is a vector of R_d. The polynomial which has the components of v_j as coefficients is a polynomial of B_d. The union $\bar{B} = \cup^{\sigma}_{d=1} B_d$ is a set of generators for I, and of course it is $F \subseteq \bar{B}$ for construction. For every polynomial f in I_d, the leading term of f must be one of the leading terms of the polynomials in B_d. In fact, if g is in B_d and H is its leading term, g is the only polynomial in B_d with T as leading term, for construction. Hence, the monoideal of all leading terms of

I is equal to the monoideal generated by the leading terms of the polynomials of $\bar{B}$. We eliminate the polynomials of $\bar{B}$ which have a multiple of a leading term of another polynomial of $\bar{B}$ as leading term. In this way we obtain the set F which is still a system of generators for I because of Lemma 1.1. □

Remark. It should be noted that F is a reduced Gröbner basis w.r.t. $<$ graduated; however, we do not use any property of Gröbner bases.

3. Computation of a minimal set of generators

Let $F = \{f_1, \ldots, f_v\}$ be the set of generators of the ideal I (of s projective points $P_1, \ldots, P_s$) computed by the Algorithm I. Therefore we constructed F by using the linear dependence among the vectors of the evaluations of the terms at the assigned points. We want to minimalize F in polynomial time; in order to achive this aim we use the results presented in [5, Cap. I Corollary 4.11]. In fact, we determine the dependent generators of degree $d+1$ by using only the generators of degree d.

We consider the monoid T ordered by a good order $<$ compatible with the multiplication of the monoid. Because of the method adopted to construct F, this set of polynomials has some properties:
1) $|F| \leq s^2(n+1)$;
2) the generators of degree d are not more then $s(n+1)$;
3) the leading term of every polynomial f_j is not a multiple of the other leading terms;
4) for every degree d, there are at most s independent terms of degree d; they are the only terms that can have a non-null coefficient in the generators of degree d together with the leading terms; moreover not all the products of the independent terms of degree d by every variable $X_0, \ldots, X_n$ can be leading terms of generators of degree $d+1$;
5) every polynomial in F is monic;
6) the maximum degree t of the degrees of polynomials in F cannot be greater than the number s of the assigned points.

Let α be the minimum degree of polynomials f_j in F. For every d such that $\alpha \leq d < t$ we construct a matrix M_{d+1} to determine the dependent polynomials of degree $d+1$. Because of property 3, we know that there are no dependent polynomials of degree α.

To construct M_{d+1} we multiply every generators of degree d by each variable $X_0, \ldots, X_n$. We obtain new polynomials originated from the generators of degree d. The first rows of M_{d+1} consist of the coefficients of these new polynomials. The last rows consist of the coefficients of generators of degree $d+1$. Every column of M_{d+1} corresponds to a term of degree $d+1$ that has a non-null coefficient in at least one polynomial, among the new polynomials and the generators of degree $d+1$.

Because of property 2, the rows of M_{d+1} are at most $s(n+1)^2+s(n+1)$, where $s(n+1)$ is the maximum number of generators of degree $d+1$ in F. Because of properties 4, terms with a non-null coefficient in at least one of the generators of a fixed degree are not more than $s(n+1)+s$: $s(n+1)$ leading terms and s independent terms. Then, after the multiplications of generators of degree d by the $n+1$ variables $X_0, \ldots, X_n$, we obtain at most $s(n+1)^2$ new polynomials in which there are at most $s(n+1)^2+s(n+1)$ terms with non-null coefficients. Of course, not more then s terms with a non-null coefficient and not more then $s(n+1)$ leading terms appear in the generators of degree $d+1$.

Note that terms that can appear in one generator of degree $d+1$, because of property 4 and by the construction, have already been considered among the $s(n+1)$ terms which are not necessarily leading terms of the new polynomials. In fact all terms of generating polynomials are multiples of the independent terms of degree d.

Because the multiplication of the monoid T is compatible with the term-ordering $<$, the multiples of the leading terms of generators of degree d are still leading terms of the new polynomials of degree $d+1$. Moreover, it can happen that some leading term of the new polynomials is one of the $s(n+1)$ multiples of the s independent terms of degree d. If the leading term of a new polynomial is not a multiple of an independent term of degree $d+1$, the corresponding column of M_{d+1} consists of elements equal to 1 or to 0, because of property 5, and because this leading term has coefficient 0 in each generators of degree $d+1$, by construction.

Let the first columns of M_{d+1} be the columns whose elements are equal to 1 or to 0. Then the first columns correspond to those leading terms of the new polynomials that are not multiples of the independent terms of degree d. Let the first rows of M_{d+1} be the rows which correspond to the new polynomials with such leading terms.

Let M'_{d+1} be the submatrix of the matrix M_{d+1} in which the first rows cross the first columns that correspond to the at most $s(n+1)^2$ terms which are not multiples of the independent terms of degree d. Now we apply the Gaussian elimination to M_{d+1}.

When we apply the Gaussian procedure to the rows of the submatrix M'_{d+1}, it is enough to make at most $s(n+1)^2 s(n+1) = s^2(n+1)^3$ subtractions among rows, in order to eliminate repetitions of elements equal to 1 in the same column. Then we apply the Gaussian elimination to the rows of the submatrix M''_{d+1} of M_{d+1} in which the remaining rows cross the remaining $s(n+1)$ columns that correspond to the multiples of the s independent terms of degree d.

It is important not to change the rows corresponding to the new polynomials with the rows corresponding to the generators of degree $d+1$, to determine the dependent generators of degree $d+1$. In fact, if a generator of degree $d+1$ is dependent, the row that corresponds to it becomes the null vector, after applying the Gaussian elimination.

Taking away the polynomials corresponding to null rows from F, we obtain a set of generators for I that is minimal according to [5, Cap. I Corollary 4.11].

ALGORITHM II
Input: a set F of generators constructed by Algorithm I and the term-ordering $<$ used in Algorithm I.
Output: a minimal set of generators.
STEP1. Set $d = \alpha$.
STEP2. $d = d + 1$. Construction of the new polynomials of degree $d + 1$, i.e., multiplication of the generators of degree d (minimalized in the previous iteration) by the variables $X_0, \ldots, X_n$.
STEP3. Construction of the matrix M_{d+1} and order of the column of the matrix in such a way that the first columns correspond to terms which are not multiples of independent terms of degree d.
STEP4. Execution of Gaussian elimination procedure to determine dependent rows of M_{d+1} which correspond to generators of degree $d + 1$.
STEP5. If $d = t$ (where t is the maximum degree of generators), then STOP, else GOTO STEP2.

4. Computational costs

Construction of the generators. To construct the matrix $\bar{G}^s_d$ we consider only the terms which are not multiples of the dependent terms of degree $d - 1$. Because it is $rk(\bar{G}^s_{d-1}) \leq s$, the terms considered are at most $s(n + 1)$.

Let every term be represented by a vector with $n + 1$ components: the hth component is the exponent of the variable X_{h-1}. Then we need at most $s(n + 1)$ additions and at most $s^2(n + 1)^3$ comparisons to construct the terms of degree d (which are multiples of the independent terms) and to order them w.r.t. the term-ordering $<$. Among the constructed terms we must identify those which are not multiples of the dependent terms. We need at most $s^2(n + 1)^3$ comparisons if we apply the same criterion used in [2].

We already mentioned in Section 2 that the degree of generators we consider cannot be higher then s. Therefore, the evaluation of terms at $P_1, \ldots, P_s$ needs at most $s^3(n + 1)$ multiplications.

Every matrix $\bar{G}^s_d$ is at most a $s \times s(n + 1)$ matrix. Thus, the computation of $\bar{\bar{G}}^s_d$ needs a number of additions and of multiplications of order at most $s^3(n + 1)$. If we consider only the operations of additions and multiplications, the proposed procedure is an $O(s^4 n)$ algorithm because we consider at most s matrices.

Computation of minimal generators. It would be useful to know the exact number of the new polynomials with a leading term that is a multiple of a dependent term of degree d. We know that these new polynomials are at most $s(n + 1)^2$; but we would like to know a bound for their number better then $s(n + 1)^2$. We cannot exclude that, for every term which is a multiple of an independent term of degree d, there are more new polynomials with this term as leading term. So, we do not find a better bound than $s(n+1)^2$ for the number of rows of the submatrix M''_{d+1} of the matrix M_{d+1}.

However, it is noteworthy that the cardinality of a minimal set of generators for the ideal I is at most sn [8]. Then, the total number of the new polynomials (that we obtain multiplying the generators by the variables $X_0, \ldots, X_n$) is at most $s(n+1)^2$. In fact, we have already carried out the minimalization procedure over the generator polynomials that we multiply in STEP2 of algorithm II by $X_0, \ldots, X_n$.

If $s \geq n+1$, we can use the bound $s^2(n+1)$ instead of $s(n+1)^2$. If $s < n+1$, when $d = 1$, we can determine at least $n+1-s$ such variables that are leading terms of generators of degree 1: such variables are the dependent terms of degree 1. So, if a polynomial in I has a leading term which is a multiple of a dependent term of degree 1, it is dependent. Then, when $s < n+1$, we construct terms of degree $d+1$ multiplying independent terms of degree d by at most s variables, not by all $n+1$ variables $X_0, \ldots, X_n$.

Finally, in both cases, we can use the bound $s^2(n+1)$ to count all new polynomials we construct to determine the matrices M_{d+1}.

Now we want to consider the block matrix M whose blocks are the matrices M_d. By construction, rows and columns that intersect in each block are different from the rows and columns that intersect in another block. Then, to execute the Gaussian elimination, we have to consider at most $s(n+1)$ columns for every block and at most $2s^2(n+1)$ rows in total. Because of property 6, the number $w = t - \alpha$ of blocks is not more than $s-1$. Let β_i be the number of rows in the ith block $M_{\alpha+i}$. The Gaussian elimination needs a number of additions and of multiplications of at most the following order:

$$\beta_1 s^2 n^2 + \beta_2 s^2 n^2 + \ldots + \beta_w s^2 n^2 =$$

$$= (\beta_1 + \beta_2 + \ldots + \beta_w) s^2 n^2 \leq$$

$$\leq (s^2 n)(s^2 n^2) =$$

$$= s^4 n^3.$$

We consider the bound $s^2(n+1)$ for the number of all new polynomials. In Section 3 we said that the application of the Gaussian elimination to the rows of a submatrix M'_{d+1} needs at most $s^2(n+1)^3$ subtractions. Now we count the total number of these subtractions by using the bound $s^2(n+1)$ for the number of all the new polynomials and we find a number of order at most $s^4(n+1)^2$, because there are $w(\leq s-1)$ blocks.

Finally we conclude with the following assertion.

Proposition 4.1. *A minimal set of generators for an ideal I of s distinct points of the projective space of dimension n over a field K is computable by an $O(s^4 n^3)$ algorithm.*

References

[1] Davis, D., Geramita, A. V., Maroscia, P., Perfect homogeneous ideals: Dubreil's theorems revisited. Bull. Sc. Math., 2e série, 108 (1984), 143-185.

[2] Faugère, J. C., Gianni, P., Lazard, D., Mora, T., Efficient computation of zero-dimensional Gröbner bases by change of ordering. Preprint, Univ. Paris VI (1989).

[3] Mora, T., Robbiano, L., Points in affine and projective spaces. Preprint, Dip. di Mat., Univ. di Genova (1992), no. 210.

[4] Orecchia, F., A polynomial algorithm for computing the conductor of points and of a curve with reduced tangent cone (1991).

[5] Ramella, I., Algoritmi di Computer Algebra relativi agli ideali di punti dello spazio proiettivo. Ph.D. Thesis, Preprint, Dip. di Mat. e Appl., Univ. di Napoli (1990), no. 26.

[6] Ramella, I., Ideals of points in generic position: a polynomial algorithm for computing a minimal set of generators. Preprint, Dip. Mat. e Appl., Univ. di Napoli (1992), no. 30.

[7] Robbiano, L., Introduzione all'Algebra Computazionale. Corso INDAM 1986/87, Roma, Preprint, Dip. di Mat., Univ. di Genova (1987), no. 7.

[8] Valla, G., Generators of ideals and multiplicities. Comm. Algebra 9 (15) (1981), 1541-1549.

Some applications of Castelnuovo theory via residuation

*Edward D. Davis**

Abstract. Two classes of results (some old, some new) are viewed as applications of Castelnuovo theory (for curves in 3-space), applied in part via residuation. Broadly viewed, the two classes are "Gherardelli-Petri type" (concerning the support of the Rao module) and "Segre type" (criteria for a subscheme to be cut out by a hypersurface).

1991 Mathematics Subject Classification: 14H50, 14F05.

Introduction

The article [GH] employs the vehicle of Gherardelli's characterization of complete intersection curves in 3–space to illustrate two basic techniques in the study of space curves: Castelnuovo Theory and Vector Bundle Methods (via the "Serre Construction"). One may view [D3] and [RV] as updatings of [GH], each working almost exclusively with one of those two techniques, which try (with some success) to escape the narrow confines of subcanonical curves and vector bundles. I had originally planned this talk as a report on the efforts to combine our methods, taking as my starting point the several proofs of Gherardelli's theorem. But a principal theme of this meeting, Castelnuovo Theory, drew me more and more toward the interplay between that theory and a third basic technique in the study of space curves, Residuation. That means studying the schemes $S' \subset S$ by passing to the scheme residual to S' in S, particularly these cases: (1) S a complete intersection (Liaison); (2) S' cut from S by a hypersurface. Those two types become one in the "second" Theorem of Gherardelli, thereby suggesting the (obvious, but apparently open) question posed at the end of §1. That question is in fact related to a question examined to some extent in §3, namely: is the support of the Rao module of every subcanonical curve necessarily an interval? But that relationship (the subject originally planned for this talk) will not be discussed here.

* Supported in part by the Consiglio Nazionale delle Ricerche as Professore Visitatore at the Politecnico di Torino. Special thanks to A. Lascu for many conversations—both recently and in years past—on the subjects discussed in this article.

Standing notation

$\mathbb{P}$ denotes $\mathbb{P}^3$ (over any algebraically closed field), $R = H^0_*\mathcal{O}_{\mathbb{P}}$, its homogeneous coordinate algebra, $\bar{\mathbb{P}}$ a general plane in $\mathbb{P}$. X denotes a curve in $\mathbb{P}$ (1–equidimensional *CM* closed subscheme), $\bar{X} = X \cap \bar{\mathbb{P}}$, $I = I_X = H^0_*\mathcal{I}_X$ its *homogeneous ideal* in R, $M = M_X = H^1_*\mathcal{I}_X$ its *Rao Module*, H its *Hilbert Function* ($H(\jmath) = H(X;\jmath) = h^0\mathcal{O}_{\mathbb{P}}(\jmath) - h^0\mathcal{I}_X(\jmath)$). $\mathcal{I}_{\bar{X}}$, $\bar{I}$, $\bar{M}$ and $\bar{H}$ denote the corresponding entities for $\bar{X}$, viewed as a subscheme of $\bar{\mathbb{P}}$. For any projective scheme S: $\alpha(S) \equiv \min\{\jmath | (I_S)_\jmath \neq 0\}$; if $\operatorname{codim} S \geq 2$, $\beta(S) \equiv \min\{\jmath | \gcd(I_S)_\jmath = 1\}$. Other notation, if not as in [AG], will be recalled from [D3]. In particular, we recall here the following.

$e = e(X) = \max\{\jmath | h^1\mathcal{O}_X(\jmath) \neq 0\}$ (*index of speciality* of X).

$s = s(X) = \sigma(\bar{X}) = \min\{\jmath | h^1\mathcal{I}_{\bar{X}}(\jmath-1) = 0\} = \max\{\jmath | h^1\mathcal{I}_{\bar{X}}(\jmath-2) \neq 0\}$.

$a = a(X) = \alpha(\bar{X})$ and $\alpha = \alpha(X)$. (Clearly: $\alpha \geq a$).

$b = b(X) = \beta(\bar{X})$ and $\beta = \beta(X)$. (Clearly: $\beta \geq b$; X integral and $\alpha \neq a \Rightarrow \alpha \geq b$.)

CI, *ACM* and *SC* are acroynms for "2–codimensional complete intersection" (in any projective space), "arithmetically Cohen–Macaulay" and "subcanonical." $CI(m,n)$ denotes a *CI* defined by hypersurfaces of degrees $m \leq n$. "X is f–*SC*" means that $\omega_X = \mathcal{O}_X(f)$. (Recall: if X is $CI(m,n)$, then it is *ACM*, e–*SC* and $e = m+n-4$; if X is reduced and *SC*, then it is necessarily e–*SC*.)

1. Gherardelli's theorem and residuation

Recall the theorem of Gherardelli–Petri ("Gherardelli's Theorem") for curves in $\mathbb{P}$: *ACM* + *SC* $\Longrightarrow$ *CI*. As indicated above [GH] gives two proofs of that result to illustrate two standard techniques of studying curves in $\mathbb{P}$. That paper would have done well to give a third proof (cf. [FL]) to illustrate yet a third standard technique (even older than the other two), namely, Residuation, and in particular, Liaison. That proof follows from these two well known facts concerning a pair of curves linked by a $CI(m,n)$: (1) the ideal sheaf on the complete intersection of either of the curves is isomorphic to the dualizing sheaf of the other twisted by $4-m-n$ (e.g., [AK, ch1(2.3)]); (2) "Gaeta's Theorem" ([G], [PS]) which says that if one of the curves (and hence the other) is *ACM*, and the homogeneous ideal of the complete intersection is generated by part of a minimal homogeneous basis for the ideal of one of the curves, then the ideal of the other curve requires fewer generators (in fact one fewer) than does the ideal of the first curve. Assume that X is f–*SC* and *ACM*, and let Y be the curve linked to X by a $CI(m,n)$, say S, containing X. Since $\mathcal{I}_{Y/S} \cong \omega_X(4-m-n)$ by (1), and $\omega_X \cong \mathcal{O}_X(f)$ by the hypothesis f–*SC*, it follows from the hypothesis *ACM* that I_Y/I_S is (R–module) isomorphic to $(R/I)(4+f-m-n)$, a cyclic module generated by an element of degree $m+n-4-f$. The upshot of that is that I_Y is generated by three forms, the two generators of I_S plus a form of degree $m+n-4-f$. So for m and n large, (2) shows that I is generated by two forms.

The above proof and the second proof in [GH] (by vector bundle methods) are valid for arbitrary X, but the most elementary of the three proofs (by Castelnuovo Theory) is valid only for integral curves. For arbitrary X, that argument proves this "numerical shadow" of Gherardelli's Theorem ([D3, (1.1)]): $ACM + QSC \iff HCI$. (X is QSC ("quasisubcanonical") provided that it is f–QSC for some f, i.e., $h^1\mathcal{O}_X(\jmath) = h^0\mathcal{O}_X(f-\jmath)$ $(\forall \jmath)$; X is *HCI* if it is *ACM* and its Hilbert function is that of a *CI*, necessarily a $CI(a, a_2)$, where a_2 is the least $\jmath$ such that $H(\jmath) \neq h^0\mathcal{O}_{\mathbb{P}}(\jmath) - h^0\mathcal{O}_{\mathbb{P}}(\jmath - a)$.) If X is HCI and not *CI*, then by [DGM, §4] (better, [D2, (3.5) & (4.4)]), $I = DJ + RG$, where: $D \in R_d = \gcd I_{b-1}$; $J = (I : D) = I_Y$, where Y is $HCI(a-d, a_2-d)$; (and obviously then) $G \in J_b$. So Y is residual in X to the subscheme X' cut out on X by the hypersurface defined by D, and moreover, X' is the *CI* defined by (D, G). Consequently, again by [AK, chI(2.3)]: $\omega_{X'} \cong \mathcal{H}om(\mathcal{O}_{X'}, \omega_X)$. It follows that X cannot be *SC*: in that event $\omega_{X'}$ would be a twist of $\mathcal{I}_{Y/X}$, whence I_Y/I would be a cyclic module—which it clearly is not. We see then that Gherardelli's theorem for arbitrary X can be recovered from its elementary version for integral X via residuation of the second kind mentioned in the introductory section above.

It is perhaps interesting to note that the "second" Theorem of Gherardelli illustrates both kinds of residuation. Here are the facts; for the details and further references the reader may consult [FL]. Suppose X is *SC*, and let Y be the curve linked to X by S, an arbitrary *CI* properly containing X. Then $Y = S \cap H$, for some hypersurface H. Conversely, suppose X is of the form $S \cap H$, with H a hypersurface and S a $CI \neq X$. Then the curve residual to X in S is *SC*. That calls to mind this Question: if X is the intersection of three hypersurfaces and is not *CI*, are the degrees of those surfaces uniquely determined? Yes, if X is integral, in which case they are the degrees of the first three forms in a "standard basis" for of I_X.

2. Greco's "Segre Theorem" for curves on integral surfaces

In $\bar{\mathbb{P}}$, let Z be a 0–dimensional subscheme of an integral curve C of degree m. For reduced Z, [S] asserts: $Z = C \cap C'$, with C' a curve of degree n if (and obviously only if) $\ell(Z) = mn$ and $\sigma(Z) \geq m+n-1$. ("ℓ" denotes "length".) But the proof in [S] implicitly assumes Z to be a divisor on C, in which case Segre's theorem becomes a special case of the principal result of [Go]. The point of view adopted in both [S] and [Go] is to make use of (variants of) the "Residue Theorem" (in the sense of [W]), which may be thought of as "liaison up to linear equivalence". On the other hand, by the characterization in [D1] of complete intersections in $\bar{\mathbb{P}}$, we have a Segre theorem for arbitrary Z and C. That characterization comes out of studying arbitrary 0–dimensional subschemes of $\bar{\mathbb{P}}$ by residuation, namely by examining the relationship among the Hilbert functions of such a scheme, a subscheme cut out on it by a curve, and the subscheme residual to that intersection. A special case of that generalized Segre theorem is this (cf. [D1, Remarks on (3.1)]).

Theorem 2.1. *Given Z and C in $\bar{\mathbb{P}}$ as above, then:*

$$\ell(Z) \leq mn \text{ and } \sigma(Z) \geq m+n-1 \Longrightarrow Z = C \cap C', \ \deg C' = n.$$

For the rest of this section we assume that our curve X lies on an integral surface S of degree m. Now if $X = S \cap S'$, where S' is a hypersurface of degree n, then $\deg X = mn$ and X is $(m+n-4)$–SC. Conversely, we have for X this theorem of Segre type [Go, (3.2)].

Theorem 2.2. *Assume:* $\deg X = mn$ *and* X *is* $(m+n-4)$–SC. *Then:*

$$X = S \cap S, \text{ where } S' \text{ is a hypersurface of degree } n.$$

Bearing in mind the definition of "quasisubcanonical" recalled in §1 above, observe:

$$f\text{–}SC \Longrightarrow f\text{–}QSC \Longrightarrow f \leq e.$$

Moreover, basic Castelnuovo theory includes (1) below (e.g., [D3, (R3e)]), and the related observation [D3, (1.2)] includes (2), a theorem of Gherardelli–Petri type.

(1) $e \leq s-3 \leq a+b-4$.

(2) $(a+b-4)$–$QSC \Longrightarrow CI$.

From those facts we can deduce this numerical shadow of (2.2).

Theorem 2.3. *Assume:* $\deg X \leq mn$ *and* X *is* f–QSC *with* $f \geq m+n-4$. *Then:*

$$X = S \cap S', \text{ where } S' \text{ is a hypersurface of degree } n.$$

Proof. Hypothesis and (1) imply: (3) $m+n-4 \leq f \leq e \leq s-3 \leq a+b-4$. Then because $\ell(\bar{X}) \leq mn$, (2.1) gives: $\bar{X}$ is $CI(a,b)$, $a = \min\{m,n\}$, $b = \max\{m,n\}$. By (2) and (3) then, X is $CI(a,b)$. Then for $F \in R_m$ a generator of I_S, take S' to be the hypersurface defined by any $F' \in I_n$ not divisible by F.

3. Gaps in the Rao modules of curves

Definitions. Hereafter, for any curve Y in $\mathbb{P}$, we employ the following notation.

$$\rho_j(Y) \equiv h^1\mathcal{I}_Y(j); \ \ \mu_0(Y) \equiv \min\{j | \rho_j(Y) \neq 0\}; \ \ \mu_1(Y) \equiv \max\{j | \rho_j(Y) \neq 0\}.$$

The notations $\mu_1(Y)$ and $\mu_1(Y)$ obviously subsume that $M_Y \neq 0$. As before, in the case of our curve X, we suppress the "X": $\rho_j \equiv \rho_j(X)$; $\mu_0 \equiv \mu_0(X)$; $\mu_1 \equiv \mu_1(X)$. $\gamma \in \mathbb{Z}$ is a *gap* the the support of the Rao module of Y (for short, a Y–*gap*) provided that: $\rho_\gamma(Y) = 0$ and $\mu_0(Y) < \gamma < \mu_1(Y)$. *NSC* (*numerically subcanonical*) holds for Y means that Y is $e(Y)$–QSC (see §1); and $NSC(j)$ (*numerically subcanonical at level* j) is defined for Y by this: $h^1\mathcal{O}_Y(i) = h^0\mathcal{O}_Y(e(Y)-i)$ $(\forall i \geq j)$. Consequently, *NSC* may also defined thusly: *NSC*(j) $(\forall j)$.

Remarks. Because $h^0\mathcal{O}_X(j)=H(j)+\rho_j$ $(\forall j)$: *NSC*$(e+1)$ holds if $\mu_0 \geq 0$; for integral X, $\mu_0 > 0$, and *NSC*(e) holds iff $h^1\mathcal{O}_X(e)=1$. For integral X, by [R, (2.2)]: *SC* iff *NSC*; *NSC*(j) iff $h^1\mathcal{O}_X(j) = h^0\mathcal{O}_X(e-j)$. Well known for integral X (see (3.10b) below): no gaps greater than $e+1$, and none greater than e when *NSC*(e) holds. [B] proves (3.1a) below (albeit under a characteristic 0 hypothesis, required by its use of "Laudal's Lemma") and shows by example that it is the best one can say about arbitrary integral curves. For integral subcanonical X, (3.1b) implies: no gaps when $e \leq 6$. At this writing there is apparently no known example of a subcanonical curve in $\mathbb{P}$ with a gap, equivalently, a rank 2 vector bundle on $\mathbb{P}$ with a gap in the support of its 1–cohomology.

Theorem 3.1. *If X is integral with γ an X–*gap, *then:*

(a) *In any case:* $e+1 \geq \gamma \geq 5$.

(b) *In case* *NSC*$(\gamma-1)$ *holds:* $e \geq \gamma \geq 6$; $e=\gamma$ *and* *NSC*(4) *fails when* $\gamma=6$.

(3.1a) follows from (3.7) below, the proof of which uses only basic numerical data concerning: the growth of $\bar{H}$, recalled in (3.2); the generation of $\bar{I}$, recalled in (3.3); Castelnuovo Theory, recalled in (3.4) and (3.6); the Rao modules and plane sections of linked curves, recalled in (3.5). (For proofs or references see [D2, §3] for (3.2) and (3.3), [D4, §0] for (3.4) and (3.5).) Our proof of (3.1b) uses those facts and certain further consequences of them developed in [D3] and recalled in (3.9).

Notation. $\Gamma(m,n)$ denotes the first difference of the Hilbert function of a *CI*(m,n) in $\bar{\mathbb{P}}$, that is, the function from $\mathbb{Z}$ to $\mathbb{N}$ defined as follows:

$$\Gamma(m,n;j)=\max\{\Lambda(j),0\}, \text{ where } \Lambda(j)=\min\{j+1,m,m+n-1-j\}.$$

Δ denotes the difference operator on maps from $\mathbb{Z}$ to $\mathbb{Z}$: $\Delta f(j) = f(j)-f(j-1)$. *The arguments below lean heavily on the behavior of* $\Delta\bar{H}$, *and are best understood by thinking in terms of the graph of that function and its relationship to the graphs of the various* $\Gamma(m,n)$ *encountered in those arguments.* For any curve Y in $\mathbb{P}$, put:

$$\delta_j(Y)\equiv\Delta\bar{H}(Y;j);\quad \mu(Y)\equiv\max\{j<s(Y)|\delta_j(Y)=\Gamma(a(Y),b(Y);j)\}.$$

And for our curve X, we do as before, namely: $\delta_j \equiv \delta_j(X)$; $\mu \equiv \mu(X)$.

Recall 3.2: on the growth of $\bar{H}$.

(a) $a \leq b \leq s \leq a+b-1$; $0 \leq \delta_j \leq \Gamma(a,b;j)$ $(\forall j)$; $\delta_j \neq 0 \iff 0 \leq j \leq s-1$.

(b) $\delta_j=\Gamma(a,b;j)$ $(\forall j \leq \mu)$; $a-1 \leq \mu$; $\Delta\delta_j \leq \Delta\Gamma(a,b;j)$ $(\forall j \leq s)$.

(c) $\Sigma\{\delta_i | i \leq j\} = \bar{H}(j) \leq \deg X$ $(\forall j)$; $\bar{H}(j)=\deg X \iff j \geq s-1$.

Recall 3.3: on the generation of $\bar{I}$ (Campanella's Estimates [C]). *Denote the dimension of the vector space* $\bar{I}_j/R_1\bar{I}_{j-1}$ *by* v_j. *Then:*

(a) *For* $j \leq a-1$ *or* $j \geq s+1$, $v_j=0$; $v_a=a+1-\delta_a$.

(b) *For* $a+1 \leq j \leq b$: $v_j \leq -\Delta\delta_j$; $v_j=a-\delta_j$ *if* $j=\mu+1$.

(c) *For* $b+1 \leq \jmath \leq s$: $v_\jmath \leq -\Delta\delta_\jmath - 1$; $v_\jmath = \Gamma(a,b;\jmath) - \delta_\jmath$ *if* $\jmath = \mu+1$.

Recall 3.4: on the "Castelnuovo Sequence." *The restriction from* $\mathbb{P}$ *to* $\bar{\mathbb{P}}$ *induces this exact sequence (of cohomology modules) in the category of graded* R*-modules.*

$$0 \to I(-1) \to I \to \bar{I} \to M(-1) \to M \to \bar{M} \to H^1_*\mathcal{O}_X(-1) \to H^1_*\mathcal{O}_X$$

The sequel denotes the Hilbert functions of $\mathrm{coker}(M(-1) \to M)$, $\ker(M(-1) \to M)$ *and* $\mathrm{coker}(I \to \bar{I})$ *by* C, K *and* $\bar{C}$, *and uses these consequences of that sequence.*

(a) *Vanishing lemmas:* $K(\jmath) = 0$ $(\forall \jmath \leq a-1)$; $C(\jmath) = 0$ $(\forall \jmath \geq s-1)$.

(b) $K = \bar{C}$. (*That is, for all* $\jmath \in \mathbb{Z}$: $\dim(\ker(M_{\jmath-1} \to M_\jmath)) = \dim(\mathrm{coker}(I_\jmath \to \bar{I}_\jmath))$.)

(c) $e = \max\{\jmath \mid \mathrm{coker}(M_{\jmath+1} \to \bar{M}_{\jmath+1}) \neq 0\} \leq s-3$.

Recall 3.5: on Liaison. *Consider a curve* Y *linked to* X *by a* $CI(m,n)$ S, *with this notation:* $\delta'_\jmath \equiv \Delta\bar{H}(Y; m+n-2-\jmath)$; $\rho'_\jmath \equiv h^1\mathcal{I}_Y(m+n-2-\jmath)$. *Then:*

(a) $\delta_\jmath + \delta'_\jmath = \delta_\jmath(S)$ $(\forall \jmath)$. *Hence:* $s(Y) = m+n-2-\mu$ $(m=a)$, $m+n-1-a$ $(m>a)$.

(b) $\rho'_\jmath = \rho_{\jmath-2}$ $(\forall \jmath)$. *Hence:* $\mu_1(Y) = m+n-4-\mu_0$; $\mu_0(Y) = m+n-4-\mu_1$.

Proposition 3.6. nonvanishing lemmas (following from the previous three items).

(a) *Suppose:* $a \leq \jmath \leq \mu_1+1$; $\Delta\delta_\imath = \Delta\Gamma(a,b;\imath)$ $(\jmath < \imath \leq s)$; $\beta = b$ *if* $\jmath < b$.
Then: $K(\jmath) \neq 0$.

(b) *For* $\mu_0 \leq \jmath \leq a-2$, $C(\jmath) \neq 0$; *if* $\mu_0 < a = \delta_a$, *then* $C(a-1) \neq 0$.

Proof. Under the hypotheses of (a), suppose $K(\jmath) = 0$. Then by (3.4b) and (3.3), $K(\imath) = 0$ $(\forall \imath \geq \jmath)$, whence the absurdity $\jmath - 1 > \mu_1$. That proves (a). For (b), take Y as in (3.5) with $m = n \gg 0$. Then by (3.5a): $\delta'_\jmath = 0$ $(\forall \jmath < a)$; $\delta'_a \geq 1$, equality iff $\delta_a = a$. By (3.5b) then, (b) follows from (a) (applied to Y) and (3.4a) (applied to both X and Y).

Proposition 3.7. *Suppose* γ *is an* X-gap. *Then:*

(a) *In any case* $a \leq \gamma \leq s-3$; *when* X *is integral* $\gamma \geq b-1$.

(b) *For* $\gamma \geq b-1$: $\beta \leq \gamma+1$; $s \leq a+b-2$; $\mu > \gamma$ *and* $\beta > b$ *when* $s = a+b-2$.

(c) *For* $\gamma = b-1$: $s \leq a+b-3$; $a < b$; $\mu < b$.

Proposition 3.8. *Assume:* $\gamma \geq b-1$ *an* X-gap; $NSC(\gamma-1)$ *holds;* $\mu_0 > 0$ (*always so when* X *is integral*). *Then:* $e \geq \gamma \geq 6$; $e = \gamma$ *and* $NSC(4)$ *fails when* $\gamma = 6$.

Proof of (3.1) *assuming* (3.7) *and* (3.8). Because $e \leq s-3$ and by "integral," $\gamma \leq e+1$ (both facts well known, e.g., (3.4c) above and (3.10b) below), (3.7) yields:

$$\max\{a, b-1\} \leq \gamma \leq \min\{e+1, s-3\}; \quad s \leq a+b-2, \text{ “} \neq \text{” if } \gamma = b-1.$$

Those data formally imply (3.1a), and show that (3.8) implies (3.1b), *a fortiori*.

Proof of (3.7a). (3.4a) gives the first assertion. Assume X integral. Since $\mathrm{im}(I_\alpha \to \bar{I}_\alpha)$ contains irreducible forms, either $\alpha = a$ and $\bar{C}(\jmath) = 0$ $(a \leq \jmath < b)$, or $\alpha > a$ and, by (3.4b), $K(\jmath) \neq 0$ $(a \leq \jmath < b)$. Therefore, since $\gamma \geq a$ and, by (3.4b), $\bar{C}(\gamma+1) = K(\gamma+1) = 0$, it follows that $\gamma \geq b-1$.

Proof of (3.7b). (3.4b) gives the first assertion. By (3.2), $b \le s \le a+b-1$ and for all $b-1 \le \jmath < s$: $\Gamma(a,b;\jmath) \ge \delta_\jmath \ge s-\jmath$; $\delta_\jmath = s-\jmath$ iff $\delta_{\imath-1} = \delta_\imath + 1$ $(\jmath < \imath \le s)$. So (3.6a) gives the the second assertion and the first conclusion of the third asssertion. To prove the second conclusion of the third assertion, assume $\beta = b$ and take Y as in (3.5) with $m = n = b$. Then by (3.5b), $2b-4-\gamma$ is a Y-gap; but that is not possible, by (3.5a) and (3.6a) (applied to Y).

Proof of (3.7c). The first two assertions follow formally from (3.7a, b). Suppose $b \le \mu$, so by (3.2), $\delta_\imath = \Gamma(a,b;\imath)$ $(\forall \imath \le b)$. Then take Y as in (3.5) with $m = n = b$. Then, as in the proof of (3.7b), by (3.5) and (3.6a), $b-3$ both is and is not a Y-gap.

The proof of (3.8) is based on (3.2)–(3.7) above and (3.9) below. It is partly organized in the lemmas (3.10)–(3.13), the first two of which have the further applications foreshadowed in the concluding remarks of this section.

Recall 3.9: on [D3, §1], the functions F and G and the formula (†).

Notation: $\tilde{\jmath} \equiv e+1-\jmath$; $\Phi(\jmath) \equiv \Sigma\{\delta_\imath | i > \jmath\}$.

Definitions: $F(\jmath) \equiv h^1\mathcal{O}_X(\jmath) - h^0\mathcal{O}_X(e-\jmath)$; $G(\jmath) \equiv \bar{H}(\jmath) + \bar{H}(\tilde{\jmath}) - \deg X$.

Observations: $G(\jmath) = G(\tilde{\jmath}) = \bar{H}(\tilde{\jmath}) - \Phi(\jmath) = \bar{H}(\jmath) - \Phi(\tilde{\jmath})$; $G(\jmath) = 0$ $(\forall \jmath \ge s-1)$.

This formula, denoted (†) *in* [D3], *is a formal consequence of the definitions.*

(a) $\Delta F(\jmath) = C(\jmath) + C(\tilde{\jmath}) + G(\jmath)$.

Next this formal consequence of the definitions and (a).

(b) $NSC(\jmath-1) \iff \Delta F(\imath) = 0$ $(\forall \imath \ge \jmath) \iff C(\imath) + C(\tilde{\imath}) = -G(\imath)$ $(\forall \imath \ge \jmath)$.

Finally this excerpt from [D3, (1.3)], *really an addendum to* (3.2) *and* (3.4c).

(c) *For* $\jmath \ge b-1$: $\Delta G(\jmath) \ge 0$, *equality iff* $G(\jmath-1) = 0$.

Lemma 3.10. *Suppose* $\gamma \ge b-2$ *is an* X-gap.

(a) *In case* $NSC(\gamma-1)$ *holds:* $C(\tilde{\gamma}) = -G(\gamma) > -G(\gamma+1) > 0$.

(b) *In case* $\mu_0 > 0$: $\gamma \le e+1$, *strict inequality if* $NSC(\gamma-1)$ *holds.*

Proof. (3.9b, c) imply (a): if $G(\gamma+1) = 0$, then $C(\imath) = G(\imath) = 0$ $(\forall \imath > \gamma)$, so $\gamma > \mu_1$. (a) implies both assertions of (b), the first because $NSC(e+1)$ necessarily holds whenever $\mu_0 > 0$. (See remarks preceding statement of (3.1).)

Lemma 3.11. For $\gamma \ge b-1$ an X-gap: $NSC(\gamma-1) \Longrightarrow \gamma > a$.

Proof. By (3.7a), $\gamma \ge a$. Suppose $\gamma = a$. Then: $\delta_a > s-a$, by (3.2b) and (3.6a) (because $a \ge b-1$); $\rho_{\tilde{a}} \ne 0$, by (3.10a). And moreover, since in any case, by (3.2a) and (3.4c), $a+b-1 \ge s \ge e+3$: $\tilde{a} \le a-2$; $\delta_a \ge \tilde{a}+3$. By (3.4a, b) and (3.6b) then:

$\rho_{\tilde{a}} \le \rho_{a-2} - (a-2-\tilde{a}) \le \rho_{a-1} - (a+1-\delta_a) = K(a) - (a+1-\delta_a) = \bar{C}(a) - (a+1-\delta_a)$.

Hence $h^0\mathcal{I}_{\bar{X}}(a) > a+1-\delta_a$. But that contradicts (3.3a).

Lemma 3.12. *For* $\gamma \ge b-1$ *an* X-gap:

(a) $\tilde{\gamma} \le a-4$, *equality iff* $\gamma \le b$ *and* $e+3 = s = a+\gamma-2$.

(b) *Assume:* $\gamma = s-3$, $\mu_0 > 0$ *and* $NSC(\gamma-1)$ *holds. Then* $\gamma = e$, *and moreover:*

$\rho_1 = \delta_{\gamma+1} + \delta_{\gamma+2} - 3 \ge 2$; $\delta_{\gamma+1} > \delta_{\gamma+2} \ge 2$.

Proof. Definitions and (3.7b, c) give (a); so we assume the hypotheses of (b). Then by (3.4c) and (3.10b), $\gamma{=}e{=}s{-}3$. By (3.9b), (3.2a, b), (a) and (3.10a) then:
$\rho_1 = -G(\gamma) = \delta_{\gamma+1}+\delta_{\gamma+2}-3 \geq 2$;
$\rho_{\gamma+1} = C(\gamma+1) = -G(\gamma+1) = \delta_{\gamma+2}-1 > 0$;
$\delta_{\gamma+1} > \delta_{\gamma+2}$.
All assertions of (b) are thereby established.

Lemma 3.13. *With* $6{\geq}\gamma{\geq}b{-}1$ *an* X–gap, *assume:* $\mu_0{>}0$ *and* $NSC(\gamma{-}1)$ *holds. Then one of the two cases described immediately below obtains.*
Case 1: $(a,b,s,e,\gamma){=}(5,6,9,6,6)$; $\delta_j{=}\Gamma(5,6;j)$ $(\forall j{<}9)$.
Case 2: *same as* Case 1, *except* $b{=}7$ *and* $4{\leq}\delta_5{\leq}5$.
Moreover, in both cases: $\rho_4{\geq}\rho_3{>}\rho_2{>}\rho_1{=}2$; $\rho_4{>}\rho_3$ *if* $\delta_5{=}5$; $\beta{=}7{\geq}\alpha{\geq}6$.

Proof. By (3.7a, b), (3.11) and (3.12a): $(a,\gamma)=(5,6)$; $s\geq 9$; $6\leq b\leq\beta\leq 7$. So in case $b{=}6$, by (3.7b): $(s,\beta){=}(9,7)$ and $\delta_j{=}\Gamma(5,6;j)$ $(\forall j{\leq}7)$. And in case $b{=}7$, by (3.7c): $(s,\beta)=(9,7)$ and $\delta_7\leq 3$. In both cases then: by (3.12b), (3.2) and (3.6b): $(\delta_7,\delta_8,\rho_1,e)=(3,2,2,6)$; $4\leq\delta_6\leq\delta_5\leq 5$; $\rho_4\geq\rho_3>\rho_2>2$; $\rho_4>\rho_3$ if $\delta_5=5$. Thus: ρ_j is as described in (3.13); δ_j is as described in (3.13) for the case $b{=}6$; and to prove to prove that for the case $b{=}7$, by (3.2), it suffices to prove $\delta_6{\neq}5$. Suppose $b{=}7$ and $\delta_6{=}5$. Then by (3.4b): $\rho_5{=}K(6){=}\bar{C}(6){\leq}3$; whence $\rho_4\leq 3{+}K(5)\leq 4$, which is absurd because $\rho_4>\rho_3\geq 4$ when $\delta_5=5$. Lastly observe that by (3.4b) again, in both cases $\alpha{>}5$: otherwise, the absurdity $2{\geq}\bar{C}(6){=}\rho_5{\geq}\rho_4{-}1{\geq}3$.

Proof of (3.8). (3.1b) gives $e\geq\gamma$. Suppose $\gamma\leq 6$. Then by (3.13), one of the two cases described there obtains. Suppose $\delta_5 = 5$ (necessarily so in Case 1). Then by (3.4b): $\rho_4\leq 5{-}C(5)$. But by (3.13): $\rho_4\geq 5$ in this case. Hence: $C(5)=0$ and $C(2) = 1$. And furthermore, direct computation shows: $G(5) = -3$. By (3.9a, b) then: $\Delta F(5){=}{-}2$; whence $NSC(4)$ fails. Likewise, if $\delta_5{=}4$ and $\alpha{=}6$, then $\rho_5{\leq}4$, and the same computation yields: $\Delta F(5) = -1$. So we are left with Case 2 and these properties: $\delta_5 = 4$; $\alpha = \beta = 7$; whence by (3.4b), $K(6) = K(5){+}3 = 5$; and by direct calculation, $G(5) = G(6){-}1 = -3$. Those data enable us to deduce an absurdity from the assumption $NSC(4)$. Under that assumption (3.9b) yields this.
$C(6){+}C(1){=}{-}G(6)$; $C(5){+}C(2){=}{-}G(5)$.
Then since $C(j){=}\rho_j{-}\rho_{j-1}{+}K(j)$, those data and those above give this.
$K(6){-}\rho_5{+}2{=}2$; $K(5){+}\rho_5{-}\rho_4{+}\rho_2{-}2{=}3$.
Consequently: $\rho_4{-}\rho_2 = 2$. The absurdity will come from making another calculation of $\rho_4{-}\rho_2$ using (3.5), and obtaining a different result. To that end take Y as in (3.5), with $m{=}n{=}7$. Then by (3.5): $s(Y){=}8$ and $(\delta'_7,\delta'_6,\delta'_5)=(3,3,2)$, so by (3.2b), $b(Y) = \beta(Y) = 7$; $(\rho'_8,\rho'_6,\rho'_5,\rho'_4) = (0,\rho_4,\rho_3,\rho_2)$. Let the functions K' and $\bar{C}'$ be the analogs for Y of the functions K and $\bar{C}$ associated to X by (3.4). Then by (3.4b) and (3.3b) (applied to Y), $\bar{C}'(5)=\bar{C}'(6)=\bar{C}'(7)=0$; whence by (3.4a) (applied to X), (3.4b) (applied to Y), and (3.5b), $\rho'_5 = \rho'_6$. Now applying (3.3c), (3.4a, b) and (3.6a) shows: $\bar{C}'(8){\leq}1$; $K'(8){\neq}0$; whence $\rho'_5{-}\rho'_4{=}1$. Hence: $\rho_4{-}\rho_2{=}\rho'_6{-}\rho'_4{=}1$. That gives our absurdity. End of proof of (3.8).

Concluding remarks: What next? For the purposes of these remarks "curve" means "curve in $\mathbb{P}$", X is an arbitrary integral curve. Consider this assertion, hereafter denoted $\mathcal{A}(n)$: $\gamma \leq n$ *an* X–gap $\Longrightarrow$ $NSC(e+4-\gamma)$ *fails*. So: "$\mathcal{A}(n)$ $(\forall n)$" implies "no gaps for subcanonical curves"; $\mathcal{A}(6)$ is a theorem, by (3.1b). My original intention for this section was to apply the methods of [D3] and [D4] to the "gap question" (with "subcanonical" at least partly in mind), expecting to arrive soon at an enlightening impass: one providing either numerical parameters for a conjectural "minimal" counterexample or clues to why none should exist. So what comes of an attempt on $\mathcal{A}(7)$ by those methods? Only these unenlightening obstructions.

(1) A plethora of "unreasonable" cases (i.e., $\mu < b-1$, as in Case 2 above).

(2) Several unyielding reasonable cases (e.g., $(a, b, s) = (5, 7, 10)$, a "Case 1"). What to do? Notice that those two obstacles vanish in characteristic 0: (1) by the uniform position of $\bar{X}$, (2) by Laudal's Lemma. What to do: retreat to characteristic 0—where integral curves are reasonable and Laudal dwells. That is the business of §4. *But there is no* §4. The reason is that (un)fortunately I have as yet neither reached that impass nor found a definitive theorem. I can report that $\mathcal{A}(12)$ holds in characteristic 0. (Consequently: in characteristic 0, no gaps for subcanonical X with $e \leq 12$.) That follows from several "general" results concerning the relationships among the members of $\{a, b, s, \alpha, \beta, \mu, e, \gamma\}$ rather than their specific values. Those results reduce the proof of $\mathcal{A}(12)$ to the consideration of a reasonably small number of numerical possibilities, most of which are straightforwardly handled by a use of (3.9) similar to its use in the proof of (3.8). In the exceptional cases the possible curves are necessarily subcanonical, and the use of (3.9) provides numerical data sufficient to show that the genus of the curve is too big for its degree, or if one prefers vector bundle methods, that the 1–cohomology of the associated vector bundle is incompatible with its Chern classes. I experimented briefly with writing a §4 which would sketch comprehensibly the results just mentioned, but found that even that would require not only an extensive elaboration of items (3.7)-(3.11), which would seriously unbalance this paper, but also a quantum leap in notation. Better to end this paper here, with the expectation of publishing, after further reflection, a more polished version of the phantom §4—very likely in the 1992/93 number of The Curves Seminar at Queen's (Volume IX).

References

[AG] Hartshorne, R., Algebraic Geometry (GTM 52). Springer, New York, 1977.

[AK] Altman, A., Kleiman, S., Introduction to Grothendieck Duality Theory (LNM 146). Springer, New York, 1970.

[B] Buraggina, A., *On the connectedness of the Hartshorne–Rao module of curves in* P^3. Preprint (1992).

[C] Campanella, G., *Standard bases of perfect homogeneous polynomial ideals of height* 2. J. Algebra **10** (1986), 47-60.

[D1] Davis, E., *0-dimensional subschemes of* P^2*: new applications of Castelnuovo's function.* Ann. Univ. Ferrara **32** (1986), 93-107.

[D2] Davis, E., *Complements to a paper of P. Dubreil.* Ric. Mat. **37** (1988), 347-357.

[D3] Davis, E., *Complements to a theorem of Gherardelli: the postulational viewpoint.* Ann. Univ. Ferrara **37** (1991), 205-219.

[D4] Davis, E., *Curves which are close to complete intersections.* The Curves Seminar at Queen's VII (A.V. Geramita, ed.) QPPAM **85** (1990).

[DGM] Davis, E., Geramita, A., Maroscia, P., *Perfect homogeneous ideals: Dubreil's theorems revisited.* Bull. Sci. Math. **108** (1984), 143-185.

[FL] Fiorentini, M., Lascu, A., *Projective embeddings and linkage.* Rend. Sem. Mat. Fis. Milano **57** (1989), 161-182.

[G] Gaeta, F., *Nuove ricerche sulle curve sghembe algebriche di residuale finito.* Ann. Mat. Pura Appl. (ser. 4) **31** (1950), 1-64.

[Go] Greco, S., *Alcune osservazioni sui sottoschemi di codimensione* 1 *di una varietà proiettiva.* Università di Bologna, Seminari di Geometria 1985, 89-100.

[GH] Griffiths, P., Harris, J., *Two proofs of a theorem concerning algebraic space curves.* Proceedings of the 8th Iranian Math Conference (1978).

[PS] Peskine, C., Szpiro, L., *Liaison des variétés algébriques.* Inv. Math. **26** (1974), 271-302.

[R] Roggero, M., *On rank* 2 *reflexive sheaves and subcanonical curves in* P^3. Comm. Alg. **16** (1988), 1770-1790.

[RV] Roggero, M., Valabrega, P., *Some vanishing properties of the intermediate cohomology of a reflexive sheaf on* P^n. Preprint (1992).

[S] Segre, B., *Sui teoremi di Bezout, Jacobi e Reiss.* Ann. Mat. Pura Appl. (4) **26** (1947), 1-26.

[W] Walker, R., Algebraic Curves, Princeton University Press, Princeton, 1950 (also reprinted by Dover Publications, New York, 1962).

On the injectivity of the nodal map

Alessandro De Paris

Abstract. In this paper we study (in characteristic 0) the nodal map $p_{d,g} : \Sigma_{d,g} \to \mathrm{Sym}^{(\delta)}(\mathbf{P}^2)$, where $\Sigma_{d,g}$ is the open set of the Severi variety, which parametrizes all curves of degree d and geometric genus g in $\mathbf{P}^2$ having at most nodes as singularities, and where $p_{d,g}$ sends a curve to the set of its nodes. In particular, we study the injectivity of $p^r_{d,g}$, the restriction of $p_{d,g}$ to the subset $\Sigma^r_{d,g}$ given by those curves, which are projections of smooth, non degenerate, irreducible curves in $\mathbf{P}^r$. We classify completely the (few) values of d, r, g for which $p^r_{d,g}$ is not injective, and obtain a simple necessary and sufficient condition for the injectivity, involving the canonical and hyperplane divisor series.

0. Introduction

The aim of the paper is to investigate some connections between the theory of algebraic curves in characteristic 0 and the study of groups of points in the plane. Starting from the idea of Brill and Noether (see [2]), who considered plane nodal models of curves, in order to obtain information on families and linear series, we deal with a "nodal map": strictly speaking, we can define this map as the function, with domain in the set of plane curves of degree d and genus g having only nodes as singularities, which sends a curve to the set of its nodes; but this map becomes really interesting when we insert it in a more general context, and this can be done both from the viewpoint of moduli and from the viewpoint of Hilbert schemes.

Let $\mathcal{M}_g$ be the moduli space of curves of genus g. Locally over $\mathcal{M}_g$, it is possible to construct varieties $\mathcal{W}^r_{d,g}$, together with morphisms $\mathcal{W}^r_{d,g} \to \mathcal{M}_g$, in such a way that for every curve C, as a point of $\mathcal{M}_g$, the fibre of the morphism parametrizes complete linear series of degree d and dimension $\geq r$. Then, we can consider the covering $\mathcal{C}^{2,r}_{d,g} \to \mathcal{W}^r_{d,g}$, such that for every point of $\mathcal{W}^r_{d,g}$, which represents a linear series on a curve C, the fibre parametrizes the subseries of dimension two and their bases. More precisely, we are interested only in those subseries which send C birationally on a plane curve having only nodes as singularities, and then we can relate $\mathcal{C}^{2,r}_{d,g}$ with $\mathrm{Sym}^\delta(\mathbf{P}^2)$ (with appropriate δ), via the nodal map.

For example, recall that the Brill-Noether conjecture, proved by several authors in the last years, says the generic fibre of the morphism $\mathcal{W}^r_{d,g} \to \mathcal{M}_g$ has dimension $\rho(d, g, r) = g - (r + 1)(g + r - d)$ (Brill-Noether number), if it is ≥ 0, otherwise the morphism is not dominant. One immediately checks that, once proved the dominancy, and after some easy calculation on the morphism $\mathcal{C}^{2,r}_{d,g} \to \mathcal{W}^r_{d,g}$, we can reduce the problem to the determination of the image and of the generic fibre of some restrictions of the nodal map (see also [5]).

As an other example, note that in the paper [1], of Arbarello and Cornalba, they obtain results on Hurwitz spaces and Severi varieties, using considerations on the sets of nodes of plane curves.

From the viewpoint of Hilbert schemes, which we shall use in this paper, we can consider the open subvariety $V_{r,d,g} \subset \mathrm{Hilb}_{r,d,g}$, given by smooth, irreducible, non-degenerate curves of degree d and genus g in $\mathbf{P}^r$; if we fix a projection $\mathbf{P}^r \to \mathbf{P}^2$, this one sends the generic curve of $V_{r,d,g}$ on a nodal curve, and one easily checks that the set of nodal plane curves obtained in this way, say $\Sigma^r_{d,g}$, is independent of the projection. So, we can obtain information about $V_{r,d,g}$ by considering the nodal map $p^r_{d,g}$: $\Sigma^r_{d,g} \to \mathrm{Sym}^\delta(\mathbf{P}^2)$, and reducing problems on components of Hilbert schemes to the study of subvarieties of $\mathrm{Sym}^\delta(\mathbf{P}^2)$. This construction, in the open set which we are interested in, is essentially equivalent to the previous one.

Here we do not consider the interesting problem of the determination of the locus inside $\mathrm{Sym}^\delta(\mathbf{P}^2)$ formed by groups of points which are nodes of curves of fixed type (and of its equations); what we are going to do, is to establish the injectivity of the nodal maps $p^r_{d,g}$.

Treger, in [15], has improved some of the results of Arbarello and Cornalba, and has proved, in the expected range, the birationality of the nodal map, defined over all the set $\Sigma_{d,g}$ of plane nodal curves of given degree and genus. These varieties were intoduced by Severi in [14]; later, Zariski in [16] has proved they are open sets of these which parametrize all plane (not necessarily nodal) curves of given degree and genus; of course, as $\Sigma^r_{d,g}$ is a closed, almost always proper subset of $\Sigma_{d,g}$, we cannot use directly Treger's results; indeed it may happen, and actually in many cases happens, that $\Sigma^r_{d,g}$ lies outside the open set where the nodal map is an isomorphism.

We show that $p^r_{d,g}$ is injective, except for a few numerical cases, which we completely classify (see table 0.1). Furthermore, we prove (corollary 2.9) that the locus in $V_{r,d,g}$, of curves for which the nodal map is not injective, is defined by the simple equation $h^0(\mathcal{O}(2K-(d-6)H)) > 0$ over the Hilbert scheme; notice that this equation involves the canonical and hyperplane series, which are in many cases the generators of the Picard group of the universal curve (see [12], or, more generally, [7] and [13]). It is also interesting to note that in some cases, this equation separates some components of Hilbert scheme, while sometimes defines proper subvarieties of components (see Example 3.3 and Example 5.1).

I thank Prof. Luca Chiantini, whose help has been essential in the preparation of this paper.

Table 0.1
(All the values of r, d, g, such that $p^r_{d,g}$ is not injective)

r	d	g
3	3	0
	4	0, 1
	5	0, 1, 2
	6	1, 2, 3, 4
	7	3, 4, 5, 6
	8	5, 6, 7, 8, 9
	9	10, 12
	10	11, 12, 15, 16
	11	23
	12	19, 24, 25
	$[13,\infty]$	$[\frac{1}{4}d^2 - d, \frac{1}{4}d^2 - d + 1]$
4	4	0
	5	0, 1
	6	1, 2
	7	3
	8	5
5	5	0
	6	1

1. Notations and recalls

All the schemes are supposed over a field k, algebraically closed of characteristic 0. If D is a divisor, we denote by $|D|$ the complete linear series of D, which is often supposed with its natural projective structure. A *curve* will mean *projective reduced scheme, pure of dimension* 1 *over* k or *divisor of* $\mathbf{P}^2$, and it will be clear from the context which meaning we are using; points will be always closed. When we deal with a singular plane curve, we consider divisors as cut on the desingularization.

Div X = divisor's group of X.

$\mathbf{C}_d = \mathbf{P}(H^0(\mathcal{O}_{\mathbf{P}^2}(d)))$.

$V_{r,d,g}$ = the set of smooth, irreducible, non-degenerate curves of degree d and genus g in $\mathbf{P}^r$ (is an open set in the Hilbert scheme).

$\Sigma_{d,g}$ = the set of irreducible curves in $\mathbf{P}^2$, of degree d and geometric genus g, having only nodes as singularities (is an open set in the Severi variety; see [14], Anhang F, and [16]).

$\Sigma^r_{d,g}$ $(r > 2)$ = the set of irreducible curves of $\Sigma_{d,g}$, which can be obtained from a curve C of $V_{r,d,g}$, by birational projection from a point $P \in \mathbf{P}^r$, $P \notin C$.

$\delta_{d,g} = \frac{(d-1)(d-2)}{2} - g$ = number of nodes of any curve of $\Sigma_{d,g}$.

$p_{d,g} : \Sigma_{d,g} \to \mathrm{Sym}^{\delta_{d,g}}(\mathbf{P}^2)$ = function which sends a curve to the set of its nodes.

$p^r_{d,g}$ = restriction of $p_{d,g}$ to $\Sigma^r_{d,g}$.

$G(r,d) = \frac{M(M-1)}{2}(r-1) + Mq$, where M and q are the quotient and the remainder of the division of $d-1$ by $r-1$

Finally, it is necessary to recall two classical results, which we shall often use in the paper.

Proposition 1.1. : *If C is a smooth, irreducible, non-degenerate curve in $\mathbf{P}^r$, of degree d and genus g, then $g \leq G(r,d)$ (Castelnuovo bound).*

Proof. See [3] and [4]. □

Proposition 1.2. *If C is a smooth, irreducible curve in $\mathbf{P}^3$ of degree d and genus g, not contained in any quadric surface, then $g \leq \frac{1}{6}d^2 - \frac{1}{2}d + 1$.*

Proof. See [8]. □

2. Preliminary results

The problem of injectivity of $p^r_{d,g}$, in many cases can be solved by easy corollaries of the propositions of Section 1 :

Proposition 2.1. *If $g < \frac{1}{4}d^2 - \frac{3}{2}d + 1$ then $p^r_{d,g}$ is injective.*

Proof. If there are $C_0, C'_0 \in \Sigma_{d,g}$, $C_0 \neq C'_0$, with $p_{d,g}(C_0) = p_{d,g}(C'_0)$, then, considering the intersection in $\mathbf{P}^2$, must be $4\delta_{d,g} \leq C_0.C'_0 = d^2$, hence, using the expression of $\delta_{d,g}$, we have $g \geq \frac{1}{4}d^2 - \frac{3}{2}d + 1$.

Hence, if $g < \frac{1}{4}d^2 - \frac{3}{2}d + 1$, $p_{d,g}$ must be injective, and so its restriction $p^r_{d,g}$.□

Corollary 2.2. *If $r \geq 6$, then $p^r_{d,g}$ is injective, for all values of d and g.*

Proof. It suffices to note that $r \geq 6$ implies $G(r,d) < \frac{1}{4}d^2 - \frac{3}{2}d + 1$; hence we have that $r \geq 6$ and $g \geq \frac{1}{4}d^2 - \frac{3}{2}d + 1$ imply $\Sigma^r_{d,g} = \emptyset$ (by Proposition 1.1), while if $g < \frac{1}{4}d^2 - \frac{3}{2}d + 1$, then $p^r_{d,g}$ is injective by Proposition 2.1. □

If we notice that $\delta_{d,g}$ fixed singular points impose $3\delta_{d,g}$ (not necessarily independent) linear conditions, we can easily find, given a curve $C_0 \in \Sigma_{d,g}$, a sufficient condition for the existence of another curve C_0', having singular points in all the nodes of C_0; but we cannot deduce from this fact that $p_{d,g}^r$ is not injective, as we need that C_0' has no other singularities out of the nodes of C_0, and that the singularities in these points are exactly nodes. We should also have that even C_0' comes, by birational projection from an external point, from a curve of $V_{r,d,g}$. We shall prove now some propositions, which allow us to solve these questions.

Proposition 2.3. *Let $C_0 \in \Sigma_{d,g}$. If there exists a curve $A \in \mathbf{C}_d$, $A \neq C_0$, having singularities in all the points of the set $p_{d,g}(C_0)$ of the nodes of C_0, then there exists a curve $C_0' \in \Sigma_{d,g}$, $C_0' \neq C_0$, with $p_{d,g}(C_0) = p_{d,g}(C_0')$.*

Proof. Let $L \subset \mathbf{C}_d$ the pencil generated by C_0 and A. The condition $p_{d,g}(C_0) = p_{d,g}(C_0')$ is equivalent to the following two open conditions:

a) C_0' is smooth outside $p_{d,g}(C_0)$.

b) Singularities in C_0' are at most nodes.

The subset of L given by those $C_0' \in L$ satisfying a) and b) is hence open and nonempty (because it contains C_0), thus in this set there are infinitely many elements among which we can find a curve distinct from C_0. □

Lemma 2.4. *Let $C_0, C_0' \in \Sigma_{d,g}$, with $p_{d,g}(C_0) = p_{d,g}(C_0')$ and with respective desingularizations C and C', let R be a line of $\mathbf{P}^2$, which cuts both C_0 and C_0' in d distinct points, and finally let $H \in$ Div C and $H' \in$ Div C' be the divisors cut by R. Then* $\dim|H| = \dim|H'|$.

Proof. Since C_0 and C_0' have the same degree and geometric genus, it suffices to show that H and H' have the same index of speciality; so, let i be the index of H and i' that one of H', and let us prove that $i' \geq i$:

If $i = 0$ it is trivial; otherwise, we find i canonical groups containing H and, consequently, i adjoint curves of C_0, say $A_1, \cdots, A_i$, of degree $d-3$, which are linearly independent on $\mathbf{P}^2$ and contain $R \cap C_0$. Since $R \cap C_0$ has d points, the adjoints, which have degree $d-3$, must have R as a component, and then they contain $R \cap C_0'$; furthermore they are adjoint of C_0' (because $p_{d,g}(C_0) = p_{d,g}(C_0')$). Hence, $A_1, \cdots, A_i$ cut i canonical groups of C', which will be independent and contain H', and this means that $i' \geq i$. Exchanging the roles of C_0 and C_0', we obtain $i \geq i'$ and then the equality. □

Lemma 2.5. *Let $C_0, C_0' \in \Sigma_{d,g}$, with $p_{d,g}(C_0) = p_{d,g}(C_0')$ and with respective desingularizations C and C'; let R be a line of $\mathbf{P}^2$, which cuts both C_0 and C_0' in a fixed common node and in other $d-2$ distinct points; finally, let $N \in$ Div C, $N' \in$ Div C' be the divisors which correspond to the node, and $H \in$ Div C, $H' \in$ Div C' the divisors cut by R. Then* $\dim|H - N| = \dim|H' - N'|$.

Proof. Similar of that of Lemma 2.4, except for the fact that now $R \cap C_0$ is a set of $d-1$ distinct points. □

Proposition 2.6. *Let $C_0 \in \Sigma^r_{d,g}$ and $C'_0 \in \Sigma_{d,g}$ such that $p_{d,g}(C_0) = p_{d,g}(C'_0)$, then $C'_0 \in \Sigma^r_{d,g}$.*

Proof. Let C and C' the respective desingularizations of C_0 and C'_0. If $|H|$ is the hyperplane section on C, relative to C_0, we must have $\dim|H| \geq r$ and $|H|$ very ample. If $|H'|$ is the hyperplane section on C', Lemma 2.4 gives that $\dim|H| \geq r$, and implies, togheter with Lemma 2.5, that $|H'|$ separates the pairs of points of C', which correspond to each node of C'_0 (indeed $\dim|H' - N'| = \dim|H - N| = \dim|H| - 2 = \dim|H'| - 2$); hence $|H'|$ is very ample.

By the above discussion, we can think of C' as embedded in $\mathbf{P}^{r'}$, with $r' \geq r$, and we have a projection of C' on $C'_0 \subset \mathbf{P}^2$ from a subspace L of $\mathbf{P}^{r'}$, of dimension $r' - 3$. The secant variety of C' only meet L in the points of the intersection of L with the secant lines which meet the nodes of C'_0, while the tangent variety does not meet L (because C'_0 has no cusps), hence it is possible to find a subspace L' of L, of dimension $r' - r - 1$ and not passing through those points of intersection. If we project C' from L' in a $\mathbf{P}^r \supset \mathbf{P}^2$ such that $L' \cap \mathbf{P}^r = \emptyset$, we have a smooth, non-degenerate, irreducible curve $C'' \subset \mathbf{P}^r$, and the projection of $\mathbf{P}^r$ from $L \cap \mathbf{P}^r$ send C'' on C_0. □

Proposition 2.7. *If $g > \frac{1}{3}d^2 - 2d + 1$ and $\Sigma^r_{d,g} \neq \emptyset$, then $p^r_{d,g}$ is not injective.*

Proof. By some easy calculations, $g > \frac{1}{3}d^2 - 2d + 1$ is equivalent to $3\delta_{d,g} < \frac{d(d+3)}{2}$; now, take a $C_0 \in \Sigma^r_{d,g}$: we have that the linear system of curves having singular points in all the nodes of C_0 has dimension $\geq \frac{d(d+3)}{2} - 3\delta_{d,g}$, which in turn (according to our hypotheses) is > 0; hence there exists a curve A, which satisfy the hypotheses of the Proposition 2.3, and then we can find a curve $C'_0 \in \Sigma_{d,g}$ such that $C'_0 \neq C_0$ and $p_{d,g}(C'_0) = p_{d,g}(C_0)$. Finally, by Proposition 2.6, C'_0 belongs to $\Sigma^r_{d,g}$. □

The previous proposition is far from providing us with a complete answer to the problem of the injectivity of $p^r_{d,g}$, because its hypothesis is too strong; however, the propositions proved till now, suggest that this problem can be solved by studying the linear series; indeed, as we have seen, the injectivity of $p^r_{d,g}$ depends on the existence, once fixed a $C_0 \in \Sigma^r_{d,g}$, of curves A, which are singular in the nodes of C_0. Such curves are called *biadjoints* (see [6]), and we have that the biadjoints of degree $2(d-3)$ cut, outside the doubled adjoint divisor (i.e. the double of the divisor corresponding to the nodes), the double of the canonical series; starting from this fact, we can state the following proposition.

Proposition 2.8. *Let $C_0 \in \Sigma_{d,g}$, with desingularization C; then there is a curve $C'_0 \in \Sigma_{d,g}$, distinct from C_0, with $p_{d,g}(C'_0) = p_{d,g}(C_0)$, if and only if $|2K-(d-6)H|$ is effective on C, where $|K|$ is the canonical series and $|H|$ is the hyperplane section relative to C_0.*

Proof. From Proposition 2.3 we have that the existence of a C'_0 which satisfies the imposed condition is equivalent to the existence of a biadjoint of C_0 of degree d. Hence, it suffices to show that the biadjoints of degree d cut, outside of $2N$, the complete series $|D| = |2K - (d-6)H|$ (here we write N for the adjoint divisor of C). This means that each biadjoint of degree d cuts a divisor linearly equivalent to $2N + D$, and for each divisor $D' \in |2K - (d-6)H|$ there is a biadjoint A which cuts $2N + D'$ on C. Now, it is clear that curves of degree d cut divisors linearly equivalent to $2N + D$; in order to prove the completeness, we observe that if we fix a divisor $D' \in |2K - (d-6)H|$, then $2N + D' = N + (N + D')$ will be cut by an adjoint A, of degree d, as we know the adjoints cut complete series outsides the nodes. But this one must be a biadjoint too: indeed, since C_0 has at most nodes as singularities, the fact that A cuts a divisor containing $2N$, implies that A has multiplicity at least 2 in each node of C_0. □

Corollary 2.9 *$p^r_{d,g}$ is not injective if and only if there is a smooth, irreducible, non-degenerate curve in $\mathbf{P}^r$, of degree d and genus g, on which $|2K - (d-6)H|$ is effective, where $|K|$ is the canonical series and $|H|$ is the hyperplane section.*

Proof. It follows immediately from Proposition 2.6, Proposition 2.8 and the fact that every smooth curve can be projected on a plane nodal curve. □

The previous assertion says that the locus of the curves in $V_{r,d,g}$ which can be projected on curves of $\Sigma^r_{d,g}$ for which $p^r_{d,g}$ is not injective, is described by the "equation" $h^0(\mathcal{O}(2K - (d-6)H)) > 0$, defined on the Hilbert scheme which contain $V_{r,d,g}$; we also notice that this equation involves the series $|K|$ and $|H|$, which in many cases are "the unique" series on the universal curve over the Hilbert scheme (i.e. the Picard group is generated by these series; see [12], or, more generally [7] and [13]). Finally we remark that this locus is independent of the chosen projection.

3. Curves in $\mathbf{P}^4$ and $\mathbf{P}^5$

In this section we solve our problem for $r = 4, 5$.

Proposition 3.1. *$p^5_{d,g}$ is not injective if and only if $(d,g) \in \{(5,0),(6,1)\}$; $p^4_{d,g}$ is not injective if and only if $(d,g) \in \{(4,0),(5,0),(5,1),(6,1),(6,2),(7,3),(8,5)\}$.*

Proof. By Propositions 1.1 and 2.1, $p^r_{d,g}$ can be not injective only if $\frac{1}{4}d^2 - \frac{3}{2}d + 1 \leq g \leq G(r,d)$. For $r = 5$ this condition holds only if $(d,g) \in \{(5,0),(6,1)\}$, and in this cases $p^r_{d,g}$ is actually not injective: indeed in the case $(d,g) = (5,0)$ it suffices to consider a plane nodal projection of the quintic rational normal curve in $\mathbf{P}^5$, and then apply Proposition 2.7, while in the case $(d,g) = (6,1)$ we can consider an elliptic smooth curve, embed it in $\mathbf{P}^5$ by a complete series $|H|$, of degree 6, and use Corollary 2.9, as in this case $|2K - (d-6)H| = 0$.

For $r = 4$, the unique pairs of values of d and g, which satisfy the condition $\frac{1}{4}d^2 - \frac{3}{2}d + 1 \leq g \leq G(r,d)$, are those listed above, and in these cases $p^4_{d,g}$ is actually not injective, indeed:

$(d,g) = (4,0)$: $\Sigma^4_{4,0} \neq \emptyset$, as there is a plane nodal projection of a quartic rational normal curve in $\mathbf{P}^4$, then $p^4_{4,0}$ is not injective by Proposition 2.7.

$(d,g) = (5,0)$: $\Sigma^4_{5,0} \neq \emptyset$, as we can obtain, projecting the quintic rational normal curve, a smooth rational quintic in $\mathbf{P}^4$; then a plane nodal projection of the last one is clearly an element of $\Sigma^4_{5,0}$. Now it suffices to apply Proposition 2.7.

$(d,g) = (5,1)$: $\Sigma^4_{5,1} \neq \emptyset$, as we can embed a smooth elliptic curve in $\mathbf{P}^4$, by any complete series of degree 5, and then as above obtain an element of $\Sigma^4_{5,1}$; again, Proposition 2.7 implies $p^4_{5,1}$ is not injective.

$(d,g) = (6,1)$: It suffices to take a smooth projection in $\mathbf{P}^4$ of the elliptic sestic in $\mathbf{P}^5$ considered in the case $r = 5$, and apply Corollary 2.9, as the condition of effectivity on $|2K - (d-6)H|$ is clearly preserved.

$(d,g) = (6,2)$: $\Sigma^4_{6,2} \neq \emptyset$, because we can take a plane nodal projection of any smooth curve in $\mathbf{P}^4$, obtained from a smooth curve of genus 2 embedded by any complete series of degree 6 (which is certainly very ample, see e.g. [10], IV, 3.2). Then, apply Proposition 2.7.

$(d,g) = (7,3)$: take a smooth curve of genus 3. A canonical divisor K has degree 4, then we can take any divisor H of degree 7 contained in $2K$, and embed by $|H|$ the curve in $\mathbf{P}^4$ (as in the above case, $|H|$ is certainly very ample). Hence $p^4_{7,3}$ is not injective, by Corollary 2.9, as the above construction assures $2K - H$ is effective.

$(d,g) = (8,5)$: $p^4_{8,5}$ is not injective by Corollary 2.9, applied to a canonical curve of genus 5 in $\mathbf{P}^4$. □

Remark 3.2. The same argument which we used in the case $(d,g) = (6,1)$, shows that if $p^r_{d,g}$ is not injective, with $r > 3$, then $p^{r-1}_{d,g}$ is not injective, and we shall use this fact in the following. We should expect that a similar property holds for the genus; indeed, once fixed the degree, we expect that it is easier that the nodal map is not injective, as the number of the nodes drops. At the end of our study (see table 0.1), by a direct cheking, we shall see that if $p^r_{d,g}$ is not injective and $\Sigma^r_{d,g+1}$ is nonempty then $p^r_{d,g+1}$ is not injective. However there exists (only) a case, where $p^r_{d,g}$ is not injective, $p^r_{d,g'}$ is injective, but $g' > g$, precisely when $r = 3$, $d = 12$, $g = 19$, $g' = 21$ (this is not in contradiction with the previous assertion, as $(d,g) = (12,20)$ is an Halphen lacuna).

Example 3.3. In the case $(d,g) = (7,3)$, there are both curves on which $|2K - (d-6)H|$ is effective and curve on which it not happens; indeed in the previous construction the effectivity depends on the choise of the seven points inside some divisor of $|2K|$. We may vary continously this choise and hence move the curve outside the locus where $p_{d,g}$ is not injective. Thus we have that the locus where the nodal map is not injective is a proper subvariety of some component. Later, in the case $(r,d,g) = (3,9,10)$, we shall see an example of an Hilbert scheme of which this locus is a whole proper component (Example 5.1).

4. Curves on a quadric surface in $\mathbf{P}^3$

When $r = 3$, there are infinitely many values of d and g satisfying the condition $\frac{1}{4}d^2 - \frac{3}{2}d + 1 \leq g \leq G(r,d)$, of the Proposition 2.1; but, by Proposition 1.2, all of these values, except a finite number, are relative to curves which must lie on a quadric surface. So in this section we shall consider curves which satisfy the condition

$$(*) \quad \frac{1}{6}d^2 - \frac{1}{2}d + 1 < g \leq G(3,d)$$

Remark 4.1. Notice that $G(3,d)$ is the integral part of $\frac{1}{4}d^2 - d + 1$ and that, $(*)$ implies $d \geq 7, g \geq 6$ (we clearly do not consider negative values); remember also that a divisor of type (a,b) on a smooth quadric surface (see [10], III, ex. 5.6), have degree $a+b$ and genus $(a-1)(b-1)$.

Proposition 4.2. *If $C \in V_{3,d,g}$ is a divisor of type (a,b) on a smooth quadric surface Q, with (d,g) satisfying $(*)$, and if $|K|$ is the canonical series and $|H|$ the plane section of C, then $|2K-(d-6)H|$ is effective if and only if $|a-b| \leq 2$.*

Proof. The canonical divisors on C are cut by divisors of type $(a-2, b-2)$ on Q, while divisors which are linearly equivalent to $|(d-6)H|$ are cut by divisors of type $(d-6, d-6)$; then divisors linearly equivalent to $\Delta = 2K - (d-6)H$, are cut by divisors of type $(a-b+2, b-a+2)$. Now, consider the exact sequence

$$(**) \quad 0 \to \mathcal{O}_Q(2-b, 2-a) \to \mathcal{O}_Q(a-b+2, b-a+2) \to \mathcal{O}_C(\Delta) \to 0.$$

First, we can assume $2-a<0$, $2-b<0$, in fact:

If $a=2$ then $d=b+2$, $g=b-1$, hence $g=d-3$, so, by easy calculations, $d-3 > \frac{1}{6}d^2 - \frac{1}{2}d + 1$ if and only if $d^2 - 9d + 24 < 0$, which is impossible; then $(*)$ excludes $a=2$. If $a=1$ then $g=0$, and again this is incompatible with $(*)$ (see Remark 4.1). Of course the case $a=0$ cannot occur.

So, let $2-a<0$, and similarly let $2-b<0$. We have $H^0(\mathcal{O}_Q(2-b,2-a)) = H^1(\mathcal{O}_Q(2-b,2-a)) = 0$ (see [10], III, ex. 5.6). Hence, from the cohomology exact sequence of $(**)$ it follows $h^0(\mathcal{O}_C(\Delta)) = h^0(\mathcal{O}_Q(a-b+2, b-a+2))$, and this is zero if and only if $|a-b| \leq 2$. □

Proposition 4.3. *If $g \geq \frac{1}{4}d^2 - d$ and $(*)$ holds, then $p^3_{d,g}$ is not injective.*

Proof. We know that for $a>0$, $b>0$, there is always a smooth curve of type (a,b) on a smooth quadric surface (see [10], III, ex. 5.6), thus we can apply Proposition 4.2 and Corollary 2.9, indeed from the relations $a+b=d$ and $(a-1)(b-1)=g$ we have that $\frac{1}{4}d^2-d\le g\le\frac{1}{4}d^2-d+1$ is equivalent to $|a-b|\le 2$. □

Proposition 4.4 *If $\frac{1}{6}d^2-\frac{1}{2}d+1<g<\frac{1}{4}d^2-d$ then $p^3_{d,g}$ is injective.*

Proof. By Corollary 2.9, it is enough to prove that there exist no smooth, irreducible, non-degenerate curves in $\mathbf{P}^3$, of degree d and genus g, with $|2K-(d-6)H|$ effective, satisfying the condition $\frac{1}{6}d^2-\frac{1}{2}d+1<g<\frac{1}{4}d^2-d$. Suppose the converse, and let C be such a curve. By Proposition 1.2, C must lie on a quadric surface, which must be smooth, as curves of degree d on a quadric cone have genus equal to the integer part of $\frac{1}{4}d^2-d+1$, and g is less than this number. Hence we can apply Proposition 4.2 and find a contradiction, since, as above, $|a-b|\le 2$ is equivalent to $\frac{1}{4}d^2-d\le g\le\frac{1}{4}d^2-d+1$. □

Proposition 4.5. *Suppose $d>12$. $p^3_{d,g}$ is not injective if and only if $\frac{1}{4}d^2-d\le g\le\frac{1}{4}d^2-d+1$.*

Proof. If $d>12$ then $\frac{1}{6}d^2-\frac{1}{2}d+1<\frac{1}{4}d^2-\frac{3}{2}d+1<\frac{1}{4}d^2-d$.

Thus, if $g<\frac{1}{4}d^2-d$ then $p^r_{d,g}$ is injective: indeed in the range $\frac{1}{6}d^2-\frac{1}{2}d+1<g<\frac{1}{4}d^2-d$ we can apply Proposition 4.4, otherwise we can apply Proposition 2.1. On the other hand, if $g>\frac{1}{4}d^2-d+1$, $\Sigma^3_{d,g}$ is empty.

Conversely, if $\frac{1}{4}d^2-d\le g\le\frac{1}{4}d^2-d+1$ then $p^3_{d,g}$ is not injective by Proposition 4.3. □

5. Curves in $\mathbf{P}^3$ of low degree

In order to complete the table 0.1, we have to consider the cases with $r=3$ and $d\le 12$. As usual, it is enough to consider the values of d and g satisfying $\frac{1}{4}d^2-\frac{3}{2}d+1\le g\le G(3,d)$. Thus $p^3_{d,g}$ can be not injective only for the following numerical values:

$d=3,\ \ g=0$
$d=4,\ \ 0\le g\le 1$
$d=5,\ \ 0\le g\le 2$
$d=6,\ \ 1\le g\le 4$
$d=7,\ \ 3\le g\le 6$
$d=8,\ \ 5\le g\le 9$
$d=9,\ \ 8\le g\le 12$
$d=10,\ \ 11\le g\le 16$
$d=11,\ \ 15\le g\le 20$
$d=12,\ \ 19\le g\le 25$

First of all, Proposition 2.7 implies that $p^3_{d,g}$ is not injective in the following cases (in which $\Sigma^3_{d,g}\ne\emptyset$; see [10], IV, 6.4.2; V, 4.13.1; V, ex. 4.14):

$d = 3,\ \ g = 0$
$d = 4,\ \ 0 \le g \le 1$
$d = 5,\ \ 0 \le g \le 2$
$d = 6,\ \ 2 \le g \le 4$
$d = 7,\ \ 4 \le g \le 6$
$d = 8,\ \ 7 \le g \le 9$
$d = 9,\ \ g = 12$

The Propositions of section 4 imply that if $(d, g) \in \{(10,15), (10,16), (11,20), (12,24), (12,25)\}$, $p^3_{d,g}$ is not injective (see Proposition 4.3), while it is injective by Proposition 4.4 if $(d, g) \in \{(9,11), (10,13), (10,14), (11,16), (11,17), (11,18), (11,19), (12,20), (12,21), (12,22), (12,23)\}$ (really in some of these cases $V_{3,d,g} = \emptyset$).

So, let us consider the remaining cases:

$(d, g) \in \{(6,1), (7,3), (8,5)\}$: $p^3_{d,g}$ is not injective by Proposition 3.1 and Remark 3.2.

$(d, g) = (8,6)$: $p^3_{8,6}$ is not injective. Indeed, take a smooth cubic surface X in $\mathbf{P}^3$; as we know, this one is isomorphic to the blowing up of $\mathbf{P}^2$ along six points in general position, so, let l be the divisor which corresponds to a line in $\mathbf{P}^2$ not passing through any of these points, and let $e_1, \cdots, e_6$ be the exceptional divisors. Consider a smooth, irreducible curve on X, linearly equivalent to $8l - 4e_1 - 3e_2 - 3e_3 - 2e_4 - 2e_5 - 2e_6$ (it exists by [10], V, 4.13), and apply Corollary 2.9. So we are done, as $|2K - (d-6)H|$ is cut by divisors equivalent to $4l - 4e_1 - 2e_2 - 2e_3$, which are effective.

$(d, g) = (9,8)$: $p^3_{9,8}$ is injective. Indeed, by Corollary 2.9, it is enough to show that there are no smooth, irreducible, non-degenerate curves in $\mathbf{P}^3$, of degree 9 and genus 8, with $|2K - 3H|$ effective.

Let C be such a curve. By Proposition 1.1, we have that $\dim|H| = 3$; furthermore letting $|L| = |K - H|$, we have $\dim|L| = 1$, by Riemann-Roch. An effective divisor equivalent to $2K - 3H$ is a point P, then $|H| = |2L - P|$ and $|K| = |3L - P|$; hence P is a base point of $|3L|$, as $\dim|3L| = \dim|K + P| = 7 = \dim|K| = \dim|3L - P|$, and then is a base point of $|L|$. Let $|L_0| = |L - P|$, so $|H| = |2L_0 + P|$. Since $|H|$ is very ample, we have that for every point Q, $\dim|H - P - Q| = 1$, $|H - P - Q| = |2L_0 - Q|$ and since $\dim|2L_0| = 2$, looking at the dimensions (see e.g. [10], IV, proof of Lemma 5.5), every effective divisor of $|2L_0|$ is a sum of a L' and a L'' both equivalent to L_0. It follows that every effective divisor of $|2L_0 - Q|$ contains a divisor of $|L_0 - Q|$; but $|L_0|$ has dimension 1, so the unique effective divisor of $|L_0 - Q|$ is formed by three base points of $|H - P - Q|$, so they must lie on the line through P and Q.

Now, project C from P: the previous discussion tells that every fiber of the projection is a group of fours aligned points, hence the image of C is a conic. Thus C lies on a quadric cone, but this is impossible, as curves of degree 9 on a quadric cone have genus 12.

$(d,g) = (9,9) : p^3_{9,9}$ is injective. Ideed, by Corollary 2.9, it is enogh to show that are no smooth, irreducible, non-degenerate curves in $\mathbf{P}^3$, of degree 9 and genus 9, with $|2K - 3H|$ effective.

Let C be such a curve, let $D \in |2K - 3H|$ (deg $D = 5$) and $L \in |K - H|$ (deg $L = 7$), so $|H| = |2L - D|$. We have that dim$|H| = 3$, by Proposition 1.1, and that dim$|L| = 2$ by Riemann-Roch. First note that $|L - D|$ cannot be effective, otherwise, taking an effective divisor $E \in |L - D|$ (deg $E = 2$), dim$|H - E| =$ dim$|L| = 2$, and this is impossible because $|H|$ is very ample of dimension 3 . It follows that D imposes exactly two conditions to $|2L|$, as dim$|2L| = 14 - 9 +$ dim$|L - D| + 1 = 5$ while dim$|2L - D| =$ dim$|H| = 3$.

Let $P \in D$ a point of $|L|$, not a base point for $|L|$, and let $Q \in D$, not a base point for $|L - P|$ (remember that dim$|L| = 2$). Looking at the dimensions, P cannot be a base point for $|L - Q|$. From the above construction and from the fact that D imposes two conditions to $|2L|$, we have that for all $D_1 \in |L - P|$ and for all $D_2 \in |L - Q|$, $D_1 + D_2 \in |2L - P - Q|$ must contain the (three) points of D out of P and Q, say P_1, P_2, P_3. These points will be base points for $|D_1|$ or $|D_2|$. Certainly P_1, P_2 and P_3 all cannot be base points of $|D_1|$ (respectively of $|D_2|$), otherwise we can impose $|D_1|$ (respectively $|D_2|$) passes through Q (respectively through P) and obtain $|\dot{D}| \subset |L|$ (and above we have excluded this fact). So, up to a change of indexes, we can suppose P_1 and P_2 are base points of $|D_1|$, and P_3 is a base point of $|D_2|$, and let $|H_1| = |D_1 - P_1 - P_2|$ and $|H_2| = |D_2 - P_3|$.

Now we have $|H| = |H_1 + H_2|$ and dim$|H_1| =$ dim$|H_2| = 1$, hence C lies on a smooth quadric surface (since the immersion of C factor through an immersion of $\mathbf{P}^1 \times \mathbf{P}^1 \to \mathbf{P}^3$), and this is impossible, as $a + b = 9$, $(a-1)(b-1) = 9$, are incompatibles on the integers (see Remark 4.1).

$(d,g) = (9,10) : p^3_{9,10}$ is not injective. It is enough to consider (see [10], II, ex. 8.4) a smooth, irreducible, non-degenerate complete intersection of two cubic surfaces in $\mathbf{P}^3$, and apply Corollary 2.9, as $|K| = |2H|$.

$(d,g) = (10,11) : p^3_{10,11}$ is not injective. Here (as usual by Corollary 2.9) it suffices to find a 2-subcanonical curve of degree 10 in $\mathbf{P}^3$. We can do this using the Serre correspondence (see [11]). Indeed if $\mathcal{E}$ is the bundle associated to an elliptic quintic curve, then it has Chern classes $c_1 = 4$ and $c_2 = 5$; so $\mathcal{E}(1)$ has $c_1 = 2$ and $c_2 = 10$, and it is generated by global sections, as $\mathcal{E}$ is. One easily see that a generic section of $\mathcal{E}(1)$ gives rise to a smooth, irreducible, 2-subcanonical curve of degree 10.

$(d,g) = (10,12) : p^3_{10,12}$ is not injective. We can mimic the proof of the case $(8,6)$, with the divisor $10l - 4e_1 - 4e_2 - 3e_3 - 3e_4 - 3e_5 - 3e_6$ on the smooth cubic surface.

$(d,g) = (11,15) : p^3_{11,15}$ is injective. Indeed, by Corollary 2.9, it is enough to show that there are no smooth, irreducible, non-degenerate curves in $\mathbf{P}^3$, of degree 11 and genus 15, with $|2K - 5H|$ effective.

Let C be such a curve, let $|L| = |K - 2H|$ (deg $L = 6$), and notice that the effective divisor of $|2K - 5K|$ is a point, say P. So $|H| = |2L - P|$ and $|K| = |5L - 2P|$.

Suppose that P is not a base point for $|L|$. By Riemann-Roch, $\dim|5L| = \dim|5L - 2P| + 1$, thus there are no divisors of $|5L|$ which contain P but not $2P$. Hence there are no divisors in $|2L|$ which contain P but not $2P$, and then $|H| = |2L - P|$ has P as a base point, but this is impossible as $|H|$ is very ample. Therefore P is a base point of $|L|$, so let $|L_0| = |L - P|$ and notice that $|2L_0| = |H - P|$, which implies $\dim|2L_0| = 2$ ($\dim|H| = 3$, by Proposition 1.1).

Now C cannot lie on a quadric surface, as in the smooth case the conditions $a + b = 11$, $(a - 1)(b - 1) = 15$ are impossible, while in a quadric cone curves of degree 11 have genus 20. So we have $\dim|2H| \geq \dim \mathbf{P}(H^0(\mathcal{O}_{\mathbf{P}^2}(2))) = 9$, which implies, by Riemann-Roch, $\dim|L| \geq 1$, hence $\dim|L_0| \geq 1$. Thus $\dim|2L_0| = 2$ implies $\dim|L_0| = 1$, so projecting C from P, we can conclude as in the end of the proof of the case $(d, g) = (9, 8)$.

$(d, g) = (12, 19) : p^3_{12,19}$ is not injective. Indeed, we can use the same argument as in the case $(9, 10)$, considering now a complete intersection of a cubic and a quartic.

Example 5.1. Notice that $\mathrm{Hilb}_{3,9,10}$ has a component of curves which lie on a quadric surface too. On this component $p^3_{9,10}$ is injective, as we can easily see, from the proof of Proposition 4.2. So, in this case we have a caracterization of those curves for which $p^3_{9,10}$ is not injective: they are the complete intersections, hence they fill one component of $\mathrm{Hilb}_{3,9,10}$.

References

[1] Arbarello, E., Cornalba, M., Footnotes to a paper of Beniamino Segre. Math. Ann. 256 (1981), 341–362.

[2] Brill, A., Noether, M., Uber die algebraischen Funktionen und ihre Anwendungen in der Geometrie. Math. Ann. 7 (1873), 269–310.

[3] Castelnuovo, G., Ricerche di geometria sulle curve algebriche. Atti R. Accad. Sci. Torino 24 (1889).

[4] Castelnuovo, G., Sui multipli di una serie lineare di gruppi di punti appartenenti ad una curva algebrica. Rend. Circ. Matem. Palermo 7 (1893).

[5] Chiantini, L., On plane curves with nodes in special position. preprint del Dipartimento di Matematica e Applicazioni "R. Caccioppoli" di Napoli 14 (1991).

[6] Enriques, F., Chisini, O., Teoria geometrica delle equazioni e delle funzioni algebriche. Zanichelli, Bologna 1924.

[7] Franchetta, A., Sulle serie lineari razionalmente determinate sulla curva a moduli generali di dato genere. Le Matematiche 11 (1954), 126–147.

[8] Gruson, L., Peskine, C., Genre des courbes de l'espace projectif. Lecture Notes in Math. 687 (1978).

[9] Harris, J., On the Severi problem. Invent. Math. 54 (1986), 445–461.

[10] Hartshorne, R., Algebraic Geometry. GTM 52, Springer Verlag, Heidelberg 1977.

[11] Hartshorne, R., Stable vector bundles of rank 2 on $\mathbf{P}^3$. Math. Ann. 238 (1978), 229–280.
[12] Mestrano, N., Conjectures de Franchetta plongées. C. R. Acad. Sc. Paris 301:1 (1985), 503–505.
[13] Mestrano, N., Conjecture de Franchetta forte, Invent. Math. 87 (1987), 365–376.
[14] Severi, F., Vorlesungen über Algebraische Geometrie. Teubner, Leipzig, 1921.
[15] Treger, R., Plane curves with nodes. Can. J. Math. 41:2 (1989), 193–212.
[16] Zariski, O., Dimension-theoretic characterization of maximal irreducible algebraic systems of plane nodal curves of a given order n and with a given number d of nodes. Amer. J. Math. 104 (1982), 209–226.

Syzygies of Points in Projective Space and Applications

*Stefan Ehbauer**

Abstract. Nowadays, there are computer programs like "Macaulay", which can compute the graded Betti numbers of a projective variety X. The question arises: Which geometric information do these numbers contain? There are several conjectures about this question. This article concerns the (M_q)-conjecture ($q = 3$).

1. Introduction

Let V be an $r+1$-dimensional vector space over the field of complex numbers $\mathbb{C}$ and let $B = \bigoplus_{q\in\mathbf{Z}} B_q$ be a graded module over the symmetric algebra $SV = \bigoplus_{p\geq 0} S^p V$. In this situation there is the Koszul complex

$$\cdots \longrightarrow \bigwedge^{p+1} V \otimes B_{q-1} \xrightarrow{d_{p+1,q-1}} \bigwedge^{p} V \otimes B_q \xrightarrow{d_{p,q}} \bigwedge^{p-1} V \otimes B_{q+1} \longrightarrow \cdots$$

The Koszul cohomology groups are defined by

$$K_{p,q}(B,V) = \frac{\ker\ d_{p,q}}{\operatorname{im}\ d_{p+1,q-1}}.$$

For a compact complex manifold X and a basepointfree line bundle L over X we put $K_{p,q}(X,L) := K_{p,q}(B,V)$ with $V = H^0(X,L)$ and $B_q = H^0(X,L^q)$. The morphism induced by L is denoted by $\varphi_L : X \longrightarrow \mathbb{P}(V) = \mathbb{P}^r$ $(h^0(X,L) = r+1)$.

In this situation one is interested in the relation between $K_{p,q}(X,L)$ and the geometry of X. The $K_{p,1}$-Theorem of M. Green gives an example, how close this relation is.

$K_{p,1}$-Theorem (M. Green, [G1]) 1.1. *Let X be a compact complex manifold, L an analytic line bundle with $h^0(X,L) = r+1$ and $\dim\varphi_L(X) = m$. Then*

(1) $K_{p,1}(X,L) = 0$ *for* $p > r-m$;

(2) $K_{r-m,1}(X,L) = 0$, *unless the variety $\varphi_L(X)$ is an m-dimensional variety of minimal degree $r-m+1$;*

(3) $K_{r-m-1,1}(X,L) = 0$, *unless* $\deg\varphi_L(X) \leq r+2-m$ *or $\varphi_L(X)$ is contained in an $m+1$-dimensional variety of minimal degree $r-m$.*

---

* Supported by a grant from DFG

The proof of this theorem is based on the "Strong Castelnuovo Lemma".

Strong Castelnuovo-Lemma (M. Green, [G1]) 1.2. *Let $P_1, ..., P_d \in \mathbb{P}^r$ be points in linear general position. Then $P_1, ..., P_d$ lie on a rational normal curve if and only if $K_{r-1,1}(P_1, ..., P_d) \neq 0$.*

In the Lemma $K_{p,q}(P_1, ..., P_d) = K_{p,q}(B, V)$ with

$$B_q := \operatorname{im}\left(H^0(\mathbb{P}^r, \mathcal{O}_{\mathbb{P}^r}(q)) \longrightarrow \bigoplus_{i=1}^{d} H^0(P_i, \mathcal{O}_{P_i}(q))\right).$$

For the curve case there is the following related conjecture concerning the existence of 1-dimensional linear series g_q^1 of degree q on the curve X and the property M_q, which is $K_{j,1}(X, L) = 0$ for $j \geq r - q$.

(M_q)-Conjecture (M. Green - R. Lazarsfeld, [GL2]) 1.3. *Let X be a smooth, irreducible curve of genus g and L a line bundle over X with $\deg L >> 2g$ and $h^0(X, L) = r + 1$. Then property M_q fails for X if and only if there exists a g_q^1 on X.*

By the theorem in [GL1] the existence of a g_q^1 on X implies $K_{r-q,1}(X, L) \neq 0$. Therefore the $K_{p,1}$-Theorem gives the M_q-conjecture for $q = 1, 2$. In this article we prove the following theorem, which verifies the M_q-conjecture for $g \geq 13$ and $q = 3$. The syzygy variety $V_{Syz}^p(X)$, which contains X, is defined in paragraph 2.

Theorem 1.4. *Let X be a smooth, irreducible curve of genus $g \geq 13$, L a line bundle on X of degree $\deg L \geq 2g + 3$ and $h^0(X, L) = r + 1$. If $K_{r-3,1}(X, L) \neq 0$, then $V_{Syz}^{r-3}(X) \subseteq \mathbb{P}^r$ is a 3-dimensional rational normal scroll of degree $r - 2$ and the 1-dimensional linear system $\{\mathbb{P}_t^2\}_{t \in \mathbb{P}^1}$ of planes on $V_{Syz}^{r-3}(X)$ cuts out a g_3^1 on X.*

Similarly to the $K_{p,1}$-Theorem the main ingredient for the proof will be a generalization of the Strong Castelnuovo-Lemma.

Generalization of the Strong Castelnuovo-Lemma 1.5. *Let $\Gamma = C \cap H$ be the intersection of a general hyperplane $H \cong \mathbb{P}^{r-1}$ and the curve $C = \varphi_L(X)$, which has degree $\deg C \geq r + 13$. If $K_{r-3,1}(X, L) \neq 0$, then $V_{Syz}^{r-3}(\Gamma) \subseteq H$ is one of the following varieties:*

- *rational normal curve*
- *irreducible curve of degree r and arithmetic genus 1*
- *irreducible surface of minimal degree $r - 2$.*

2. The Syzygy Variety

The polynomial ring over the field of complex numbers $\mathbb{C}$ in $r+1$ variables is denoted by $S = \mathbb{C}[X_0, ..., X_r]$ and we identify its graded components S_q with the graded components S^qV of the symmetric algebra SV over the $r+1$-dimensional vector space $V = H^0(\mathbb{P}^r, \mathcal{O}_{\mathbb{P}^r}(1))$. If a line bundle L over a complex manifold X induces a morphism $\varphi_L : X \longrightarrow \mathbb{P}^r$, then V and $H^0(X, L)$ will be identified. Furthermore, for a projective variety $Y \subseteq \mathbb{P}^r$ let $I_Y = \bigoplus_{n \in \mathbf{Z}} H^0(\mathbb{P}^r, \mathcal{I}_Y(n))$ be the homogeneous ideal with graded components $(I_Y)_q = H^0(\mathbb{P}^r, \mathcal{I}_Y(q))$ and let $S_Y = \bigoplus_{q \in \mathbf{Z}} (S_Y)_q$ be the homogeneous coordinate ring. For this ring there exists a minimal free resolution

$$0 \longleftarrow S_Y \xleftarrow{\delta_0} F_0 \xleftarrow{\delta_1} F_1 \xleftarrow{\delta_2} \dots$$

with

$$F_i \cong \bigoplus_{j \in \mathbf{Z}} S(-j)^{\beta_{ij}(Y)} \qquad (\beta_{ij}(Y) \in \mathbb{N}).$$

The numbers $\beta_{ij}(Y)$ are called *graded syzygy numbers* or *graded Betti numbers.* They equal the dimensions

$$\beta_{i,j}(Y) = \dim_{\mathbb{C}} \mathrm{Tor}_i^S(S_Y, \mathbb{C})_j.$$

Suppose we have a complex manifold X and a very ample line bundle L over X defining an embedding $\varphi_L : X \longrightarrow \mathbb{P}^r$. As it was mentioned in the introduction there are the Koszul complexes

$$\dots \longrightarrow \bigwedge^{p+1} V \otimes B_{q-1} \xrightarrow{d_{p+1,q-1}} \bigwedge^{p} V \otimes B_q \xrightarrow{d_{p,q}} \bigwedge^{p-1} V \otimes B_{q+1} \longrightarrow \dots$$

with $B_q = H^0(X, L^q)$. In terms of a basis $X_0, ..., X_r$ of V and an element $b_q \in B_q$ the differentials $d_{p,q}$ are given by

$$\begin{array}{llll} d_{p,q} : & \bigwedge^p V \otimes B_q & \longrightarrow \bigwedge^{p-1} V \otimes B_{q+1} \\ & X_{i_1} \wedge ... \wedge X_{i_p} \otimes b_q & \longrightarrow \sum_{l=1}^p (-1)^{l+1} X_{i_1} \wedge ... \wedge \widehat{X}_{i_l} \wedge ... \wedge X_{i_p} \otimes X_{i_l} b_q \end{array}$$

The Koszul cohomology groups are defined by

$$K_{p,q}(B, V) = K_{p,q}(X, L) = \frac{\ker\, d_{p,q}}{\mathrm{im}\, d_{p+1,q-1}}.$$

If $\varphi_L(X) = Y \subseteq \mathbb{P}^r$ is projectively normal, i. e. $H^1(\mathbb{P}^r, \mathcal{I}_X(n)) = 0$ for $n \geq 0$, then

$$K_{p,q}(X, L) \cong \mathrm{Tor}_p^S(S_Y, \mathbb{C})_{p+q}$$

holds, therefore $\beta_{p,p+q}(Y) = \dim_{\mathbb{C}} K_{p,q}(X, L)$.

We come to the definition of the syzygy variety. For a non-degenerate variety $Y \subseteq \mathbb{P}^r$ $((S_Y)_1 = V)$ there is the following commutative diagram with the canonical map ϵ_Y :

$$\begin{array}{ccccc}
\bigwedge^{p+1} V & \xrightarrow{d_{p+1,0}} & \bigwedge^{p} V \otimes V & \xrightarrow{d_{p,1}} & \bigwedge^{p-1} V \otimes S_2 \\
\| & & \| & & \downarrow \epsilon_Y \\
\bigwedge^{p+1} V & \xrightarrow{d_{p+1,0}} & \bigwedge^{p} V \otimes V & \xrightarrow{\epsilon_Y \circ d_{p,1}} & \bigwedge^{p-1} V \otimes (S_Y)_2
\end{array}$$

We get

$$\mathrm{Tor}_p^S(S_Y, \mathbb{C})_{p+1} = \frac{\ker\,(\epsilon_Y \circ d_{p,1})}{\mathrm{im}\ d_{p+1,0}}.$$

Remark 2.1. Let $Z \subseteq Y$ be a nondegenerate subvariety. Then the diagram immediately implies

$$\mathrm{Tor}_p^S(S_Y, \mathbb{C})_{p+1} \subseteq \mathrm{Tor}_p^S(S_Z, \mathbb{C})_{p+1}.$$

In particular $\beta_{p,p+1}(Z) \geq \beta_{p,p+1}(Y)$ holds.

For $\alpha \in \ker\,(\epsilon_Y \circ d_{p,1})$ the eqivalence class in

$$\mathrm{Tor}_p^S(S_Y, \mathbb{C})_{p+1} \cong \frac{\ker\,(\epsilon_Y \circ d_{p,1})}{\mathrm{im}\ d_{p+1,0}}$$

is denoted by $[\alpha]$. Let

$$\alpha = \sum_{0 \leq i_1 < \ldots < i_p \leq r} \sum_{j=0,\ldots,r} X_{i_1} \wedge \ldots \wedge X_{i_p} \otimes c^{[\alpha]}_{j i_1 \ldots i_p} X_j \in \ker\,(\epsilon_Y \circ d_{p,1}),$$

then the definition implies

$$d_{p,1}(\alpha) = \sum_{0 \leq i_1 < \ldots < i_{p-1} \leq r} \sum_{j=0,\ldots,r} X_{i_1} \wedge \ldots \wedge X_{i_{p-1}} \otimes Q^{[\alpha]}_{i_1 \ldots i_{p-1}} \in \bigwedge^{p-1} V \otimes (I_Y)_2.$$

Definition 2.2. For $[\alpha] \in \mathrm{Tor}_p^S(S_Y, \mathbb{C})_{p+1}, \neq 0$ we define the ideal

$$I([\alpha]) := (Q^{[\alpha]}_{i_1 \ldots i_{p-1}} \mid 0 \leq i_1 < \ldots < i_{p-1} \leq r) \text{ and } V([\alpha]) := V(I([\alpha]))$$

as the zero set of the ideal $I([\alpha])$. For a nondegenerate variety $Y \subseteq \mathbb{P}^r$ we call

$$V^{(p)}_{Syz}(Y) := \bigcap_{[\alpha] \in \mathrm{Tor}_p^S(S_Y, \mathbb{C})_{p+1}} V([\alpha])$$

the *p-th syzygy variety of* Y. For the case of a complex manifold X and a very ample line bundle L over X with $Y = \varphi_L(X)$ we also write $V^p_{Syz}(X)$.

Remarks 2.3.

a) If $Y \subseteq \mathbb{P}^r$ contains the "coordinate points"

$$P_0 := [1 : 0 : \ldots : 0], \ldots, P_r := [0 : \ldots : 0 : 1]$$

of $\mathbb{P}^r$, then $c^{[\alpha]}_{ji_1...i_p} = 0$ for $j = i_1, ..., i_p$ and $Q^{[\alpha]}_{i_1...i_{p-1}}$ is a sum of degree 2 monomials in the variables X_i, $i \in \{0, ..., r\} \setminus \{i_1, ..., i_p\}$, where the monomials X_i^2, $i = 0, ..., r$ do not occur.

b) $Q^{[\alpha]}_{i_1...i_{p-1}} = 0$ $(0 \le i_1 < ... < i_{p-1} \le r)$ implies $[\alpha] = 0$.

c) It holds $Y \subseteq V^p_{Syz}(Y)$ and $V^{p-1}_{Syz}(Y) \subseteq V^p_{Syz}(Y)$. In particular, $\beta_{p-1,p}(Y) = 0$ implies $\beta_{p,p+1}(Y) = 0$.

d) For a non-degenerate subvariety $Z \subseteq Y \subseteq \mathbb{P}^r$ we have either $\beta_{p,p+1}(Z) = \beta_{p,p+1}(Y)$ or $\beta_{p,p+1}(Z) > \beta_{p,p+1}(Y)$, which gives the existence of a quadric in $(I_Z)_2 \setminus (I_Y)_2$.

The following example describes the quadrics arising by a syzygy.

Example 2.4. Let $Y \subseteq \mathbb{P}^4$ be a variety containing the coordinate points of $\mathbb{P}^4$. A syzygy $[\alpha] \in \mathrm{Tor}^S_2(S_Y, \mathbb{C})_3$ induces the following quadrics

$$\begin{aligned}
Q^{[\alpha]}_0 :=& (-c^{[\alpha]}_{102} - c^{[\alpha]}_{201})X_1X_2 + (-c^{[\alpha]}_{103} - c^{[\alpha]}_{301})X_1X_3 + (-c^{[\alpha]}_{104} - c^{[\alpha]}_{401})X_1X_4 + \\
& (-c^{[\alpha]}_{203} - c^{[\alpha]}_{302})X_2X_3 + (-c^{[\alpha]}_{204} - c^{[\alpha]}_{402})X_2X_4 + (-c^{[\alpha]}_{304} - c^{[\alpha]}_{403})X_3X_4 \\
Q^{[\alpha]}_1 :=& (-c^{[\alpha]}_{012} + c^{[\alpha]}_{201})X_0X_2 + (-c^{[\alpha]}_{013} + c^{[\alpha]}_{301})X_0X_3 + (-c^{[\alpha]}_{014} + c^{[\alpha]}_{401})X_0X_4 + \\
& (-c^{[\alpha]}_{213} - c^{[\alpha]}_{312})X_2X_3 + (-c^{[\alpha]}_{214} - c^{[\alpha]}_{412})X_2X_4 + (-c^{[\alpha]}_{314} - c^{[\alpha]}_{413})X_3X_4 \\
Q^{[\alpha]}_2 :=& (c^{[\alpha]}_{012} + c^{[\alpha]}_{102})X_0X_1 + (-c^{[\alpha]}_{023} + c^{[\alpha]}_{302})X_0X_3 + (-c^{[\alpha]}_{024} + c^{[\alpha]}_{402})X_0X_4 + \\
& (-c^{[\alpha]}_{123} + c^{[\alpha]}_{312})X_1X_3 + (-c^{[\alpha]}_{124} + c^{[\alpha]}_{412})X_1X_4 + (-c^{[\alpha]}_{324} - c^{[\alpha]}_{423})X_3X_4 \\
Q^{[\alpha]}_3 :=& (c^{[\alpha]}_{013} + c^{[\alpha]}_{103})X_0X_1 + (c^{[\alpha]}_{023} + c^{[\alpha]}_{203})X_0X_2 + (-c^{[\alpha]}_{034} + c^{[\alpha]}_{403})X_0X_4 + \\
& (c^{[\alpha]}_{123} + c^{[\alpha]}_{213})X_1X_2 + (-c^{[\alpha]}_{134} + c^{[\alpha]}_{413})X_1X_4 + (-c^{[\alpha]}_{234} + c^{[\alpha]}_{423})X_2X_4 \\
Q^{[\alpha]}_4 :=& (c^{[\alpha]}_{014} + c^{[\alpha]}_{104})X_0X_1 + (c^{[\alpha]}_{024} + c^{[\alpha]}_{204})X_0X_2 + (c^{[\alpha]}_{034} + c^{[\alpha]}_{304})X_0X_3 + \\
& (c^{[\alpha]}_{124} + c^{[\alpha]}_{214})X_1X_2 + (c^{[\alpha]}_{134} + c^{[\alpha]}_{314})X_1X_3 + (c^{[\alpha]}_{234} + c^{[\alpha]}_{324})X_2X_3
\end{aligned}$$

The next proposition is an easy consequence of a lemma given in [Mat] (p 140) and the long cohomology sequence.

Proposition 2.5. *Let $Y \subseteq \mathbb{P}^r$ be a non-degenerate, irreducible, projectively normal variety and $H \cong \mathbb{P}^{r-1}$ a general hyperplane. Then there is an isomorphism*

$$\mathrm{Tor}^S_i(S_Y, \mathbb{C})_j \cong \mathrm{Tor}^{S_H}_i(S_{Y\cap H}, \mathbb{C})_j.$$

If Y is a non-degenerate, irreducible, linearly normal variety, i. e. $H^1(\mathbb{P}^r, \mathcal{I}_Y(1)) = 0$, then there is an isomorphism

$$\mathrm{Tor}^S_i(S_Y, \mathbb{C})_{i+1} \cong \mathrm{Tor}^{S_H}_i(S_{Y\cap H}, \mathbb{C})_{i+1}.$$

Corollary 2.6. *Let $Y \subseteq \mathbb{P}^r$ be a non-degenerate, irreducible, linearly normal variety and $H \cong \mathbb{P}^{r-1}$ a general hyperplane. Then*

$$V^p_{Syz}(Y) \cap H = V^p_{Syz}(Y \cap H).$$

3. The Uniform Position Principle

The following ideas are essentially due to J. Harris. The uniform position principle applied to a set $\Gamma = \{P_1, ..., P_d\} \subseteq \mathbb{P}^r$ of points means that these points are indistinguishable.

Definition 3.1. A set $\Gamma = \{P_1, ..., P_d\} \subseteq \mathbb{P}^r$ of points are in linear general position, if each $r+1$ points of Γ span $\mathbb{P}^r$.

We will see in a moment why it is necessary to consider the uniform position property. The next example shows that in the "Strong Castelnuovo Lemma" the linear general position of the points must be assumed.
Suppose we have $d = r+8$ points. $r-k+4$ of these lie on $W_{r-k} \cong \mathbb{P}^{r-k}$ and the remaining $k+4$ points lie on $W_k \cong \mathbb{P}^k$, where W_k intersects W_{r-k} in one point. Then $\beta_{r-1,r}(\Gamma) \neq 0$ ($\Gamma = \{P_1, ..., P_d\}$), but Γ is not contained in a rational normal curve.

The next example shows that the "Generalization of the Strong Castelnuovo Lemma" given in the introduction needs a stronger uniform position property.
Let $E \subseteq \mathbb{P}^4$ be an irreducible, non-degenerate curve of degree 5 and $\Gamma = \{P_1, ..., P_{12}\}$ twelve points in linear general position, where $P_1, ..., P_{11}$ lie on E and P_{12} is neither contained in E nor in the secant variety $\operatorname{Sec} E$. According [BE] the curve E has the following minimal free resolution

$$0 \longleftarrow S_E \longleftarrow S \longleftarrow S(-2)^5 \stackrel{\varphi}{\longleftarrow} S(-3)^5 \longleftarrow S(-4) \longleftarrow 0.$$

φ is a skewsymmetric matrix whose five pfaffians generate the ideal of E. This resolution gives $\beta_{2,3}(\Gamma) = 1$. So Γ can neither lie on a rational normal curve $C \subseteq \mathbb{P}^4$ ($\beta_{2,3}(C) = 8$) nor on an irreducible surface $F \subseteq \mathbb{P}^4$ of minimal degree 3 ($\beta_{2,3}(F) = 2$). Therefore Γ does not fulfill the assumption of the "Generalization of the Strong Castelnuovo Lemma". But here it is a set of points lying somewhere and in the Lemma it is a general hyperplane section of a curve. In fact we will see that a general hyperplane section of a curve has this powerful uniform position property, which is needed.

Let $C \subseteq \mathbb{P}^{r+1}$ be an irreducible, non-degenerate curve of degree d and let $U \subseteq \left(\mathbb{P}^{r+1}\right)^*$ be the open set of hyperplanes intersecting C transversally in d smooth points. There are the following incidence varieties

$$I := \{(P, H) \in \mathbb{P}^{r+1} \times U \mid P \in H\}$$

$$I(m) := \{(P_1, ..., P_m; H) \in C \times ... \times C \times U \mid P_1, ..., P_m \in H \cap C; P_i \neq P_j (i \neq j)\}.$$

The projection $\pi_2 : I(1) \longrightarrow U$ is a d-sheeted topological covering. Hence for a base point $H_0 \in U$ and $\Gamma_0 = H_0 \cap C$ there is the monodromy map

$$\pi_1(U, H_0) \longrightarrow Aut(\Gamma_0)$$

of the fundamental group $\pi_1(U, H_0)$ into the automorphism group $Aut(\Gamma_0)$. The image of this map is called the *monodromy group* of π_2.

Monodromylemma (J. Harris, ([ACGH] p 111)) 3.2. *The monodromy group of π_2 is the full symmetric group $\mathcal{S}_d$.*

Remark 3.3. The statement of the Monodromylemma is equivalent to the irreducibility of $I(m)$ $(m \leq d)$.

Let $W = \{(P_1, ..., P_d) \in \mathbb{P}^r \times ... \times \mathbb{P}^r \mid P_i \neq P_j (i \neq j)\}$. The automorphism group $PGl(r)$ of $\mathbb{P}^r$ operates on $\mathbb{P}^r \times ... \times \mathbb{P}^r$:

$$PGl(r) \longrightarrow Aut(\mathbb{P}^r \times ... \times \mathbb{P}^r)$$
$$A \longrightarrow \begin{cases} A : \mathbb{P}^r \times ... \times \mathbb{P}^r \longrightarrow \mathbb{P}^r \times ... \times \mathbb{P}^r \\ \qquad (P_1, ..., P_d) \longrightarrow (A(P_1), ..., A(P_d)) \end{cases}$$

Definition 3.4. Let V be a closed, $PGl(r)$-invariant (i. e. $A(V) \subseteq V$ $(A \in PGl(r))$) subspace of W. Then d points $P_1, ..., P_d \in \mathbb{P}^r$ are in *uniform position with respect to* V if for $(P_1, ..., P_d) \in V$ the d-tuple $(P_{\sigma(1)}, ..., P_{\sigma(d)})$ is an element of V for every permutation $\sigma \in \mathcal{S}_d$.

Proposition 3.5. *Let $C \subseteq \mathbb{P}^{r+1}$ be an irreducible, non-degenerate curve of degree d and let $\varphi_H : H \xrightarrow{\cong} \mathbb{P}^r$ be a general hyperplane. Let $V_1, ..., V_n \subseteq W$ $(n \in \mathbb{N})$ be finitely many closed, $PGl(r)$-invariant subspaces of W. Then for $H \cap C = \{P_1, ..., P_d\}$ the points $\varphi_H(P_1), ..., \varphi_H(P_d)$ are in uniform position with respect to $V_1, ..., V_n$.*

Proof. It is sufficient to consider the case $n = 1$ where $V := V_1$. The incidence variety $\pi : I \longrightarrow U$ is a $\mathbb{P}^r$-bundle over U. Let

$$\Phi_i : I_{U_i} := \pi^{-1}(U_i) \xrightarrow{\cong} \mathbb{P}^r \times U_i \qquad (U = \bigcup_i U_i)$$

be trivializations with $\Phi_i(P, H) = \varphi_i(P) \times \{H\}$ $(H \in U_i)$. We define

$$X_i := \{(\varphi_i(P_1), ..., \varphi_i(P_d); H) \in \mathbb{P}^r \times ... \times \mathbb{P}^r \times U_i \mid \{P_1, ..., P_d\} = H \cap C\}$$

and

$$X := \bigcup_i X_i.$$

There is the projection

$$\begin{array}{cccc} \Phi : & X & \longrightarrow & W \\ & (\varphi_i(P_1), ..., \varphi_i(P_d), H) & \longrightarrow & (\varphi_i(P_1), ..., \varphi_i(P_d)) \ (H \in U_i) . \end{array}$$

For $H \in U_i \cap U_j$ the d-tuple $(\varphi_i(P_1), ..., \varphi_i(P_d))$ is in V if and only if $(\varphi_j(P_1), ..., \varphi_j(P_d))$ is an element of V, since V is $PGl(r)$-invariant. Now, $\Phi^{-1}(V) \subseteq X$ is closed and

$$\begin{array}{rcl}\Psi: & X & \longrightarrow U \\ & (\varphi_i(P_1),...,\varphi_i(P_d),H) & \longrightarrow H\end{array}$$

is a $(d!:1)$ topological covering. Since $X \cong I(d)$ is irreducible, there are the two possibilities $\Phi^{-1}(V) = X$ or $\dim \Phi^{-1}(V) < \dim X$. In the first case $\varphi_H(P_1),...,$ $\varphi_H(P_d)$ is in uniform position with respect to V for every $H \in U$. In the second case this holds for every H in the Zariski open set $U \setminus \Psi(\Phi^{-1}(V))$. □

Remark 3.6. Suppose $C \subseteq \mathbb{P}^5$ is a non-degenerate, irreducible, linearly normal curve of degree d. Let $[\alpha] \in \mathrm{Tor}_2^S(S_C, \mathbb{C})_3$, $\neq 0$ be a syzygy. Then there exixts a Zariski open subset $U([\alpha]) \subseteq (\mathbb{P}^5)^*$ of hyperplanes in $\mathbb{P}^5$ such that for each $H \in U([\alpha])$ the following holds: We identify H with $\mathbb{P}^4$ and assume that $C \cap H$ contains the coordinate points $P_0 = [1:0:...:0],...,P_4 = [0:...:0:1]$ of $\mathbb{P}^4$. The syzygy $[\alpha]$ induces a syzygy $[\alpha\,|_H] \in \mathrm{Tor}_2^{S_H}(S_{C\cap H}, \mathbb{C})_3$ by the isomorphism

$$\mathrm{Tor}_2^S(S_C, \mathbb{C})_3 \cong \mathrm{Tor}_2^{S_H}(S_{C\cap H}, \mathbb{C})_3.$$

According to paragraph 2 we get quadrics

$$Q_k^{[\alpha|_H]} = \sum_{\substack{0\leq i<j\leq 4\\ i,j\neq k}} b_{ijk}^{[\alpha|_H]} X_i X_j \qquad (k = 0,...,4).$$

Then all coefficients $b_{ijk}^{[\alpha|_H]}$ are different from zero. Suppose this does not hold. Then we consider the path γ_{ijk} in $\pi_1(U([\alpha]), H)$ corresponding to the permutation $\sigma \in Aut(X)$ which exchanges P_i and P_j $(0 \leq i < j \leq 4; i,j \neq k)$. The quadric $Q_i^{[\alpha|_H]}$ will become the quadric $Q_j^{[\alpha|_H]}$ and $b_{hik}^{[\alpha|_H]}$ will become $b_{hjk}^{[\alpha|_H]}$ $(h \in \{0,...4\}, \neq k)$, where $b_{hik}^{[\alpha|_H]} := b_{ihk}^{[\alpha|_H]}$ for $i < h$. This implies that all coefficients vanish, which means $[\alpha\,|_H] = 0$, a contradiction.

Corollary 3.7. *With the assumptions of Remark* 3.6. *we have*

$$Q_k^{[\alpha|_H]} \neq 0 \qquad (k = 0,...,4).$$

For a set $\{P_1,...,P_d\}$ of points in $\mathbb{P}^r$ and for $I = \{i_1,...,i_s\} \subseteq \{1,...,d\}$ the projective space spanned by P_i $(i \in I)$ is denoted by $\mathbb{P}^I$. If $\{P_1,...,P_d\}$ is in linear general position and $s \leq r$, then there are points P_j $(j \in J \subseteq \{1,...,d\} \setminus I)$ defining a projection

$$\pi^{(I)}: \mathbb{P}^r \setminus \mathbb{P}^J \longrightarrow \mathbb{P}^I.$$

We put $Z^{(I)} := \pi^{(I)}(Z)$ for a variety $Z \subseteq \mathbb{P}^r$ (see also paragraph 6).

Definition 3.8. Let $V_{k,s} \subseteq W$ be the set of all d-tuples $(P_1,...,P_d)$ of points in $\mathbb{P}^r$ in linear general position, such that for an index set $I_0 := \{i_1,...,i_s\} \subseteq \{1,...,d\}$ each subset of $\Gamma^{(I_0)}$ $(\Gamma = \{P_1,...,P_d\})$ consisting of m points imposes the same number of conditions on hypersurfaces of degree k in $\mathbb{P}^{I_0}$. We call a set $\{P_1,...,P_d\}$

in $\mathbb{P}^r$ in *uniform position with respect to quadrics* if this set is in uniform position with respect to $V_{2,1}, ..., V_{2,d}$.

Remark 3.9. A general hyperplane section of an irreducible, non-degenerate, projective curve $C \subseteq \mathbb{P}^{r+1}$ is in uniform position with respect to quadrics.

4. Syzygies and Chern Classes

Let $Y \subseteq \mathbb{P}^r$ be a non-degenerate, projective variety and let $X_0, ..., X_r$ be the basis of the vector space $V = H^0(\mathbb{P}^r, \mathcal{O}_{\mathbb{P}^r}(1))$. As it is known, the p-th exterior power $\Omega^p_{\mathbb{P}^r}$ of the sheaf of differential forms is the kernel of the Koszul differential

$$d_p : \bigwedge^p V \otimes \mathcal{O}_{\mathbb{P}^r}(-p) \longrightarrow \bigwedge^{p-1} V \otimes \mathcal{O}_{\mathbb{P}^r}(-p+1).$$

This induces the exact sequence

$$0 \longrightarrow H^0(\mathbb{P}^r, \Omega^{p-1}_{\mathbb{P}^r}(p+1) \otimes \mathcal{I}_Y) \longrightarrow \bigwedge^{p-1} V \otimes (I_Y)_2 \longrightarrow \bigwedge^{p-2} V \otimes (I_Y)_3$$

and therefore

$$H^0(\mathbb{P}^r, \Omega^{p-1}_{\mathbb{P}^r}(p+1) \otimes \mathcal{I}_Y) \cong \mathrm{Tor}^S_{p-1}(I_Y, \mathbb{C})_{p+1} \cong \mathrm{Tor}^S_p(S_Y, \mathbb{C})_{p+1}$$

An element $[\alpha] \in \mathrm{Tor}^S_p(S_Y, \mathbb{C})_{p+1}$ with

$$d_{p,1}(\alpha) = \sum_{0 \leq i_1 < ... < i_{p-1} \leq r} X_{i_1} \wedge ... \wedge X_{i_{p-1}} \otimes Q^{[\alpha]}_{i_1 ... i_{p-1}} \in \bigwedge^{p-1} V \otimes (I_Y)_2$$

corresponds to the global section

$$s^{[\alpha]} := (Q^{[\alpha]}_{i_1 ... i_{p-1}})_{0 \leq i_1 < ... < i_{p-1} \leq r} \in H^0(\mathbb{P}^r, \Omega^{p-1}_{\mathbb{P}^r}(p+1) \otimes \mathcal{I}_Y)$$
$$\subseteq (I_Y)_2^{\binom{r+1}{p-1}} \cong \bigwedge^{p-1} V \otimes (I_Y)_2.$$

We identify $H^{2n}(\mathbb{P}^r, \mathbb{Z})$ $(n \leq r)$ with $\mathbb{Z}$ and denote by $c_n(\mathcal{F})$ the n-th Chern class of a sheaf $\mathcal{F}$. We will apply the following proposition ([Gro], [H] p 429 ff) to a global section induced by a syzygy.

Proposition 4.1. *Let $\mathcal{F}$ be a locally free sheaf of rank n on $\mathbb{P}^r$ and let $s \in H^0(\mathbb{P}^r, \mathcal{F})$ be a global section. Let Y be the zero scheme, which is defined by the exact sequence*

$$\mathcal{F}^\vee \xrightarrow{s^\vee} \mathcal{O}_{\mathbb{P}^r} \longrightarrow \mathcal{O}_Y.$$

If codim $(Y, \mathbb{P}^r) = n$, *then* $c_n(\mathcal{F}) = \deg Y$.

Remark 4.2. According to the sequence

$$0 \longrightarrow \Omega^1_{\mathbb{P}^r}(3) \longrightarrow V \otimes \mathcal{O}_{\mathbb{P}^r}(2) \longrightarrow \mathcal{O}_{\mathbb{P}^r}(3) \longrightarrow 0$$

and ([H], appendix) we get the following Chern classes:

$$c_r(\Omega^1_{\mathbb{P}^r}(3)) = \sum_{l=0}^{r} \binom{r+1}{l} 2^l (-3)^{r-l} .$$

So $c_4(\Omega^1_{\mathbb{P}^4}(3)) = 11$, $c_3(\Omega^1_{\mathbb{P}^3}(3)) = 5$ and $c_2(\Omega^1_{\mathbb{P}^2}(3)) = 3$.

For the rest of this paragraph we consider the case $r = 4$ and $p = 2$.
Let $Y = \{P_0, ..., P_{4+g}\} \subseteq \mathbb{P}^4$ be a set of $5 + g$ points in linear general position, where $P_0, ..., P_4$ are the coordinate points of $\mathbb{P}^4$. For $[\alpha] \in \mathrm{Tor}_2^S(S_Y, \mathbb{C})_3$ we want to describe the zero scheme $V([\alpha])$ given by $s^{[\alpha]} \in H^0(\mathbb{P}^4, \Omega^1_{\mathbb{P}^4}(3))$. In particular, we are interested in $\mathrm{codim}\,(V([\alpha]), \mathbb{P}^4)$. For $[\alpha] \in \mathrm{Tor}_2^S(S_Y, \mathbb{C})_3, \neq 0$ we have

$$d_{2,1}(\alpha) = \sum_{0 \leq i \leq 4} X_i \otimes Q_i^{[\alpha]} \in V \otimes (I_Y)_2.$$

The zero scheme $V([\alpha])$ is the intersection of the five quadrics $Q_0^{[\alpha]}, ..., Q_4^{[\alpha]}$. Because of $d_{1,2} \circ d_{2,1} = 0$ we get a relation

$$\sum_{i=0}^{4} X_i Q_i^{[\alpha]} = 0.$$

This relation and the linear general position of Y prove the following Lemma.

Lemma 4.3. *If Y consists of more than 7 points, then* $\dim_{\mathbb{C}} (Q_0^{[\alpha]}, ..., Q_4^{[\alpha]})_2 \geq 3$. *If Y contains more than 8 points, then the five quadrics $Q_0^{[\alpha]}, ..., Q_4^{[\alpha]}$ are irreducible and* $\mathrm{codim}_{\mathbb{C}}(V([\alpha]), \mathbb{P}^4) \geq 2$.

By Proposition 4.1. and $c_4(\Omega^1_{\mathbb{P}^4}(3)) = 11$ we get

Lemma 4.4. *If Y contains more than 11 points, then* $2 \leq \mathrm{codim}\,(V([\alpha]), \mathbb{P}^4) \leq 3$.

5. Generalization of the Strong Castelnuovo Lemma in P^4

We apply the previous results to $r = 4$. Let Y be a general hyperplane section of an irreducible, non-degenerate, linearly normal curve $C \subseteq \mathbb{P}^5$ of degree $d \geq 18$ and genus g. So we assume that $Y = \{P_0, ..., P_{4+g}\}$ is a set of $d = 5 + g \geq 18$ points in linear general position in $\mathbb{P}^4$, where $P_0, ..., P_4$ are the coordinate points of $\mathbb{P}^4$. We will prove the following proposition:

Proposition 5.1. *If $\beta_{2,3}(Y) \neq 0$, then the 2nd syzygy variety $V^2_{Syz}(Y) \subseteq \mathbb{P}^4$ is either*
- *a rational normal curve in $\mathbb{P}^4$ or*
- *an irreducible curve $E \subseteq \mathbb{P}^4$ of degree 5 and arithmetic genus 1 or*
- *an irreducible surface $F \subseteq \mathbb{P}^4$ of minimal degree 3.*

First we describe $V([\alpha])$ for $[\alpha] \in \mathrm{Tor}_2^S(S_Y, \mathbb{C})_3, \neq 0$ with

$$d_{2,1}(\alpha) = \sum_{0 \leq i \leq 4} X_i \otimes Q_i^{[\alpha]} \in H^0(\mathbb{P}^4, \mathcal{O}_{\mathbb{P}^4}(1)) \otimes (I_Y)_2 \text{ and } \sum_{0 \leq i \leq 4} X_i Q_i^{[\alpha]} = 0.$$

By Lemma 4.4. $V([\alpha])$ is a curve plus eventually isolated points or a surface plus eventually curves and isolated points. For a general hyperplane $H \cong \mathbb{P}^3$ there is an isomorphism

$$\Omega^1_{\mathbb{P}^4}(3) \otimes \mathcal{O}_H \cong \Omega^1_H(3) \oplus \mathcal{O}_H(2).$$

Considering $[\alpha]$ as an element of $H^0(\mathbb{P}^4, \Omega^1_{\mathbb{P}^4}(3))$ we denote by α_H the corresponding element in $H^0(H, \Omega^1_H(3))$ and by Q_H the quadric in the second summand $H^0(H, \mathcal{O}_H(2))$. To be more precise, there exists a basis $\{X_{i_1}, ..., X_{i_4}, X_H\}$ of $H^0(\mathbb{P}^4, \mathcal{O}_{\mathbb{P}^4}(1))$ and a quadric $Q_H \in (Q_0^{[\alpha]}, \ldots, Q_4^{[\alpha]})_2$ with

$$\sum_{1 \leq k \leq 4} X_{i_k} Q_{i_k}^{[\alpha]} + X_H Q_H = 0.$$

We get $\alpha_H = (Q_{i_1}^{[\alpha]}|_H, ..., Q_{i_4}^{[\alpha]}|_H)$ and $Q_H|_H \in H^0(H, \mathcal{O}_H(2))$. Let $V(\alpha_H)$ be the zero scheme of α_H and let $V(\alpha^H) := V(Q_{i_1}, ..., Q_{i_4})$. Then

$$V([\alpha]) \cap H = V(\alpha_H) \cap V(Q_H|_H) \text{ and } V(\alpha^H) \cap H = V(\alpha_H).$$

In $\{Q_0^{[\alpha]}, ..., Q_4^{[\alpha]}\}$ there exists a regular sequence of length $\mathrm{codim}\,(V([\alpha]), \mathbb{P}^4)$. Without loss of generality let $Q_0^{[\alpha]}, Q_1^{[\alpha]}$ be a regular sequence of length 2 and let the quadrics $Q_0^{[\alpha]}, Q_1^{[\alpha]}, Q_2^{[\alpha]}$ be linearly independent according to Lemma 4.3.

Suppose $V(Q_0^{[\alpha]}, Q_1^{[\alpha]})$ is irreducible. Then $V(Q_0^{[\alpha]}, Q_1^{[\alpha]}, Q_2^{[\alpha]})$ is a curve of degree 8. Let $H \cong \mathbb{P}^3$ be a general element of the pencil $\{H_\lambda\}_{\lambda \in \mathbb{P}^1}$ defined by

$$H_{[\lambda_3:\lambda_4]} := \{[X_0 : ... : X_4] \in \mathbb{P}^4 \mid \lambda_3 X_3 + \lambda_4 X_4 = 0\} \qquad ([\lambda_3 : \lambda_4] \in \mathbb{P}^1).$$

A 1-dimensional irreducible component of $V(Q_0^{[\alpha]}, Q_1^{[\alpha]}, Q_2^{[\alpha]}) \cap H$ could only be contained in the base locus $V(X_3, X_4)$ of the pencil $\{H_\lambda\}_{\lambda \in \mathbb{P}^1}$. The Remark 3.6. and the form of the quadrics $Q_0^{[\alpha]}, ..., Q_4^{[\alpha]}$ imply

$$V(Q_0^{[\alpha]}, Q_1^{[\alpha]}, Q_2^{[\alpha]}) \cap V(X_3, X_4) = V(X_0X_1, X_0X_2, X_1X_2) = \{P_0, P_1, P_2\}$$

and therefore

$$\mathrm{codim}\,(V(Q_0^{[\alpha]}, Q_1^{[\alpha]}, Q_2^{[\alpha]}) \cap H, H) = 3.$$

In particular, we get $\operatorname{codim}(V(\alpha_H), H) = 3$ for $[\alpha] \in \operatorname{Tor}_2^S(S_Y, \mathbb{C})_3$ and a general hyperplane $H \cong \mathbb{P}^3$. Proposition 4.1. implies $\deg V(\alpha_H) = c_3(\Omega_H^1(3)) = 5$ and because of $V([\alpha]) \subseteq V(\alpha^H)$ we have $\deg V([\alpha]) \leq 5$.

Suppose $V(Q_0^{[\alpha]}, Q_1^{[\alpha]}) = F_1 \cup ... \cup F_k$, where $F_1, ..., F_k$ are 2-dimensional irreducible components, $2 \leq k \leq \deg V(Q_0^{[\alpha]}, Q_1^{[\alpha]}) = 4$.
For $k = 4$ we have $V(Q_0^{[\alpha]}, Q_1^{[\alpha]}) \cong \mathbb{P}^2 \cup \mathbb{P}^2 \cup \mathbb{P}^2 \cup \mathbb{P}^2$. By the linear general position, Y could consist of at most 12 points contradicting $\deg Y \geq 18$.
Similarly for $k = 3$ there is an isomorphism $V(Q_0^{[\alpha]}, Q_1^{[\alpha]}) \cong F \cup \mathbb{P}^2 \cup \mathbb{P}^2$ with an irreducible surface F of degree 2 which is degenerate. Again, the linear general position would imply $\deg Y \leq 10$, which is a contradiction to $\deg Y \geq 18$.
We get $k = 2$ and $V(Q_0^{[\alpha]}, Q_1^{[\alpha]}) = F_1 \cup F_2$ with two surfaces F_1, F_2. For $\deg F_1 = \deg F_2 = 2$ the surfaces would be degenerate and $\deg Y \leq 8$ would hold contradicting $\deg Y \geq 18$. It remains $V(Q_0^{[\alpha]}, Q_1^{[\alpha]}) \cong F \cup \mathbb{P}^2$ with an irreducible surface F of minimal degree 3. This surface contains at least 15 points of Y because of the linear general position.
We proved the following Lemma:

Lemma 5.2. *Let $Q_{i_0}^{[\alpha]}, Q_{i_1}^{[\alpha]}$ be a regular sequence of length 2 in $\{Q_0^{[\alpha]}, ..., Q_4^{[\alpha]}\}$.*

a) *If $V(Q_{i_0}^{[\alpha]}, Q_{i_1}^{[\alpha]})$ is irreducible, then $V([\alpha])$ is a curve plus eventually isolated points and $\deg V([\alpha]) \leq 5$ holds.*

b) *If $V(Q_{i_0}^{[\alpha]}, Q_{i_1}^{[\alpha]})$ is reducible, then this variety is isomorphic to $F \cup \mathbb{P}^2$ with an irreducible surface F of minimal degree 3.*

Lemma 5.3. *Let $F \subseteq \mathbb{P}^4$ be an irreducible surface of minimal degree 3 containing at least 13 points $P_1, ..., P_{13}$ of Y. Then $Y \subseteq F$.*

Proof. Each $m \geq 9$ points of Y impose at least 10 conditions on quadrics, i. e., if 10 points of Y are contained in 6 linearly independent quadrics $Q_1, ..., Q_6$, then $Y \subseteq V(Q_1, ..., Q_6)$.
If 11 points of Y lie on 5 linearly independent quadrics $Q_1, ..., Q_5$, then the 11 points lie on a 6th quadric, so we have the first case, or the uniform position principle gives $Y \subseteq V(Q_1, ..., Q_5)$ (see Remark 3.9.).
Similarly, if 12 points of Y lie on 4 linearly independent quadrics $Q_1, ..., Q_4$, then one of the cases before applies or by the uniform position principle $Y \subseteq V(Q_1, ..., Q_4)$ holds.
By assumption there exist 3 linearly independent quadrics Q_1, Q_2, Q_3 defining the ideal of F and containing at least 13 points of Y. So by the cases above or by the uniform position principle $Y \subseteq F$ holds. □

Let $V([\alpha])$ be a curve plus eventually isolated points and $\deg V([\alpha]) \leq 5$.

Lemma 5.4. *Let $C \subseteq \mathbb{P}^r$ be a curve plus eventually isolated points and* $\deg C \leq 5$. *Let the ideal of C be generated by quadrics. Then an irreducible, non-degenerate curve $E \subseteq C$ of degree 5 has arithmetic genus $p_a(E) = 1$.*

Proof. According to the Theorem of Castelnuovo ([GH]) we have $p_a(E) \leq 1$. Suppose $p_a(E) = 0$. Let $\tilde{E}$ be the normalisation of E, $\tilde{\mathcal{O}}_P$ $(P \in E)$ the integral closure of the local ring $\mathcal{O}_P$ in the quotient field and δ_P = length $\tilde{\mathcal{O}}_P/\mathcal{O}_P$ $(P \in E)$. Because of

$$p_a(E) = p_a(\tilde{E}) + \sum_{P \in E} \delta_P.$$

we get $E \cong \tilde{E} \cong \mathbb{P}^1$. Furthermore, we get the following commutative diagram, where l_{5H} is the 5-tuple embedding, π the projection from a point R not contained in $l_{5H}(\mathbb{P}^1)$ and f the canonical map.

$$\begin{array}{ccccc} \mathbb{P}^1 & \xrightarrow{l_{5H}} & l_{5H}(\mathbb{P}^1) & \subseteq & \mathbb{P}^5 \\ \| & & \pi \downarrow & & \pi \downarrow \\ \mathbb{P}^1 & \xrightarrow{f} & E & \subseteq & \mathbb{P}^4 \end{array}$$

We show the existence of a 2-dimensional linear projective space $W \cong \mathbb{P}^2$ through R containing three points A_1, A_2, A_3 of $l_{5H}(\mathbb{P}^1)$. The divisors of degree 3 on $l_{5H}(\mathbb{P}^1) \cong \mathbb{P}^1$ are parametrized by $\mathbb{P}(H^0(\mathbb{P}^1, \mathcal{O}_{\mathbb{P}^1}(3))) \cong \mathbb{P}^3$. For $s \in H^0(\mathbb{P}^1, \mathcal{O}_{\mathbb{P}^1}(3))$ the class in $\mathbb{P}(H^0(\mathbb{P}^1, \mathcal{O}_{\mathbb{P}^1}(3)))$ is denoted by $[s]$. We define

$$I := \{(A, [s]) \in \mathbb{P}^5 \times \mathbb{P}(H^0(\mathbb{P}^1, \mathcal{O}_{\mathbb{P}^1}(3))) \mid A \in \overline{(s)}\},$$

where $\overline{(s)}$ denotes the span of the divisor (s). There are the two projections

$$\pi_1 : I \longrightarrow \mathbb{P}^5$$
$$\pi_2 : I \longrightarrow \mathbb{P}(H^0(\mathbb{P}^1, \mathcal{O}_{\mathbb{P}^1}(3))) \cong \mathbb{P}^3.$$

The variety I is a $\mathbb{P}^2$-bundle over $\mathbb{P}^3$ and has therefore dimension 5. In particular, I is a closed set. We show the birationality of π_1. Let us take

$$[s], [s'] \in \mathbb{P}(H^0(\mathbb{P}^1, \mathcal{O}_{\mathbb{P}^1}(3)))$$

and suppose there exists a point $A \in \overline{(s)} \cap \overline{(s')}$. Then $s+s'$ is a divisor of degree 6 on $l_{5H}(\mathbb{P}^1)$, whose span $\overline{(s+s')}$ is at most 4-dimensional. Since the rational normal curve is non-degenerate, we get $\overline{(s)} \cap \overline{(s')} = \emptyset$. Hence the general fiber of π_1 is 0-dimensional, i. e., π_1 is surjective. This proves that for the point $R \in \mathbb{P}^5 \setminus l_{5H}(\mathbb{P}^1)$ there is a linear space $W \cong \mathbb{P}^2$ intersecting $l_{5H}(\mathbb{P}^1)$ in three points.
The line $\pi(W)$ is a 3-secant of E and by Bezout's Theorem we have $\pi(W) \subseteq C$. The fact that C contains the curve $\pi_P(W) \cup E$ of degree 6 contradicts $\deg C \leq 5$.
□

Lemma 5.5. *Let $V([\alpha])$ be a curve plus eventually isolated points. Let* $\deg V([\alpha]) \leq 5$. *Then $V([\alpha])$ has an irreducible component, which contains Y and which is either a rational normal curve or an irreducible curve of degree* 5 *and arithmetic genus* 1.

Proof. Suppose $V([\alpha]) = Y_1 \cup \ldots \cup Y_k \cup \Gamma$ with 1-dimensional irreducible components $Y_1, \ldots, Y_k$ and a set Γ of isolated points. In $\{Q_0^{[\alpha]}, \ldots, Q_4^{[\alpha]}\}$ we choose a regular sequence of length 3. Without loss of generality we can take $\{Q_0^{[\alpha]}, Q_1^{[\alpha]}, Q_2^{[\alpha]}\}$ to be such a sequence. Then $V(Q_0^{[\alpha]}, Q_1^{[\alpha]}, Q_2^{[\alpha]}) = Y_1 \cup \ldots Y_k \cup Z$ is a 1-dimensional complete intersection of degree 8 with an 1-dimensional irreducible component Z of degree $(8 - \sum_{i=1}^k \deg Y_i)$.

For a generic quadric $Q \in (Q_0^{[\alpha]}, \ldots, Q_4^{[\alpha]})_2 \setminus (Q_0^{[\alpha]}, Q_1^{[\alpha]}, Q_2^{[\alpha]})_2$ we have that codim $(V(Q) \cap Z, \mathbb{P}^4) = 4$ and $\Gamma \subseteq V(Q) \cap Z$. Bezout's Theorem implies $\deg \Gamma \leq 2(8 - \sum_{i=1}^k \deg Y_i)$.
By $\deg V([\alpha]) \leq 5$ we have $1 \leq k \leq 5$. Without loss of generality $\deg Y_1 \geq \ldots \geq \deg Y_k$. We consider all possible cases $(\deg Y_1, ..., \deg Y_k)$ $(k = 1, ..., 5)$. Recall that the linear general position implies that on a curve in $\mathbb{P}^3$ there are at most 4, in $\mathbb{P}^2$ at most 3 and on $\mathbb{P}^1$ at most 2 points of Y.

- Let $k \in \{1, ..., 5\}$ and $\deg Y_1 = ... = \deg Y_k = 1$, i. e. $Y_1 \cong ... \cong Y_k \cong \mathbb{P}^1$. Then $Y_1 \cup ... \cup Y_k$ contains at most $2k$ points of Y and Γ contains at most $2(8-k)$ points of Y. But Y consists of more than $2k + 2(8-k) = 16$ points.
- Let $\deg Y_1 = ... = \deg Y_k = 2$ inducing $k \in \{1, 2\}$. Y_1, Y_k are plane conics and $Y_1 \cup Y_k$ contains at most $3k$ points of Y. Γ contains at most $2(8-2k)$ points of Y. This contradicts $\deg Y > 2(8-2k) + 3k = 16 - k$.
  For $\deg Y_1 = \deg Y_l = 2$ $(l \in \{1, 2\})$; $\deg Y_{l+1} = ... = \deg Y_k = 1 (k \in \{l+1, 3\})$ the variety $Y_1 \cup ... \cup Y_k$ could contain at most $3l + 2(k-l)$ points of Y and in Γ there could lie at most $2(8 - 2l - (k-l))$ points of Y contradicting $\deg Y \geq 18 > 2(8 - l - k) + 2k + l = 16 - l$.
- For $\deg Y_1 = 3$ we have $k \in \{1, 2, 3\}$. Y_1 is either a plane curve or a twisted cubic in $\mathbb{P}^3$ and contains at most 3 or 4 points of Y. For $k = 2$, Y_2 is either a plane conic or a line containing at most 3 or 2 points of Y. For $k = 3$, Y_2 and Y_3 are lines. So on $Y_1 \cup Y_2 \cup Y_3$ there lie at most $4+2+2 = 8$ or $3+2+2 = 7$ points of Y. Γ contains for $k = 1$ at most 10, for $k = 2, \deg Y_2 = 2$ at most 6, for $k = 2, \deg Y_2 = 1$ at most 8 and for $k = 3$ at most 6 points of Y. In each case Y could not consist of more than 14 points.
- If $\deg Y_1 = 4$, we have $k \leq 2$. Suppose $k = 2, \deg Y_1 = 4, \deg Y_2 = 1$. Γ contains at most 6 and Y_2 at most 2 points of Y. Since Y_1 is non-degenerate, as Y consists of more than $4 + 2 + 6 = 12$ points, Y_1 is a rational normal curve in $\mathbb{P}^4$ and each $m \geq 9$ points of Y_1 impose 9 conditions on quadrics. Y consists of at least 18 points implying that at least 10 of them lie on Y_1. The uniform position principle implies that each 10 points of Y impose 9 conditions on quadrics. Consequently, Y must be contained in Y_1.
  For $k = 1$, Γ contains at most $2(8-4) = 8$ points of Y and Y_1 contains at least 10 points of Y. Analogous to the case $k = 2$, $Y \subseteq Y_1$ holds.
- For $\deg Y_1 = 5$, Γ consists of at most 6 points. Since Y_1 is non-degenerate, as Y consists of more than $4+6 = 10$ points, Y_1 is a non-degenerate, irreducible curve of degree 5 contained in $V([\alpha])$. By Lemma 5.4. $p_a(Y_1) = 1$ holds. Furthermore, Y_1 is arithmetically Cohen-Macaulay and because of $\omega_{Y_1} \cong \mathcal{O}_{Y_1}$ arithmetically Gorenstein (ω_{Y_1} denotes the canonical sheaf). By [BE] S_{Y_1} has a minimal free resolution of the following form:

$$0 \longleftarrow S_{Y_1} \longleftarrow S \longleftarrow S(-2)^5 \xleftarrow{\varphi} S(-3)^5 \longleftarrow S(-4) \longleftarrow 0.$$

φ is a skew-symmetric matrix, whose five pfaffians generate the ideal of Y_1. Therefore each $m \geq 10$ points of Y_1 impose 10 conditions on quadrics. Since Y_1 contains at least 11 points of Y, the uniform position principle implies that each 11 points of Y impose 10 conditions on quadrics. This induces $Y \subseteq Y_1$. The following table shows the various cases again.

k	$(d_1, ..., d_k)$	m	description	n	m+n
1	1	2	Y_1 line	14	16
1	2	3	Y_1 plane conic	12	15
1	3	3	Y_1 plane cubic	10	13
1	3	4	Y_1 twisted cubic	10	14
1	4	3	Y_1 plane quartic	8	11
1	4	4	$Y_1 \subseteq \mathbb{P}^3$ non-degenerate	8	12
1	4	10	$Y \subseteq Y_1, Y_1$ rational normal curve	8	18
1	5	3	Y_1 plane quintic	6	9
1	5	4	$Y_1 \subseteq \mathbb{P}^3$ non-degenerate	6	10
1	5	11	$Y \subseteq Y_1, p_a(Y_1) = 1$	6	17

k	$(d_1, ..., d_k)$	m	description	n	m+n
2	(1,1)	4	Y_1, Y_2 lines	12	16
2	(2,1)	5	Y_1 plane conic, Y_2 line	10	15
2	(3,1)	5	Y_1 plane cubic, Y_2 line	8	13
2	(3,1)	6	$Y_1 \subseteq \mathbb{P}^3$ non-degenerate	8	14
2	(4,1)	5	Y_1 plane quartic	6	11
2	(4,1)	6	$Y_1 \subseteq \mathbb{P}^3$ non-degenerate	6	12
2	(4,1)	10	$Y \subseteq Y_1, Y_1$ rational normal curve	6	16
2	(2,2)	6	Y_1, Y_2 plane conics	8	14
2	(3,2)	6	Y_1 plane cubic	6	12
2	(3,2)	7	$Y_1 \subseteq \mathbb{P}^3$ twisted cubic	6	13
3	(1,1,1)	6	Y_1, Y_2, Y_3 lines	10	16
3	(2,1,1)	7	Y_1 plane conic	8	15
3	(3,1,1)	7	Y_1 plane cubic	6	13
3	(3,1,1)	8	Y_1 twisted cubic	6	14
3	(2,2,1)	8	Y_1, Y_2 plane conics, Y_3 line	6	14
4	(1,1,1,1)	8	$Y_1, ..., Y_4$ lines	8	16
4	(2,1,1,1)	9	Y_1 plane conic	6	15
5	(1,1,1,1,1)	10	$Y_1, ..., Y_5$ lines	6	16

$d_i := \deg Y_i$ $(i = 1, ..., 5)$,
m = maximal number of points in Y lying on $Y_1 \cup ... \cup Y_k$
n = maximal number of points in Γ. □

Remark 5.6. Let $Z \subseteq \mathbb{P}^r$ be one of the following varieties:
- rational normal curve
- irreducible curve of degree $r + 1$ and arithmeic genus 1

- irreducible surface of minimal degree $r-1$.
  Then $V^{r-2}_{Syz}(Z) = Z$ holds.

Proof. The minimal free resolutions of these varieties are known and for each quadric $Q \in (I_Z)_2$ there exists a syzygy $[\alpha] \in \mathrm{Tor}^S_{r-2}(S_Z, \mathbb{C})_{r-1}$ with $Q \in I([\alpha])$. This implies $I_Z \subseteq I_{V^{r-2}_{Syz}(Z)}$ and $Z = V^{r-2}_{Syz}(Z)$. □

Proof of Proposition 5.1.:
The Eagon-Northcott complex gives $\beta_{2,3}(F) = 2$ for a 2-dimensional rational normal scroll F and $\beta_{2,3}(C) = 8$ for a rational normal curve C. The minimal free resolution of an irreducible, non-degenerate curve E with $\deg E = 5$ and $p_a(E) = 1$ gives $\beta_{2,3}(E) = 5$.
By the previous Lemmas Y is contained in one of the varieties given in the statement of Proposition 5.1. In particular we have $\beta_{2,3}(Y) \geq 2$. For $\beta_{23}(Y) = 2$, Y must be contained in an irreducible surface F of minimal degree 3, but can neither be contained in a rational normal curve nor in an irreducible curve of degree 5 and arithmetic genus 1 according to Remark 2.3. $\mathrm{Tor}^S_2(S_Y, \mathbb{C})_3 = \mathrm{Tor}^S_2(S_F, \mathbb{C})_3$ implies $V^2_{Syz}(Y) = V^2_{Syz}(F) = F$.
Suppose $\beta_{2,3}(Y) > 2$ and suppose Y is contained in an irreducible surface F of minimal degree 3, but neither in a rational normal curve nor in an irreducible curve of degree 5 and arithmetic genus 1. By Lemma 5.2. we may assume the existence of a syzygy $[\alpha]$ in $\mathrm{Tor}^S_2(S_Y, \mathbb{C})_3$ and a quadric $Q_F \in S_2$, such that the ideal I_F of F is generated by the quadrics $Q^{[\alpha]}_0, Q^{[\alpha]}_1, Q_F$. Without loss of generality we may assume that $Q^{[\alpha]}_0, Q^{[\alpha]}_1$ is a regular sequence of length 2, for which $V(Q^{[\alpha]}_0, Q^{[\alpha]}_1) \cong F \cup \mathbb{P}^2$ holds. By Remark 2.3. there exists a quadric $Q \in (I_Y)_2 \setminus (I_F)_2$ induced by an $[\alpha] \in \mathrm{Tor}^S_2(S_Y, \mathbb{C})_3$. $V(Q) \cap F$ is a divisor on F of type $2H - D$, where H is the general hyperplane section and D an effective divisor.
For $\deg D = 0$ the variety $V(Q) \cap F$ is a curve of degree 6. The mapping cone

$$\begin{array}{ccccccccccc}
0 \longleftarrow & S_F(-2) & \longleftarrow & S(-2) & \longleftarrow & S(-4)^3 & \longleftarrow & S(-5)^2 & \longleftarrow 0 \\
 & Q\downarrow & & Q\downarrow & & Q\downarrow & & Q\downarrow & \\
0 \longleftarrow & S_F & \longleftarrow & S & \longleftarrow & S(-2)^3 & \longleftarrow & S(-3)^2 & \longleftarrow 0 \\
 & \downarrow & & & & & & & \\
 & S_{V(Q)\cap F} & & & & & & &
\end{array}$$

gives the following graded betti numbers

$\beta_{i,i+j}(V(Q) \cap F)$ $i =$	0	1	2	3
$j = 0$	1	–	–	–
1	–	4	2	–
2	–	–	3	2

Because of $\beta_{2,3}(V(Q) \cap F) < \beta_{2,3}(Y)$ there is a further quadric $Q' \in (I_Y)_2 \setminus (I_{V(Q)\cap F})_2$.
For $V(Q) \cap F$ irreducible $V(Q') \cap V(Q) \cap F$ consists of at most 12 points contradicting $\deg Y \geq 18$.
Suppose $V(Q) \cap F = C_1 \cup \ldots \cup C_k \cup \Gamma$ with 1-dimensional irreducible components

$C_1, ..., C_k$ $(2 \leq k \leq 6)$ and a set Γ of isolated points. In $\{Q, Q_0^{[\alpha]}, Q_1^{[\alpha]}, Q_F\}$ there exists a regular sequence of length 3 A completely analogous discussion as in the proof of Lemma 5.4. shows that Γ consists of at most 4 points and $V(Q) \cap F$ has an irreducible component, which is either a rational normal curve containing Y, which contradicts our assumption, or which is an irreducible curve E of degree 5 containing at least 12 points of Y. The quadric Q' is an element of I_E, since otherwise $V(Q')$ would intersect E in at most 10 points, but $V(Q') \cap E$ contains Y. Therefore, the ideal I_E is generated by the quadrics $Q', Q, Q_0^{[\alpha]}, Q_1^{[\alpha]}, Q_F$ and Lemma 5.4. implies $p_a(E) = 1$. The uniform position principle gives $Y \subseteq E$ contradicting our assumption.

So $\deg D > 0$ holds and $V(Q) \cap F$ is a curve of degree ≤ 5. Without loss of generality $Q = Q_2^{[\alpha]}$ may be assumed. $Q_0^{[\alpha]}, Q_1^{[\alpha]}, Q$ is a regular sequence of length 3 and completely analogous to the proof of Lemma 5.5. $V(Q) \cap F$ has an irreducible component, which contains Y and is either a rational normal curve or an irreducible curve of degree 5 and arithmetic genus 1. This contradicts our assumption. We have proved that for $\beta_{2,3}(Y) > 2$ Y is contained in a rational normal curve or in an irreducible curve of degree 5 and arithmetic genus 1. In particular, $\beta_{2,3}(Y) > 2$ implies $\beta_{2,3}(Y) \geq 5$.

For $\beta_{2,3}(Y) = 5$, Y can only be contained in an irreducible curve E of degree 5 and $p_a(E) = 1$. $\mathrm{Tor}_2^S(S_Y, \mathbb{C})_3 \cong \mathrm{Tor}_2^S(S_E, \mathbb{C})_3$ implies $V_{Syz}^2(Y) = V_{Syz}^2(E) = E$.

Suppose $\beta_{2,3}(Y) > 5$ and Y is not contained in a rational normal curve, but Y lies on an irreducible curve E with $\deg E = 5$ and $p_a(E) = 1$. Then there exists a quadric Q'' in $(I_Y)_2 \setminus (I_E)_2$. But $V(Q'') \cap E$ consists of at most 10 points contradicting $\deg Y \geq 18$. Hence for $\beta_{2,3}(Y) > 5$ Y is contained in a rational normal curve C, which implies $\beta_{2,3}(Y) \geq 8$.

Suppose $\beta_{2,3}(Y) > 8$. Then there exists a quadric $Q''' \in (I_Y)_2 \setminus (I_C)_2$, such that $V(Q''') \cap C$ consists of at most 8 points contradicting $\deg Y \geq 18$.

Therefore, Y is contained in a rational normal curve C if and only if $\beta_{2,3}(Y) = \beta_{2,3}(C) = 8$. As before, this implies $V_{Syz}^2(Y) = V_{Syz}^2(C) = C$. □

6. Generalization of the Strong Castelnuovo Lemma in P^r

Let $Y \subseteq \mathbb{P}^r$ be a variety containing the points $P_0 = [1 : 0 : ... : 0], ..., P_r = [0 : ...0 : 1]$. For $i \in \{0, ..., r\}$ and $I_i := \{0, ..., r\} \setminus \{i\}$ let

$$\begin{aligned} \pi^{(I_i)} : \quad & \mathbb{P}^r \setminus P_i \longrightarrow \mathbb{P}^{I_i} \\ & [X_0 : ... : X_r] \longrightarrow [X_0 : ... : \widehat{X_i} : ... : X_r] \end{aligned}$$

be the projection from the coordinate point P_i. We denote by $^{(I_i)}$ the projected objects:

$S^{(I_i)} := \mathbb{C}[X_0, ..., \widehat{X_i}, ..., X_r]$, $P^{(I_i)} := \pi^{(I_i)}(P)$ $(P \in \mathbb{P}^r \setminus \{P_i\})$, $Z^{(I_i)} = \pi^{(I_i)}(Y \setminus \{P_i\})$ for a projective variety $Z \subseteq \mathbb{P}^r$ and $V^{(I_i)} := H^0(\mathbb{P}^{I_i}, \mathcal{O}_{\mathbb{P}^{I_i}}(1))$. $V^{(I_i)}$ has the

basis $\{X_0, ..., \widehat{X_i}, ..., X_r\}$. For $I = \{0, ..., r\} \setminus \{i_1, ..., i_p\}$ we define $\pi^{(I)} := \pi^{(I_{i_p})} \circ ... \circ \pi^{(I_{i_1})}$. Again by ${}^{(I)}$ we denote the projected objects:
$S^{(I)} := \mathbb{C}[X_i \mid i \in I]$, $Z^{(I)} := \pi^{(I)}(Z \setminus \{P_{i_1}, ..., P_{i_p}\})$ for a projective variety $Z \subseteq \mathbb{P}^r$, $V^{(I)} = H^0(\mathbb{P}^I, \mathcal{O}_{\mathbb{P}^I}(1))$. Again let $\{X_0, ..., X_r\}$ resp. $\{X_j \mid j \in I\}$ be a basis of the vector space $V := H^0(\mathbb{P}^r, \mathcal{O}_{\mathbb{P}^r}(1))$ resp. $V^{(I)}$. We define the projection of syzygies:

$$\begin{array}{rl} \tilde{\pi}^{(I_i)} : \operatorname{Tor}_p^S(S_Y, \mathbb{C})_{p+1} & \longrightarrow \operatorname{Tor}_{p-1}^S(S_{X^{(I_i)}}^{(I_i)}, \mathbb{C})_p \\ [\alpha] & \longrightarrow \tilde{\pi}^{(I_i)}([\alpha]) \end{array}$$

Let $[\alpha] \in \operatorname{Tor}_p^S(S_Y, \mathbb{C})_{p+1}$ with

$$\alpha = \sum_{\substack{0 \le i_1 < ... < i_p \le r \\ j \ne i_1, ..., i_p}} X_{i_1} \wedge ... \wedge X_{i_p} \otimes c_{j i_1 ... i_p}^{[\alpha]} X_j \in \bigwedge^p V \otimes V$$

and

$$d_{p,1}(\alpha) = \sum_{0 \le i_1 < ... < i_{p-1} \le r} X_{i_1} \wedge ... \wedge X_{i_{p-1}} \otimes Q_{i_1 ... i_{p-1}}^{[\alpha]} \in \bigwedge^{p-1} V \otimes (I_Y)_2.$$

We define

$$c_{j i_1 ... i_{p-1}}^{([\alpha], i)} := \begin{cases} c_{j i i_1 ... i_{p-1}}^{[\alpha]} & \text{for } 0 \le i < i_1 \\ c_{j i_1 ... i_k\, i\, i_{k+1} ... i_{p-1}}^{[\alpha]} & \text{for } i_k < i < i_{k+1} \ \ (k \in \{1, ..., p-1\}) \\ c_{j i_1 ... i_{p-1}\, i}^{[\alpha]} & \text{for } i_{p-1} < i \le r \end{cases}$$

and

$$Q_{i_1 ... i_{p-2}}^{([\alpha], i)} := \begin{cases} Q_{i i_1 ... i_{p-2}}^{[\alpha]} & \text{for } 0 \le i < i_1 \\ Q_{i_1 ... i_k\, i\, i_{k+1} ... i_{p-2}}^{[\alpha]} & \text{for } i_k < i < i_{k+1} \ \ (k \in \{1, ..., p-2) \\ Q_{i_1 ... i_{p-2}\, i}^{[\alpha]} & \text{for } i_{p-1} < i \le r \end{cases}$$

Then $\tilde{\pi}^{(I_i)}([\alpha]) = [\beta]$ with

$$\beta := \sum_{\substack{0 \le i_1 < ... < i_{p-1} \le r \\ i_1, ..., i_{p-1} \ne i}} \sum_{\substack{0 \le j \le r \\ j \ne i, i_1, ..., i_{p-1}}} \sum_{l=1}^{p-1} \frac{i - i_l}{\mid i - i_l \mid} X_{i_1} \wedge ... \wedge X_{i_{p-1}} \otimes c_{j i_1 ... i_{p-1}}^{([\alpha], i)} X_j.$$

Remark 6.1. The map $\tilde{\pi}^{(I_i)}$ is welldefined, since the Koszul differential

$$\begin{array}{rl} d_{p-1,1}^{(I_i)} : \bigwedge^{p-1} V^{(I_i)} \otimes V^{(I_i)} & \rightarrow \bigwedge^{p-2} V^{(I_i)} \otimes S_2^{(I_i)} \\ X_{i_1} \wedge ... \wedge X_{i_{p-1}} \otimes X_j & \rightarrow \sum_{l=1}^{p-1} (-1)^{l+1} X_{i_1} \wedge ... \wedge \widehat{X_{i_l}} \wedge ... \wedge X_{i_{p-1}} \otimes X_{i_l} X_j \end{array}$$

applied to β gives

$$d_{p-1,1}^{(I_i)}(\beta) = \sum_{\substack{0 \le i_1 < \ldots < i_{p-2} \le r \\ i_1,\ldots,i_{p-2} \neq i}} X_{i_1} \wedge \ldots \wedge X_{i_{p-2}} \otimes Q_{i_1 \ldots i_{p-2}}^{([\alpha],i)} \in \bigwedge^{p-2} V^{(I_i)} \otimes (I_{Y^{(I_i)}})_2.$$

For $I = \{0, \ldots, r\} \setminus \{i_1, \ldots, i_s\}$ we define

$$\tilde{\pi}^{(I)} : \mathrm{Tor}_p^S(S_Y, \mathbb{C})_{p+1} \longrightarrow \mathrm{Tor}_{p-s}^{S^{(I)}}(S_{Y^{(I)}}^{(I)}, \mathbb{C})_{p-s+1}.$$

by $\tilde{\pi}^{(I)} = \tilde{\pi}^{(I_{i_s})} \circ \ldots \circ \tilde{\pi}^{(I_{i_1})}$

Examples 6.2. Let $[\alpha] \in \mathrm{Tor}_{r-2}^S(S_Y, \mathbb{C})_{r-1}$ and $I = \{0, \ldots, 4\}$. Then $\tilde{\pi}^{(I)}([\alpha]) = [\beta]$ with

$$\beta = \sum_{0 \le i_1 < i_2 \le 4} \sum_{\substack{0 \le j \le 4 \\ j \neq i_1, i_2}} X_{i_1} \wedge X_{i_2} \otimes c_{j i_1 i_2 5 \ldots r}^{[\alpha]} X_j.$$

The Koszul differential

$$d_{p,q}^I : \bigwedge^p V^{(I)} \otimes S_q^{(I)} \longrightarrow \bigwedge^{p-1} V^{(I)} \otimes S_{q+1}^{(I)}$$

gives

$$d_{2,1}^{(I)}(\beta) = \sum_{0 \le k \le 4} X_k \otimes Q_{k5\ldots r}^{[\alpha]}.$$

Because of $d_{1,2}^{(I)} \circ d_{2,1}^{(I)}(\beta) = 0$ we get

$$\sum_{0 \le k \le 4} X_k Q_{k5\ldots r}^{[\alpha]} = 0.$$

The quadrics $Q_{k5\ldots r}^{[\alpha]}$ are of the same form as the quadrics $Q_k^{[\alpha]}$ in example 2.4. Here the coefficients are denoted by $c_{j i_1 i_2 5 \ldots r}^{[\alpha]}$ instead of $c_{j i_1 i_2}^{[\alpha]}$ there.

Let $Y \subseteq H \cong \mathbb{P}^r$ be a general hyperplane section of an irreducible, non-degenerate, linearly normal curve $C \subseteq \mathbb{P}^{r+1}$ of degree $d \ge r + 14$ and genus g. We identify H with $\mathbb{P}^r$. Then $Y \subseteq \mathbb{P}^r$ is a set of d points in linear general position and in uniform position with respect to sufficiently many closed, $PGl(r)$-invariant subspaces $V_1, \ldots, V_n$ $(n \in \mathbb{N})$ of $W = \{(P_1, \ldots, P_d) \in \mathbb{P}^r \times \ldots \times \mathbb{P}^r \mid P_i \neq P_j \ (i \neq j)\}$. The main result of this article is the following Theorem:

Theorem 6.3. *Suppose $\beta_{r-2,r-1}(Y) \neq 0$. Then $V_{Syz}^{r-2}(Y) \subseteq \mathbb{P}^r$ is one of the following varieties:*

- *rational normal curve*
- *irreducible curve of degree $r + 1$ and arithmetic genus 1*
- *irreducible surface F of minimal degree $r - 1$.*

We assume $\beta_{r-2,r-1}(Y) \neq 0$. The following Lemmas will prove the proposition.

Lemma 6.4. *Let $I_0 := \{j_0, ..., j_4\} \subseteq \{0, ..., r\}$ be fixed. If $V^2_{Syz}(Y^{(I_0)})$ is a rational normal curve, then $V^2_{Syz}(Y^{(I)})$ is a rational normal curve for every $I = \{i_0, ..., i_4\} \subseteq \{0, ..., r\}$. If $V^2_{Syz}(Y^{(I_0)})$ is an irreducible curve of degree 5 and arithmetic genus 1, then $V^2_{Syz}(Y^{(I)})$ is such a curve for every $I = \{i_0, ..., i_4\} \subseteq \{0, ..., r\}$. If $V^2_{Syz}(Y^{(I_0)})$ is an irreducible surface of minimal degree 3, then $V^2_{Syz}(Y^{(I)})$ is such a surface for every $I = \{i_0, ..., i_4\} \subseteq \{0, ..., r\}$.*

Proof. By the uniform position principle $Q^{[\alpha]}_{i_1 \dots i_{p-1}} \neq 0$ for $(0 \leq i_1 < ... < i_{p-1} \leq r)$. The projection of syzygies implies $\beta_{2,3}(Y^{(I)}) \neq 0$ for every $I = \{i_0, ..., i_4\}$. By Proposition 5.1. $V^2_{Syz}(Y^{(I)}) \subseteq \mathbb{P}^4$ is either a rational normal curve or an irreducible curve of degree 5 and arithmetic genus 1 or an irreducible surface of minimal degree 3. Suppose $Y^{(I)} \subseteq \mathbb{P}^4$ is contained in a rational normal curve. Then $Y^{(I)}$ imposes 9 conditions on quadrics in $\mathbb{P}^4$. If $Y^{(I)}$ is contained in an irreducible curve of degree 5 and arithmetic genus 1, then Y^I imposes exactly 10 conditions on quadrics in $\mathbb{P}^4$. We show: If $Y^{(I)}$ is contained in an irreducible surface F of minimal degree 3 and neither in a rational normal curve nor in an irreducible curve of degree 5 and arithmetic genus 1, then $Y^{(I)}$ imposes exactly 12 conditions on quadrics in $\mathbb{P}^4$. Suppose there exists a quadric $Q \in (I_{Y^{(I)}})_2$ not contained in the ideal I_F of F. Then as in the proof of Proposition 5.1. $V(Q) \cap F$ is a divisor on F of type $2H - D$ and completely analogous as in the proof of propostion 5.1. we get a contradiction.
The uniform position principle implies that $Y^{(I)}$ imposes the same number of conditions on quadrics in $\mathbb{P}^4$. This proves the statement. □

We put $I_0 := \{0, ..., 4\}$.

Lemma 6.5. *If $V^2_{Syz}(Y^{(I_0)})$ is a rational normal curve in $\mathbb{P}^4$, then $V^{r-2}_{Syz}(Y)$ is a rational normal curve in $\mathbb{P}^r$.*

Proof. We recall $Y = \{P_0, ..., P_{r+g}\}$, where $P_0, ..., P_r$ are the coordinate points of $\mathbb{P}^r$ and $P_{r+k} = [a_0^{(k)} : ... : a_r^{(k)}]$ $(k = 1, ..., g)$. The linear general position implies $a_i^{(k)} \neq 0$. By Lemma 6.4. $V^2_{Syz}(Y^{(I)})$ is a rational normal curve in $\mathbb{P}^I$ for every $I = \{i_0, ..., i_4\}$. By a Lemma of Castelnuovo (see [GH]) $\pi^{(I)}(P_{i_0}), ..., \pi^{(I)}(P_{i_4})$, $\pi^{(I)}(P_{r+k})$ $(k = 1, ..., g)$ are contained in a unique rational normal curve $C^{(I)} \subseteq \mathbb{P}^4$ and by the Steiner Construction ([GH]) the ideal $I_{C^{(I)}}$ is generated by the quadrics

$$\det \begin{pmatrix} X_{i_0}X_{i_m} & X_{i_0}X_{i_n} & X_{i_m}X_{i_n} \\ a^{(1)}_{i_0}a^{(1)}_{i_m} & a^{(1)}_{i_0}a^{(1)}_{i_n} & a^{(1)}_{i_m}a^{(1)}_{i_n} \\ a^{(2)}_{i_0}a^{(2)}_{i_m} & a^{(2)}_{i_0}a^{(2)}_{i_n} & a^{(2)}_{i_m}a^{(2)}_{i_n} \end{pmatrix} \qquad (1 \leq m < n \leq 4).$$

Again by the Steiner Construction the quadrics

$$\det\begin{pmatrix} X_0X_m & X_0X_n & X_mX_n \\ a_0^{(1)}a_m^{(1)} & a_0^{(1)}a_n^{(1)} & a_m^{(1)}a_n^{(1)} \\ a_0^{(2)}a_m^{(2)} & a_0^{(2)}a_n^{(2)} & a_m^{(2)}a_n^{(2)} \end{pmatrix} \qquad (1 \le m < n \le r)$$

define a rational normal curve C in $\mathbb{P}^r$ containing Y. We have $\beta_{r-2,r-1}(Y) \ge \beta_{r-2r-1}(C)$.
If $\beta_{r-2,r-1}(Y) > \beta_{r-2r-1}(C)$, there would exist a quadric $Q^{[\alpha]}_{i_1\dots i_{r-3}} \in (I_Y)_2 \setminus (I_C)_2$ induced by a syzygy $[\alpha] \in \operatorname{Tor}^S_{r-2}(S_Y,\mathbb{C})_{r-1}$. For $\{0,\dots,r\} \setminus I_0 \subseteq \{i_1,\dots,i_{r-3}\}$ $Y^{(I_0)}$ would be contained in $V(Q^{[\alpha]}_{i_1\dots i_{r-3}})^{(I_0)} \cap C^{(I_0)}$. This variety consists of at most 8 points, which contradicts $\deg Y^{(I_0)} \ge 18$.
$\beta_{r-2,r-1}(Y) = \beta_{r-2r-1}(C)$ implies $V^{r-2}_{Syz}(Y) = V^{r-2}_{Syz}(C) = C$. □

For an analogous statement to Lemma 6.5. applied to the remaining cases we need the lifting of syzygies. The following remark is well known and essential for the lifting of syzygies.

Remark 6.6. Let $I = (f_1,\dots,f_s) \subseteq S$ be an ideal generated by quadrics $f_1,\dots,f_s$. Then $X_0,\dots,X_n$ $(n \le r)$ is a regular sequence in S/I if and only if every relation $\overline{R}$ among the quadrics $f_1 \bmod (X_0,\dots,X_n),\dots,f_s \bmod (X_0,\dots,X_n)$ in $S/(X_0,\dots,X_n)$ can be lifted to a relation $\tilde{R} = R + \sum_{i=1}^{s} l_i f_i$ with polynomials $l_1,\dots,l_s$ in $X_0,\dots,X_n$, i. e. $\tilde{R} \bmod (X_0,\dots,X_n) = \overline{R}$.

Lemma 6.7. *If* $V^2_{Syz}(Y^{(I_0)}) \subseteq \mathbb{P}^4$ *is an irreducible curve of degree* 5 *and arithmetic genus* 1, *then* $V^{r-2}_{Syz}(Y) \subseteq \mathbb{P}^r$ *is an irreducible curve of degree* $r+1$ *and arithmetic genus* 1.

Proof. By Lemma 6.4. $V^2_{Syz}(Y^{(I)}) \subseteq \mathbb{P}^4$ is an irreducible curve of degree 5 and arithmetic genus 1 for every $I = \{i_0,\dots,i_4\} \subseteq \{0,\dots,r\}$. By the Theorem of Castelnuovo (see [GH]) such a curve is arithmetically Cohen-Macaulay and because of $\omega_{V^2_{Syz}(Y^{(I)})} \cong \mathcal{O}_{V^2_{Syz}(Y^{(I)})}$ arithmetically Gorenstein. By [BE] $S_{V^2_{Syz}(Y^{(I)})}$ has a minimal free resolution

$$0 \longleftarrow S^{(I)}_{V^2_{Syz}(Y^{(I)})} \longleftarrow S^{(I)} \longleftarrow S^{(I)}(-2)^5 \xleftarrow{\ \varphi\ } S^{(I)}(-3)^5 \longleftarrow S^{(I)}(-4) \longleftarrow 0.$$

φ is a skewsymmetric matrix and the five pfaffians $q_0^{(I)},\dots,q_4^{(I)} \in S_2^{(I)}$ generate $I_{V^2_{Syz}(Y^{(I)})}$. Let $R_0^{(I)},\dots,R_4^{(I)}$ be the relations among $q_0^{(I)},\dots,q_4^{(I)}$.
A general hyperplane $H \cong \mathbb{P}^3$ in $\mathbb{P}^4$ intersects $V^2_{Syz}(Y^{(I)})$ in 5 points. Without loss of generality we may assume that the hyperplane H_0 with equation $X_{i_0} = 0$ does so. We have $H_0 \cap V^2_{Syz}(Y^{(I)}) = \{P^{(I)}_{i_1},\dots,P^{(I)}_{i_4},P^{(I)}\}$ and

$$(q_0^{(I)},\dots,q_4^{(I)},X_{i_0}) = (f^{(I)}_{i_1i_3}, f^{(I)}_{i_1i_4}, f^{(I)}_{i_2i_3}, f^{(I)}_{i_2i_4}, f^{(I)}_{i_3i_4}, X_{i_0})$$

where $f^{(I)}_{i_k i_l} := X_{i_k}X_{i_l} - c^{(I)}_{i_k i_l}X_{i_1 i_2}$ $(1 \le k < l \le 4; (k,l) \ne (1,2); c^{(I)}_{i_k i_l} \in \mathbb{C} \setminus \{0\})$. The numbers $c^{(I)}_{i_k i_l}$ are determined by the coordinates of $P^{(I)}$.
The Standardbasisalgorithm ([S2] appendix) shows that there are five linear relations $\overline{R}^{(I)}_0, ..., \overline{R}^{(I)}_4$ among $f^{(I)}_{i_1 i_3}, f^{(I)}_{i_1 i_4}, f^{(I)}_{i_2 i_3}, f^{(I)}_{i_2 i_4}, f^{(I)}_{i_3 i_4}$ and that these relations can be lifted to

$$\tilde{R}_l = \sum_{0 \le k \le 4} b_{k,l}\overline{R}^{(I)}_k q^{(I)}_k + X_{i_0} q^{(I)}_k \qquad (l = 0, ..., 4; b_{k,l} \in \mathbb{C})$$

in the sense of Remark 6.6. Let J be the ideal generated by the quadrics

$$q^{ij}_k := q^{(I)}_k \qquad (I = \{0,1,2,i,j\}; 3 \le i < j \le r; k = 0, ..., 4).$$

We show that $V(J)$ is an irreducible curve of degree $r+1$ and arithmetic genus 1. For $I = \{0,1,2,i,j\}$ we put $f_{i_k i_l} := f^{(I)}_{i_k i_l}$ $(1 \le k < l \le 4; (k,l) \ne (1,2))$ and $\overline{R}^{ij}_k := \overline{R}^{(I)}_k$ $(k = 0, ..., 4)$. We have

$$J \bmod (X_0) = (f_{kl} \mid 1 \le k < l \le r; (k,l) \ne (1,2)) \bmod (X_0)$$

and

$$V(J + (X_0)) = \{P_1, ..., P_r, P\}.$$

The relations among q^{ij}_k $(3 \le i < j \le r; k = 0, ..., 4)$ are generated by the relations R^{ij}_k $(3 \le i < j \le r; k = 0, ..., 4)$ according the Standardbasisalgorithm. The relations among f_{kl} $(1 \le k < l \le r, (k,l) \ne (1,2))$ are generated by $\overline{R}^{ij}_k$ $(k = 0, ..., 4; 3 \le i < j \le r)$. This proves that every relation among f_{kl} $(1 \le k < l \le r; (k,l) \ne (1,2))$ can be lifted in the sense of Remark 6.6. $V(J)$ is a curve of degree $r+1$. It is irreducible and of arithmetic genus 1, since it is mapped to $V^2_{Syz}(Y^{(I)})$ by the projection

$$\begin{array}{llll} \pi^{(I)}: & \mathbb{P}^r \setminus \mathbb{P}^{I^c} & \longrightarrow & \mathbb{P}^I \\ & [X_0 : ... : X_r] & \longrightarrow & [X_{i_0} : ... : X_{i_4}] \end{array} \qquad (I = \{i_0, ..., i_4\} = \{0, ..., r\} \setminus I^c)$$

We have $\beta_{r-2,r-1}(Y) \ge \beta_{r-2,r-1}(V(J))$. Suppose $\beta_{r-2,r-1}(Y) > \beta_{r-2,r-1}(V(J))$. Then a syzygy $[\alpha] \in \mathrm{Tor}^S_{r-2}(S_X, \mathbb{C})$ induces a quadric $Q^{[\alpha]}_{i_1 ... i_{r-3}} \in (I_Y)_2 \setminus (I_{V(J)})_2$. Without loss of generality $\{0, ..., r\} \setminus I_0 \subseteq \{i_1, ..., i_{r-3}\}$. Then $V(Q^{[\alpha]}_{i_1 ... i_{r-3}})^{(I_0)} \cap V(J)^{(I_0)}$ consists of at most 10 points contradicting $\deg Y^{(I_0)} \ge 18$. Therefore $\beta_{r-2,r-1}(Y) = \beta_{r-2r-1}(V(J))$ must hold. This implies $V^{r-2}_{Syz}(Y) = V^{r-2}_{Syz}(V(J)) = V(J)$. □

The surface case is proved in the same way as the last Lemma.

Lemma 6.8. *If $V^2_{Syz}(Y^{(I_0)})$ is an irreducible surface of minimal degree 3 in $\mathbb{P}^4$, then $V^{r-2}_{Syz}(Y)$ is an irreducible surface of minimal degree $r-1$ in $\mathbb{P}^r$.*

Proof. By Lemma 6.4. the variety $V^2_{Syz}(Y^{(I)})$ is an irreducible surface of degree 3 for every $I = \{i_0, ..., i_4\}$. Such a surface has a minimal free resolution of the following form:

$$0 \longleftarrow S^{(I)}_{V^2_{Syz}(Y^{(I)})} \longleftarrow S^{(I)} \longleftarrow S^{(I)}(-2)^3 \longleftarrow S^{(I)}(-3)^2 \longleftarrow 0.$$

The ideal $I_{V^2_{Syz}(Y^{(I)})}$ is generated by three quadrics $q_0^{(I)}, q_1^{(I)}, q_2^{(I)} \in S_2^{(I)}$. Among these there are 2 linear relations $R_1^{(I)}, R_2^{(I)}$. Let $W \cong \mathbb{P}^2$ be a general 2-dimensional linear subspace of $\mathbb{P}^4$ intersecting $V^2_{Syz}(Y^{(I)})$ in 3 different points. Without loss of generality we may assume the existence of a syzygy $[\alpha^{(I)}] \in \mathrm{Tor}_2^{S^{(I)}}(S^{(I)}_{Y^{(I)}}, \mathbb{C})_3$ with

$$q_0^{(I)}, q_1^{(I)}, q_2^{(I)} \in \{Q_0^{[\alpha^{(I)}]}, ..., Q_4^{[\alpha^{(I)}]}\}.$$

The projective space $W_0 \cong \mathbb{P}^2$ given by the equations $X_{i_0} = X_{i_1} = 0$ intersects $V^2_{Syz}(Y^{(I)})$ in $\{P_{i_2}^{(I)}, P_{i_3}^{(I)}, P_{i_4}^{(I)}\}$ and we get

$$(q_0^{(I)}, q_1^{(I)}, q_2^{(I)}, X_{i_0}, X_{i_1}) = (X_{i_2}X_{i_3}, X_{i_2}X_{i_4}, X_{i_3}X_{i_4}, X_{i_0}, X_{i_1}).$$

We may assume $X_{i_2}X_{i_3}, X_{i_2}X_{i_4}, X_{i_3}X_{i_4}$ are the leading forms of $q_0^{(I)}, q_1^{(I)}, q_2^{(I)}$ with respect to the inverse lexicographic order on $S^{(I)}$. Among $X_{i_2}X_{i_3}, X_{i_2}X_{i_4}, X_{i_3}X_{i_4}$ there are 2 linear syzygies $\overline{R}_1^{(I)}, \overline{R}_2^{(I)}$, which can be lifted to relations

$$\tilde{R}_k^{(I)} = b_{1,k}^{(I)} R_1^{(I)} + b_{2,k}^{(I)} R_2^{(I)} + \sum_{j=0}^{2} l_{j,k}^{(I)} q_j^{(I)} \qquad (k = 1, 2)$$

with $l_{j,k}^{(I)} \in \mathbb{C}[X_{i_0}, X_{i_1}]$ according the Standardbasisalgorithm.
Let J be the ideal generated by $q_0^{(I)}, q_1^{(I)}, q_2^{(I)}$ with $0, 1 \in I$ ($i_0 = 0, i_1 = 1$). Then

$$J + (X_0, X_1) = (X_i X_j \mid 2 \leq i < j \leq r) + (X_0, X_1)$$

and

$$V(J + (X_0, X_1)) = \{P_2, ..., P_r\}.$$

The relations among $X_i X_j$ $(2 \leq i < j \leq r)$ are generated by $\overline{R}_1^{(I)}, \overline{R}_2^{(I)}$ $(I = \{0, 1, k, l, m\}$ $(2 \leq k < l < m \leq r))$. Therefore every relation among these quadrics can be lifted in the sense of Remark 6.6. $V(J)$ is a surface of degree $r - 1$. It is irreducible, since it is mapped to $V^2_{Syz}(Y^{(I)})$ by the projection

$$\begin{array}{llcll} \pi^{(I)}: & \mathbb{P}^r \setminus \mathbb{P}^{I^c} & \longrightarrow & \mathbb{P}^I & \qquad (I = \{i_0, ..., i_4\} = \{0, ..., r\} \setminus I^c) \\ & [X_0 : ... : X_r] & \longrightarrow & [X_{i_0} : ... : X_{i_4}] & \end{array}$$

We get $\beta_{r-2,r-1}(Y) \geq \beta_{r-2r-1}(V(J))$. Suppose $\beta_{r-2,r-1}(Y) > \beta_{r-2r-1}(V(J))$. Then $[\alpha] \in \mathrm{Tor}^S_{r-2}(S_Y, \mathbb{C})_{r-1} \setminus \mathrm{Tor}^S_{r-2}(V(J), \mathbb{C})_{r-1}$ induces a quadric $Q^{[\alpha]}_{i_1 \ldots i_{r-3}} \in (I_Y)_2 \setminus (I_{V(J)})_2$. Without loss of generality $\{0, ..., r\} \setminus I_0 \subseteq \{i_1, ..., i_{r-3}\}$.
The syzygy $\tilde{\pi}^{(I_0)}([\alpha]) \in \mathrm{Tor}_2^{S^{(I_0)}}(S^{(I_0)}_{Y^{(I_0)}}, \mathbb{C})_3$ is not in $\mathrm{Tor}_2^{S^{(I_0)}}(V(J)^{(I_0)}, \mathbb{C})_3$. This contradicts $\beta_{2,3}(Y^{(I_0)}) = \beta_{2,3}(V^2_{Syz}(Y^{(I_0)})) = \beta_{2,3}(V(J)^{(I_0)})$.

Therefore $\beta_{r-2,r-1}(Y) = \beta_{r-2,r-1}(V(J))$ and $V^{r-2}_{Syz}(Y) = V^{r-2}_{Syz}(V(J)) = V(J)$ hold. □

7. Syzygies of Curves

We apply the results of the previous paragraphs to curves.

Theorem 7.1. *Let X be a smooth, irreducible curve of genus g and let L be a very ample line bundle over X of degree $d \geq r+13$ ($h^0(X,L) = r+1$). Let $C := \varphi_L(X) \subseteq \mathbb{P}^r$ be linearly normal. Then for $\beta_{r-3,r-2}(C) \neq 0$ the syzygy variety $V^{r-3}_{Syz}(C) \subseteq \mathbb{P}^r$ is one of the following varieties:*
- *irreducible surface of minimal degree $r-1$*
- *irreducible surface of degree r*
- *irreducible 3-dimensional variety of minimal degree $r-2$.*

Proof. Since C is linearly normal, we have an isomorphism

$$\mathrm{Tor}^S_p(S_C, \mathbb{C})_{p+1} \cong \mathrm{Tor}^{S_H}_p(S_{C\cap H}, \mathbb{C})_{p+1},$$

where $H \cong \mathbb{P}^{r-1}$ is a general hyperplane. This isomorphism implies

$$V^p_{Syz}(C \cap H) = V^p_{Syz}(C) \cap H.$$

By Theorem 6.3. $V^{r-3}_{Syz}(C) \cap H \subseteq H$ is a rational normal curve or an irreducible curve of degree $r-1$ or a surface of minimal degree $r-2$. □

We prove the conjecture 1.3. for $q = 3$ and $g \geq 13$.

Theorem 7.2. *Let X be a smooth, irreducible, non-hyperelliptic curve of genus $g \geq 13$ and let L be a line bundle over X of degree $d \geq 2g+3$. Let $C = \varphi_L(X) \subseteq \mathbb{P}^r = \mathbb{P}(H^0(X,L))$ $(h^0(X,L) = r+1)$. If $\beta_{r-3,r-2}(C) \neq 0$, then $V^{r-3}_{Syz}(C) \subseteq \mathbb{P}^r$ is a 3-dimensional rational normal scroll of degree $r-2$ and the pencil $\{\mathbb{P}^2_t\}_{t\in\mathbb{P}^1}$ of planes $\mathbb{P}^2_t \cong \mathbb{P}^2$ cuts out a g^1_3 on C.*

We need the classification of rational surfaces due to M. Nagata.

Lemma ([N1], [N2]) 7.3. *A non-degenerate surface in $\mathbb{P}^r$ of degree $r \geq 10$ is either the image of a surface in $\mathbb{P}^{r+1}$ of minimal degree under the projection or a cone over a smooth elliptic curve.*

Lemma 7.4. *For $g \geq 13$ and $\deg L = d \geq 2g+3$ the curve $C \subseteq \mathbb{P}^r$ can neither be contained in a surface $F \subseteq \mathbb{P}^r$ of degree r, which is the image of a surface in $\mathbb{P}^{r+1}$ of minimal degree r under projection, nor in a cone over a smooth elliptic curve.*

Proof. Since C is linearly normal, the first case is excluded.
Suppose C is contained in a 2-dimensional cone Z over a smooth, elliptic curve E. Let P be the vertex and L_Q the line determined by P and $Q \in E$. If the line L_Q intersects C in one point, we would get $g = 1$ contradicting the assumption. Hence L_Q intersects C in at least 2 points. The plane $\overline{PQT}$ $(Q, T \in E)$ spanned by L_Q and L_T intersects C in at least four points $P_1, ..., P_4$. Identifying $P_1, ..., P_4$ with its preimages in X we get the special line bundle $L(-P_1 - P_2 - P_3 - P_4)$, i. e.

$$\deg L(-P_1 - P_2 - P_3 - P_4) = \deg L - 4 \leq \deg K_C = 2g - 2$$

contradicting $\deg L \geq 2g + 3$. □

Proof of Theorem 7.2:
By Riemann-Roch we have $r + 1 = d - g + 1 \geq g + 4 \geq 17$, in particular $r \geq 10$. Then by Lemma 7.4. and Theorem 7.1. $V_{Syz}^{r-3}(C) \subseteq \mathbb{P}^r$ is a 2-dimensional or 3-dimensional rational normal scroll. Suppose $V_{Syz}^{r-3}(C)$ is a 2-dimensional rational normal scroll of degree $r - 1$. Then $\beta_{r-2,r-1}(X) \geq \beta_{r-2,r-1}(V_{Syz}^{r-3}(X)) \neq 0$ holds and X would be hyperelliptic by the $K_{p,1}$-Theorem. Therefore $V_{Syz}^{r-3}(C)$ is a 3-dimensional rational normal scroll, which has a pencil $\{\mathbb{P}_t^2\}_{t \in \mathbb{P}^1}$ of planes $\mathbb{P}_t^2 \cong \mathbb{P}^2$. Suppose $\mathbb{P}_t^2$ intersects C in at least 4 points $P_1, ..., P_4$, then as before $L(-P_1 - P_2 - P_3 - P_4)$ would be a special line bundle. contradicting $d \geq 2g + 3$. $\mathbb{P}_t^2 \cap C$ consists of more than 2 points, since C is neither a rational normal curve $(g \geq 13)$ nor hyperelliptic. Therefore $\{\mathbb{P}_t^2 \cap C\}_{t \in \mathbb{P}^1}$ is a g_3^1 on C. □

If one applies Theorem 7.1. to the canonical bundle $L = K_X$ and uses the geometric version of Riemann-Roch, then one would get the main results of [S2] and [V] under the additional assumption $g \geq 14$.

References

[ACGH] E. Arbarello, M. Cornalba, P. A. Griffith, J. Harris, Geometry of algebraic curves, Vol I. Heidelberg-Berlin-New York, 1985.

[BE] D. Buchsbaum, D. Eisenbud, Algebra Structures for Finite Free Resolutions and Some Structure Theorems for Ideals of Codimension 3. Amer. J. Math. 99 (1977), 447–485.

[G1] M. Green, Koszul-homology and the geometry of projective varieties I. J. Diff. Geom. 19 (1984), 125–171.

[G2] M. Green, Koszul-homology and the geometry of projective varieties II. J. Diff. Geom. 20 (1984), 279–289.

[GL1] M. Green, R. Lazarsfeld, The non-vanishing of certain Koszul-cohomology groups. J. Diff. Geom. 19 (1984), 168–170.

[GL2] M. Green, R. Lazarsfeld, On the projective normality of complete linear series on an algebraic curve. Inv. Math. 83 (1986), 73–90.

[GL3] M. Green, R. Lazarsfeld, Some results on the syzygies of finite sets and algebraic curves. Comp. Math. 67 (1988), 301–304.

[GH] P. Griffith, J. Harris, Principles of algebraic geometry. New York-Chichester-Brisbane-Toronto, 1978.

[Gro] A. Grothendieck, La Theorie des Classes de Chern. Bull. Soc. Math. France 86 (1958), 137–154.

[Ha1] J. Harris, A bound on the genus of the projective varieties. Ann. Scuola Norm. Imp. (4) 8 (1981), 35–68.

[Ha2] J. Harris, Curves in projective space. Sem. de Mathematiques Supèrieures Université de Montreal, 1982.

[H] R. Hartshorne, Algebraic Geometry. Heidelberg-Berlin-New York, 1977.

[Mat] H. Matsumura, Commutative ring theory. Cambridge, Cambridge University Press, 1990.

[MS] G. Martens, F. O. Schreyer, Line bundles and syzygies of trigonal curves. Abh. Math. Sem. Univ. Hamburg 56 (1986), 169–189.

[N1] M. Nagata, On rational surfaces I. Memoirs of the College of Science University of Kyoto 32 (3), (1960), 351–370.

[N2] M. Nagata, On rational surfaces II. Memoirs of the College of Science University of Kyoto 33 (2) (1960), 271–293.

[S1] F.-O. Schreyer, Syzygies of canonical curves and special linear series. Math. Ann. 275 (1986), 105–137.

[S2] F.-O. Schreyer, A standard basis approach to syzygies of canonical curves. J. reine angew. Math. 421 (1991), 83–123.

[V] C. Voisin, Courbes tètragonal et cohomologie de Koszul. J. reine angew. Math. 387 (1988), 111–121.

The zero dimensional scheme defined by the intersection of two curve singularities

Juan Elias

Abstract. In this paper the author gives an upper bound of the contact of a pair of curve singularities in terms of the degrees of a minimal standard bases of the ideals, generalizing Noether's formula to space curve singularities.

Mathematics Subject Classification: 14H20.

1. Introduction

The purpose of this paper is to study the length of the zero dimensional scheme defined by the intersection of a pair of curve singularities, so-called contact between the pair of curve singularities.

In the case of plane curve singularities the contact is the intersection multiplicity of the pair. Noether's formula express the multiplicity of intersection in terms of the products of the multiplicities of the shared infinitely near points.

In this paper we give an upper bound of the contact of a pair of curve singularities in terms of the degrees of a minimal standard bases of the ideals, generalizing Noether's formula to space curve singularities.

In the first section we prove some results on the Hilbert function of rank one modules free of torsion over a one-dimensional Cohen-Macaulay ring. We related the Hilbert function of an ideal as module and the Hilbert function of the ground ring with respect the ideal. Using these results in the second section we will generalize Noether's formula. As corollary we prove that if X, Y is a pair of curve singularities such that their contact achieves the bound then the scheme $X \cap Y$ is an hypersurface of X.

2. Torsion free modules of rank one

Throughout this section A will be a one-dimensional Cohen-Macaulay local ring with maximal ideal m. We assume that $\mathbf{k} = A/m$ is an infinite field.

We will recall some basic facts about the Hilbert functions. Let I be an open ideal of A, the Hilbert function of A with respect to I is $h_I^{1*}(n) := \text{length}_A(A/I^n)$. It is well known that for $n >> 0$,

$$h_I^{1*}(n) := e_0^*(I)n - e_1^*(I)$$

for some integers $e_0^*(I), e_1^*(I)$.

Let tot(A) be the total ring of quotients of A and let $\overline{A}$ the integral closure of A in tot(A). Recall that the blowing-up of A centered in I is

$$BL_I(A) = \sum_n (I^n : I^n)_{\overline{A}}.$$

The blowing-up enable us to compute the integer $e_1^*(I)$:

$$e_1^*(I) = \text{length}_A(Bl_I(A)/A)$$

([12], Theorem 1.5). Let x be a transversal element of I, i.e. $x \in I$ and $I^{n+1} = xI^n$ for $n >> 0$. Then ([12], Theorem 1.9)

$$e_1^*(I) = \text{length}_A(E/Ex)$$

for all $E \subset \overline{A}$ sub$-A-$module finitely generated.

Let M be a finitely generated $A-$module, we will denote by h_M^1 the Hilbert function of M, i.e. $h_M^1(n) := \text{length}_A(M/Mm^n)$. It is well known that there are integers $e_0(M), e_1(M)$ such that for all $n >> 0$,

$$h_M^1(n) := e_0(M)n - e_1(M).$$

Remark. Let I be an open ideal of A. Since I is in particular an $A-$module notice that we have two Hilbert functions associated to I: $h_I^{1*}(n) = \text{length}_A(A/I^n)$, $h_I^1(n) = \text{length}_A(I/Im^n)$.

Proposition. *Let M be a finitely generated rank one $A-$module free of torsion. Then $e_0(M) = e_0(A)$ and $e_1(M) \leq e_1(A)$. If $M = I$ is an open ideal of A then*

$$e_1(A) - e_1(I) = e_0^*(I) - \text{length}_A(A/I).$$

Proof. Since M is a finitely generated module rank one free of torsion we may assume that $M = I$ is an ideal of A. Since I has no torsion, I contains a non-zero divisor of A. Hence for $n >> 0$ we have $m^n \subset I$ and we can consider the sequence

$$0 \longrightarrow m^n/m^nI \longrightarrow I/m^nI \longrightarrow I/m^n \longrightarrow 0.$$

From this sequence we obtain

$$\begin{aligned} h_I^1(n) &= \text{length}_A(m^n/m^nI) + \text{length}_A(A/m^n) - \text{length}_A(A/I) \\ &= \text{length}_A(m^n/m^nI) + e_0(A)n - e_1(A) - \text{length}(A/I). \end{aligned}$$

On the other hand recall that m^n is an $\overline{A}$-module for $n >> 0$, in particular m^n is a $Bl_I(A)-$module. Hence we have $Bl_I(A)m^n = m^n$, and then

$$\text{length}_A(m^n/m^nI) = \text{length}_A(m^n/m^nIBl_I(A)).$$

Let ω be a transversal element of I, then $IBl_I(A) = \omega Bl_I(A)$ and we get

$$\text{length}_A(m^n/m^nI) = \text{length}_A(m^n/m^n\omega) = e_0^*(I).$$

Therefore, for $n >> 0$, we have

$$h_I^1(n) = e_0(A)n - e_1(A) + \text{length}_A(A/I) - e_0^*(I).$$

From this equality we obtain $e_0(I) = e_0(A)$ and

$$e_1(I) = e_1(A) - e_0^*(I) + \text{length}_A(A/I).$$

Since $e_0^*(I) \geq \text{length}_A(A/I)$ ([15], Lemma 1), we obtain $e_1(A) \geq e_1(I)$.

Proposition. *Let M be a finitely generated rank one $A-$module free of torsion. Then*

$$h_M^1(n) \geq h_A^1(n)$$

for all $n \geq 0$.

Remark. The moduli space of torsion free rank one modules over a curve singularity has been studied by Greuel and Pfister in [G-Pf].

3. A generalization of Noether's formula

A curve singularity of $(\mathbf{k}^N, 0) = \text{Spec}(R)$ is a one-dimensional, reduced and closed subscheme X of $(\mathbf{k}^N, 0)$, i.e., $X = \text{Spec}(R/I)$ where $I = I(X)$ is a reduced height N-1 ideal of $R = \mathbf{k}[[X_1, ..., X_N]]$; we put $\mathcal{O}_X = R/I$.

Let X be a curve singularity and let Q be an infinitely near point of X, see [6], [16]. It is known that there exists a unique sequence $Q_{i=0,...,t}$ of infinitely near points of X such that $Q_0 = 0, ..., Q_t = Q$, and that Q_{i+1} belongs to the first neighborhood of Q_i for $i = 0, ..., t-1$. We denote by (X, Q) the union of the irreducible components through Q of the proper transform of X by the composition of the blowing-up centered at Q_i for $i = 0, ..., t-1$. We denote by $e_0(X, Q)n - e_1(X, Q)$ the Hilbert polynomial of the local ring $\mathcal{O}_{(X,Q)}$. We will denote by $\text{Inf}(X)$ be the set of infinitely near points of X.

Definition. Let X, Y be curves of $\mathbf{k}$. We denote by $(X.Y)$ the contact between X and Y, i.e., $(X.Y) = \dim_{\mathbf{k}}(R/I(X) + I(Y))$ ([H]).

Remark. Notice that $(X.Y)$ is not an intersection multiplicity. Let us consider X_1, X_2, X_3 curve singularities such that $T_{X_i,0} \cap T_{X_j,0} =$ for $i \neq j$. Then $(X_i, X_j) =$

1 for all $i \neq j$, so $(X_1, X_2 \cup X_3) = 1$ and $(X_1, X_2) + (X_1, X_3) = 1$, i.e., we have no additivity.

Proposition (Noether's formula). *Let X, Y be plane curve singularities, then*

$$(X.Y) = \sum_{P \in \mathrm{Inf}(X) \cap \mathrm{Inf}(Y)} e_0(X, P) e_0(Y, P).$$

Proof. see [13], or [16].

We will generalize Noether's formula for space curve singularities.

Proposition. *Let X, Y be curve singularities, then we have*

$$(X.Y) = \sum_{P \in \mathrm{Inf}(X) \cap \mathrm{Inf}(Y)} (e_1(X \cup Y, P) - e_1(X, P) - e_1(Y, P)).$$

Proof. Let us consider the following diagram:

$$\begin{array}{ccccccccc} 0 & \longrightarrow & \mathcal{O}_{X \cup Y} & \longrightarrow & \mathcal{O}_X \bigoplus \mathcal{O}_Y & \longrightarrow & \mathcal{O}_{X \cap Y} & \longrightarrow & 0 \\ & & \downarrow \varphi_1 & & \downarrow \varphi_2 & & \downarrow \varphi_3 & & \\ 0 & \longrightarrow & \mathcal{O}_{Bl_p(X \cup Y)} & \longrightarrow & \mathcal{O}_{Bl_p(X)} \bigoplus \mathcal{O}_{Bl_p(Y)} & \longrightarrow & \mathcal{O}_{Bl_p(X) \cap Bl_p(Y)} & \longrightarrow & 0 \end{array}$$

where the vertical arrows are the morphisms induced by

$$Bl_P(Z) \longrightarrow Z$$

and Z is X, Y or $X \cup Y$.

From the exact sequence below we deduce

$$(X.Y) - (Bl_P(X).Bl_P(Y)) = \dim_{\mathbf{k}}(\mathrm{Coker}(\varphi_1)) - \dim_{\mathbf{k}}(\mathrm{Coker}(\varphi_2)),$$

by [14] or [1] we obtain

$$(X.Y) - (Bl_P(X).Bl_P(Y)) = e_1(X \cup Y, P) - e_1(X, P) - e_1(Y, P).$$

By induction on the cardinality of $\mathrm{Inf}(X) \cap \mathrm{Inf}(Y)$ we obtain the claim.

Remark. Recall that if Z is a plane curve singularity then

$$e_1(Z) = e_0(Z)(e_0(Z) - 1)/2,$$

so from the last result we obtain Noether's formula.

Remark. From [14] and the last result we deduce Hironaka's formula ([8]): let $Y_1, ..., Y_r$ be the branches of a curve singularity Y, then

$$\delta(Y) = \sum_{i=1}^{r} \delta(Y_i) + \sum_{i=1}^{r-1} (Y_i.(\bigcup_{j=i+1}^{r} Y_j)),$$

where $\delta(Y)$ is the singularity order of Y, i.e. $\delta(Y) = \dim_{\mathbf{k}} \overline{\mathcal{O}_Y}/\mathcal{O}_Y$.

Definition. From now on we will write

$$[X.Y]_P = e_1(X \cup Y, P) - e_1(X, P) - e_1(Y, P)$$

for the contribution of P to $(X.Y)$.

Remark. Notice that $[X.Y]_P = 0$ if and only if $P \notin \mathrm{Inf}(X) \cap \mathrm{Inf}(Y)$. If we denote by $T_{Z,P}$ the tangent space to a curve singularity Z, then we have $[X.Y]_P = 1$ if and only if $T_{X,P} \cap T_{Y,P} = \{P\}$.

Remark. Very often we can compute the Hilbert polynomial of a curve singularity from the knowledge of its singularity order and the contact of their branches, see [E-1,2,3].

Definition. Let X be a curve singularity. The regularity index of X is the integer

$$i(X) = \mathrm{Min}\{n \mid h^1_X(n) = e_0(X)n - e_1(X)\}.$$

Notice that $i(X) \leq e_0(X) - 1$ ([5]).

Let $I(X)^*$ be the homogeneous ideal defined by the initial forms of the elements of $I(X)$. Recall that in [5] we prove that if $f_1, .., f_s$ is a minimal system of generators of $I(X)^*$ then $\deg(f_i) \leq i(X) + 1$ for $i = 1, ..., s$. We will denote by $\sigma(X) = \mathrm{Max}\{\deg(f_1), ..., \deg(f_s)\}$, see [9] Chapter III, section 1.

Proposition. *Let X, Y be curve singularities, then*

$$[X.Y]_0 \leq \mathrm{Min}\{e_0(X)\sigma(Y), \sigma(X)e_0(Y)\}.$$

Proof. We may assume that the coset of X_1 in $\mathcal{O}_X$ is a degree one superficial element. Let $f_1, .., f_s$ be a standard basis of $I = I(Y)$ such that their initial forms are a minimal basis of I^*. From [9], Chapter III, Lemma 6, we have

$$I' = (\frac{f_1}{X_1^{d_1}}, ..., \frac{f_s}{X_1^{d_s}}) \subset \mathcal{O}_{Bl_0(\mathbf{k}^n,0)} = R[X_2/X_1, ..., X_n/X_1]$$

is the strict transform of I, where $d_i = \mathrm{order}(f_i)$, $i = 1, ..., s$.

Let J be the ideal defined by I' in $\Lambda = \mathcal{O}_{Bl_0(X)}$. Since X_1 is a non-zero divisor we have

$$\begin{aligned}\dim_{\mathbf{k}} \frac{\Lambda}{X_1^{\sigma(Y)}J} &= \dim_{\mathbf{k}} \frac{\Lambda}{J} + \dim_{\mathbf{k}} \frac{J}{X_1^{\sigma(Y)}J} \\ &= (Bl_0(X).Bl_0(Y)) + e_0(X)\sigma(Y).\end{aligned}$$

Again since X_1 is a non-zero divisor we obtain

$$\frac{\Lambda}{X_1^{\sigma(Y)}J} \cong \frac{X_1^n\Lambda}{X_1^n\Lambda X_1^{\sigma(Y)}J}.$$

On the other hand there is α such that $X_1\Lambda = m^{\alpha}$ ([11], Theorem 12.1), so we have $X_1^n\Lambda = m^{\alpha n}$ for $n >> 0$, and then

$$\dim_{\mathbf{k}} \frac{\Lambda}{X_1^{\sigma(Y)}J} = \dim_{\mathbf{k}} \frac{m^{\alpha n}}{m^{\alpha n}X_1^{\sigma(Y)}J}.$$

It is easy to prove that

$$m^{\alpha n}X_1^{\sigma(Y)}J \subset m^{\alpha n}I\Lambda = m^{\alpha n}I.$$

Hence we obtain

$$\begin{aligned}\dim_{\mathbf{k}} \frac{\Lambda}{X_1^{\sigma(Y)}J} &\geq \dim_{\mathbf{k}} \frac{m^{\alpha n}}{m^{\alpha n}I} \\ &= \dim_{\mathbf{k}} \frac{I}{m^{\alpha n}I} - \dim_{\mathbf{k}} \frac{A}{m^{n\alpha}} + \dim_{\mathbf{k}} \frac{A}{I} \\ &= e_0(A)(\alpha n) - e_1(I) - e_0(A)(\alpha n) - e_1(A) + \dim_{\mathbf{k}} \frac{A}{I} \\ &= e_1(A) - e_1(I) + \dim_{\mathbf{k}} \frac{A}{I}.\end{aligned}$$

By Theorem 2.1

$$\dim_{\mathbf{k}} \frac{\Lambda}{X_1^{\sigma(Y)}J} \geq e_0^*(I).$$

Hence from [15] we have

$$\dim_{\mathbf{k}} \frac{\Lambda}{X_1^{\sigma(Y)}J} \geq e_0^*(I) \geq \dim_{\mathbf{k}} \frac{A}{I} = (X.Y). \tag{1}$$

From this we deduce

$$[X.Y]_0 = (X.Y) - (Bl_0(X).Bl_0(Y)) \leq e_0(X)\sigma(Y)$$

By the symmetry of the problem we obtain the claim.

From this result we can give the generalization of Noerther's formula for curve singularities:

Corollary. *Let X, Y be curve singularities, then*

$$\begin{aligned}(X.Y) &\leq \sum_{P\in \mathrm{Inf}(X)\cap \mathrm{Inf}(Y)} \mathrm{Min}\{e_0(X,P)\sigma(Y,P), \sigma(X,P)e_0(Y,P)\} \\ &\leq \sum_{P\in \mathrm{Inf}(X)\cap \mathrm{Inf}(Y)} \mathrm{Min}\{e_0(X,P)(i(Y,P)+1), (i(X,P)+1)e_0(Y,P)\} \\ &\leq \sum_{P\in \mathrm{Inf}(X)\cap \mathrm{Inf}(Y)} e_0(X,P)e_0(Y,P).\end{aligned}$$

Proof. From the Proposition and [3], we have

$$\begin{aligned}[X.Y]_0 &\leq \mathrm{Min}\{e_0(X)\sigma(Y), \sigma(X)e_0(Y)\} \\ &\leq \mathrm{Min}\{e_0(X)(i(Y)+1), (i(X)+1)e_0(Y)\}\end{aligned}.$$

By [11], Theorem 12.11, we obtain

$$e_0(X)(i(Y)+1), \quad (i(X)+1)e_0(Y) \leq e_0(X)e_0(Y),$$

summing up we get the claim.

Corollary. *Let X, Y be curve singularities and assume that $[X.Y]_0 = e_0(X)\sigma(Y)$. Then there exists a hypersurface $H \subset (\mathbf{k}^n, 0)$ such that $X \cap Y = X \cap H$.*

Proof. We put $I = I(Y)\mathcal{O}_X$. From the proof of Proposition 3.3 (see (1)) we deduce $e_0^*(I) = \dim_{\mathbf{k}}(\mathcal{O}_X/I)$, so $I = (F)$ ([15], Corollary 1). Let H be the hypersurface of $(\mathbf{k}^n, 0)$ defined by a representative of $F \in A$, then we get the claim.

Remark. Notice that the last result means that if the bound of the main result is achieved then $X \cap Y$ is a hypersurface of X. See [10] for results on the Hilbert scheme of closed subchemes of a curve singularity.

References

[1] Casas, E., Sobre el cálculo efectivo del género de las curvas algebraicas. Collectanea Math. 25 (1973), 3-11.

[2] Elias, J., Characterization of the Hilbert-Samuel polynomials of curve singularities. Compositio Math. 74(2) (1990), 135-155.

[3] Elias, J., The conjecture of Sally on the Hilbert function for curve singularities. To appear in J. Algebra.

[4] Elias, J., An algorithm for computing curve singularities with given Hilbert-Samuel polynomial. Preprint, 1991.

[5] Elias, J., On the analytic equivalence of curves. Math. Proc. Camb. Phil. Soc. 100, 57-64 (1986)

[6] Enriques, F. and Chisini, O., Teoria geometrica delle equazione e delle funzione algebriche. Nicola Zanichelli, Bologna 1918.

[7] Greuel, G. M. and Pfister, G., Moduli spaces for torsion free modules an curve singularities I. Preprint 1991.
[8] Hironaka, H., Certain numerical characters of singularities. J. Math. Kyoto Univ. 10 (1970), 151-187.
[9] Hironaka, H., Resolution of singularities of an algebraic variety over a field of characteristic zero: II. Annals of Math. vol 79, 2.
[10] Iarrobino, A., Hilbert Scheme of Points: Overview of Last Ten Years. Proceedings of Symposia in Pure Mathematics. Vol 40 (1987), 297-320.
[11] Matlis, E., 1-Dimensional Cohen-Macaulay rings. L.N.M. 327, Springer Verlag 1977.
[12] Lipman, J., Stable ideals and arf rings. Amer. J. of Math. 93 (1971), 649-685
[13] Noether, M., Rationale Ausführung der Operationen in der Theorie der algebraischen Funktionen. Math. Ann. 23 (1884), 311-358.
[14] Northcott, D.G., The reduction number of a one-dimensional local ring. Mathematika 6 (1959), 87-90.
[15] Northcott, D.G., A note on the coefficients of the abstract Hilbert function. J. London Math. Soc. 35 (1960), 209-214.
[16] Van der Waerden, B.L., Infinitely near points. Indagationes Math. 12 (1950), 401-411.

Sharp Bounds for the p-th Module of Syzygies of Certain Rational Surfaces

A.V. Geramita (*), A. Gimigliano, R.M. Miró-Roig (**)

0. Introduction

Let $Z = \{P_1, \cdots, P_s\}$ be a set of s distinct points in $\mathbb{P}^2 = \mathbb{P}^2_k$ (k an algebraically closed field) and let $P_i \longleftrightarrow \wp_i = (L_{i1}, L_{i2}) \subset k[x,y,z] = R$. Let $J = \wp_1^{\alpha_1} \cap \cdots \cap \wp_s^{\alpha_s}$ ($\alpha_i \geq 1$). Then J_d describes the (incomplete) linear system of plane curves of degree d which have P_i as a point of multiplicity at least α_i. Let $\mathbb{P}^2(Z)$ be the surface obtained from $\mathbb{P}^2$ by blowing up the points of Z. Then J_d corresponds to a complete linear system on $\mathbb{P}^2(Z)$. If $\dim_k J_d = N + 1$ then J_d provides a morphism $\mathbb{P}^2(Z) \longrightarrow \mathbb{P}^N$ which, if the linear system on $\mathbb{P}^2(Z)$ is very ample, gives an embedding of $\mathbb{P}^2(Z)$ into $\mathbb{P}^N$. Earlier work in [7], [4], and [6] has studied the defining ideals of the variety in $\mathbb{P}^N$ thus obtained and, in some cases, the degrees in a minimal generating set of these ideals has been obtained. The goal of this work is to expand that study to a study of all graded Betti numbers in a free resolution of the defining ideal of such varieties. Such a study has been sucessful in only two cases so far. In [7], where White surfaces are studied, a complete description of the minimal free resolution of the defining ideal is given. In [4] a complete description is also available (but in a more theoretical form) for the case of Room surfaces.

To be more precise, we give free (not necessarily minimal) resolutions of the corresponding ideal sheaves and sharp bounds for the degrees of the generators of the p-th module of syzygies of the homogeneous ideals and lower and upper bounds for all graded Betti numbers. We conclude this work with several illustrative examples. Our main tools will be Beilinson's spectral sequence and a generalized mapping cone construction.

---

* Partially supported by a grant from the Natural Sciences and Engineering Research Council of Canada

** Partially supported by DGICYT PB88-0224.

Notation and preliminaries.

In this section, we fix our notation and recall some of the basic facts we will need.

Let Z' be the subscheme of $\mathbb{P}^2$ defined by J. Then Z' is a scheme of multiplicity $e = \sum_{i=1}^{s} \binom{\alpha_i+1}{2}$ which is supported on Z. It is called a scheme of "fat points" on Z and will be denoted $(Z; \alpha_1, \ldots \alpha_s)$.

We summarize some well-known facts about the Hilbert function $H(Z', \ )$ of Z'. By definition $H(Z',t) = \dim_k(k[x,y,z]/J)_t$. The first difference of $H(Z', \ )$ is defined by $\Delta H(Z',t) = H(Z',t) - H(Z',t-1)$. If we let α be the least degree of a curve in $\mathbb{P}^2$ which contains Z' we have (see e.g.[3]):

(1) $\Delta H(Z',t) \geq 0$ for all t;

(2) $\Delta H(Z',t) \leq t+1$ for all t with equality if and only if $0 \leq t \leq \alpha - 1$;

(3) $\Delta H(Z',t) \geq \Delta H(Z',t+1)$ for all $t \geq \alpha$;

(4) $\sum_{i=0}^{\infty} \Delta H(Z',i) = \sum_{i=1}^{s} \binom{\alpha_i+1}{2} = e$.

If we let σ be the least integer t for which $\Delta H(Z',t) = 0$ then the ideal J can always be generated by forms of degree $\leq \sigma$ ([1]).

Let $E_1, \cdots, E_s$ be the divisor classes in $\mathrm{Pic}(\mathbb{P}^2(Z))$ which contain the divisors corresponding to the exceptional lines on $\mathbb{P}^2(Z)$ i.e. the lines which are the blow ups of the points $P_1, \cdots, P_s$, respectively. Further, let E_0 be the divisor class in $\mathrm{Pic}(\mathbb{P}^2(Z))$ which contains the divisor corresponding to the transform of a line in $\mathbb{P}^2$ which misses all the points of Z. Then

$$\mathrm{Pic}(\mathbb{P}^2(Z)) \cong \mathbb{Z}^{s+1} \cong < E_0, E_1, \cdots, E_s >$$

where $E_0^2 = 1 = -E_1^2 = \cdots = -E_s^2$ and $E_iE_j = 0$ if $i \neq j$.

The complete linear system on $\mathbb{P}^2(Z)$ corresponding to J_d consists of all the effective divisors in the divisor class $D_d = dE_0 - \sum_{i=1}^{s} \alpha_i E_i$. It is well known that D_d is very ample when $d \geq \sigma + 1$ and that D_σ is very ample if and only if no line of $\mathbb{P}^2$ meets Z' in multiplicity $\geq \sigma$. From now on we will always assume this measure of generality about Z'. Therefore, all the divisors D_t, $t \geq \sigma$ on $\mathbb{P}^2(Z)$ are very ample and they define morphisms:

$$\varphi_{t,Z'} : \mathbb{P}^2(Z) \longrightarrow \mathbb{P}^N$$

where $N = h^0(D_t) = \binom{t+2}{2} - e - 1$. These morphisms embed $\mathbb{P}^2(Z)$ into $\mathbb{P}^N$ and we will denote the image of $\varphi_{t,Z'}$ by $V_{t,Z'}$.

Since the Veronese surface in $\mathbb{P}^r$, $r = \binom{t+2}{2} - 1$ is the image of $\mathbb{P}^2$ in $\mathbb{P}^r$ obtained by using the complete linear system R_t and the surfaces $V_{t,Z'}$ are projections of that surface, the surfaces $V_{t,Z'}$ are often called *Veronesean surfaces.* Moreover, the surfaces $V_{t,Z'}$ are all smooth and arithmetically Cohen-Macaulay of degree $t^2 - e$ (see [7] in cases Z' is reduced and [5] for the more general case).

1. Free resolutions and sharp bounds for the generators of the p-th module of syzygies

We start with the following fact: let $X \subset \mathbb{P}^n$ be a closed subscheme with ideal sheaf I_X and let F, G be coherent sheaves on $\mathbb{P}^n$. It is well known that if we are given (minimal) free resolutions of F and G (see below),

$$\begin{array}{ccccccccccccc}
 & & & & & & & & & & 0 & & \\
 & & & & & & & & & & \downarrow & & \\
0 & \to & G_m & \to & \cdots & \to & G_2 & \to & G_1 & \to & G & \to & 0 \\
 & & \downarrow & & & & \downarrow & & \downarrow & & \downarrow & & \\
0 & \to & F_m & \to & \cdots & \to & F_2 & \to & F_1 & \to & F & \to & 0 \\
 & & & & & & & & & & \downarrow & & \\
 & & & & & & & & & & I_X & & \\
 & & & & & & & & & & \downarrow & & \\
 & & & & & & & & & & 0 & &
\end{array}$$

we obtain a (not necessarily minimal) free resolution of I_X, via the so-called "mapping cone" construction:

$$0 \longrightarrow G_m \longrightarrow F_m \oplus G_{m-1} \longrightarrow \ldots \longrightarrow F_2 \oplus G_1 \longrightarrow F_1 \longrightarrow I_X \longrightarrow 0,$$

where, $F_i = \oplus_j \mathcal{O}(-d_{ij})$ and $G_i = \oplus_j \mathcal{O}(-r_{ij})$.

Proposition 1.1 (Generalized mapping cone). *Let $X \subset \mathbb{P}^n$ be a closed subscheme with ideal sheaf I_X. Assume that we have an exact sequence:*

$$0 \longrightarrow H_r \longrightarrow H_{r-1} \longrightarrow \ldots \longrightarrow H_1 \longrightarrow I_X \longrightarrow 0 \qquad (*)$$

with H_i coherent $\mathcal{O}_{\mathbb{P}^n}$-modules. Let

$$0 \longrightarrow L_t^{n_t} \longrightarrow L_t^{n_t-1} \longrightarrow \ldots \longrightarrow L_t^1 \longrightarrow H_t \longrightarrow 0, \quad L_t^i = \oplus_j \mathcal{O}(-n_t^{ij}),$$

be a free resolution of H_t, $1 \leq t \leq r$. Set $s = \max\{n_t\}$ and $L_t^p = 0$ if $n_t < p \leq s$. Then,

$$0 \longrightarrow L_r^s \longrightarrow L_r^{s-1} \oplus L_{r-1}^s \longrightarrow \ldots \longrightarrow L_r^{s-r+1} \oplus L_{r-1}^{s-r+2} \oplus \ldots \oplus L_2^{s-1} \oplus L_1^s \longrightarrow \ldots$$

$$\longrightarrow L_3^1 \oplus L_2^2 \oplus L_1^3 \longrightarrow L_2^1 \oplus L_1^2 \longrightarrow L_1^1 \longrightarrow I_X \longrightarrow 0$$

is a (not necessarily minimal) free resolution of I_X.

Proof. First of all, we cut (*) into short exact sequences:

$$\begin{array}{ccccccccccccc}
0 & \to & H_r & \to & H_{r-1} & & \to & & H_{r-2} & & \to & & \cdots \\
 & & & & & \searrow & & \nearrow & & \searrow & & & \\
 & & & & & & K_{r-2} & & & & K_{r-3} & & \cdots \\
 & & & & & \nearrow & & \searrow & & \nearrow & & & \\
 & & & & 0 & & & & 0 & & & &
\end{array}$$

$$\begin{array}{ccccccccccc}
\cdots & & \rightarrow & & H_2 & & \rightarrow & & H_1 & \rightarrow & I_X & \rightarrow & 0 \\
& \searrow & & \nearrow & & \searrow & & \nearrow & & & & & \\
\cdots & & K_2 & & & & K_1 & & & & & & \\
& \nearrow & & \searrow & & \nearrow & & \searrow & & & & & \\
& & & & 0 & & & & 0 & & & &
\end{array}$$

The mapping cone between the given free resolutions of H_{r-1} and H_r

$$\begin{array}{ccccccccccccc}
 & & & & & & & & & & 0 & & \\
 & & & & & & & & & & \downarrow & & \\
0 & \rightarrow & L_r^s & \rightarrow & \cdots & \rightarrow & L_r^2 & \rightarrow & L_r^1 & \rightarrow & H_r & \rightarrow & 0 \\
 & & \downarrow & & & & \downarrow & & \downarrow & & \downarrow & & \\
0 & \rightarrow & L_{r-1}^s & \rightarrow & \cdots & \rightarrow & L_{r-1}^2 & \rightarrow & L_{r-1}^1 & \rightarrow & H_{r-1} & \rightarrow & 0 \\
 & & & & & & & & & & \downarrow & & \\
 & & & & & & & & & & K_{r-2} & & \\
 & & & & & & & & & & \downarrow & & \\
 & & & & & & & & & & 0 & &
\end{array}$$

gives the free resolution of K_{r-2}:

$$0 \longrightarrow L_r^s \longrightarrow L_r^{s-1} \oplus L_{r-1}^s \longrightarrow L_r^{s-2} \oplus L_{r-1}^{s-1} \longrightarrow \cdots$$

$$\longrightarrow L_{r-1}^2 \oplus L_r^1 \longrightarrow L_{r-1}^1 \longrightarrow K_{r-2} \longrightarrow 0.$$

Repeating the process gives us the stated free resolution of I_X.

Beilinson's Theorem ([8]). *Let E be a coherent sheaf on $\mathbb{P}^n$. There is a spectral sequence $E_r^{p,q}$ with E_1-term*

$$E_1^{p,q} = H^q(\mathbb{P}^n, E(p)) \otimes \Omega^{-p}(-p)$$

which converges to

$$E^i = \begin{cases} E & \text{for } i = 0; \\ 0 & \text{otherwise} \end{cases}$$

i.e., $E_\infty^{p,q} = 0$ for $p + q \neq 0$ and $\oplus_{p=0}^n E_\infty^{-p,p}$ is the associated graded sheaf of a filtration of E.

Here $\Omega^p(p)$ is the p-th exterior power of the twisted cotangent bundle $\Omega^1(1)$. The differential maps

$$d_1^{p,q} \in \operatorname{Hom}(E_1^{p,q}, E_1^{p+1,q}) \cong \operatorname{Hom}(H^0\mathcal{O}(1) \otimes H^q(\mathbb{P}^n, E(p)), H^q(\mathbb{P}^n, E(p+1)))$$

coincide with the natural multiplication maps, they are determined by the module structure of $\oplus_p H^q E(p)$.

The following well-known fact will be very useful.

Lemma 1.2. *The Koszul resolution of $\mathbb{P}^n$ gives us the following minimal free resolution of Ω^p, $1 \leq p \leq n$:*

$$0 \longrightarrow \mathcal{O}(-n-1) \longrightarrow (n+1)\mathcal{O}(-n) \longrightarrow \ldots. \longrightarrow \binom{n+1}{p+2}\mathcal{O}(-p-2) \longrightarrow$$

$$\longrightarrow \binom{n+1}{p+1}\mathcal{O}(-p-1) \longrightarrow \Omega^p \longrightarrow 0.$$

Proposition 1.3. *Let $I_{V_{t,Z'}}$ be the ideal sheaf of $V_{t,Z'} \subset \mathbb{P}^N$. Then, $I_{V_{t,Z'}}$ has a locally free resolution of the following type:*

$$0 \longrightarrow \beta_N \Omega^N(N) \longrightarrow \beta_{N-1}\Omega^{N-1}(N-1) \longrightarrow \ldots \longrightarrow \beta_4\Omega^4(4) \longrightarrow$$

$$\longrightarrow \beta_3\Omega^3(3) \oplus \alpha\mathcal{O} \longrightarrow I_{V_{t,Z'}}(2) \longrightarrow 0$$

where $\alpha = h^0(\mathbb{P}^N, I_{V_{t,Z'}}(2))$ and $\beta_i = h^3(\mathbb{P}^N, I_{V_{t,Z'}}(2-i))$, $3 \le i \le N$.

Proof. We apply Beilinson's spectral sequence to $E = I_{V_{t,Z'}}(2)$. The E_1^{pq}-term looks as follows:

$$\begin{array}{ccccccc}
0 & 0 & \cdots & 0 & 0 & 0 & 0 \\
\cdot & \cdot & \cdots & \cdot & \cdot & \cdot & \cdot \\
\cdot & \cdot & \cdots & \cdot & \cdot & \cdot & \cdot \\
\cdot & \cdot & \cdots & \cdot & \cdot & \cdot & \cdot \\
0 & 0 & \cdots & 0 & 0 & 0 & 0 \\
\beta_N\Omega^N(N) & \beta_{N-1}\Omega^{N-1}(N-1) & \cdots & \beta_3\Omega^3(3) & 0 & 0 & 0 \\
0 & 0 & \cdots & 0 & 0 & 0 & 0 \\
0 & 0 & \cdots & 0 & 0 & 0 & 0 \\
0 & 0 & \cdots & 0 & 0 & 0 & \alpha\mathcal{O}
\end{array}$$

Therefore, $E_2^{pq} = E_\infty$ and $E = I_{V_{t,Z'}}(2)$ has a locally free resolution of the following type:

$$0 \longrightarrow \beta_N \Omega^N(N) \longrightarrow \beta_{N-1}\Omega^{N-1}(N-1) \longrightarrow \ldots \longrightarrow \beta_4\Omega^4(4) \longrightarrow$$

$$\longrightarrow \beta_3\Omega^3(3) \oplus \alpha\mathcal{O} \longrightarrow I_{V_{t,Z'}}(2) \longrightarrow 0.$$

Remark 1.3.1. We can give the explicit values of the β_i's in terms of the degree d of $V_{t,Z'}$ and the sectional genus g of $V_{t,Z'}$. In fact, using the Riemann-Roch Theorem we get:

$\beta_3 = g-1$, and
$\beta_i = \beta_{i-1} + g - 1 + (i-3)d$, for $4 \le i \le N$.

Hence, $\beta_i = (i-2)(g-1) + \binom{i-2}{2}d$, for $3 \le i \le N$.

Theorem 1.4. *The ideal sheaf $I_{V_{t,Z'}}$ of $V_{t,Z'} \subset \mathbb{P}^N$ has a (not necessarily minimal) free resolution of the following type:*

$$0 \longrightarrow \beta_N\mathcal{O}(-1) \oplus \beta_{N-1}\mathcal{O}(-2) \oplus \ldots \oplus \beta_3\mathcal{O}(-N+2) \longrightarrow$$

$$(N+1)\beta_{N-1}\mathcal{O}(-1) \oplus (N+1)\beta_{N-2}\mathcal{O}(-2) \oplus \ldots \oplus (N+1)\beta_3\mathcal{O}(-N+3) \longrightarrow$$

$$\binom{N+1}{2}\beta_{N-2}\mathcal{O}(-1) \oplus \binom{N+1}{2}\beta_{N-3}\mathcal{O}(-2) \oplus \ldots \oplus \binom{N+1}{2}\beta_3\mathcal{O}(-N+4) \longrightarrow \cdots$$

$$\longrightarrow \binom{N+1}{5}\beta_4\mathcal{O}(-1) \oplus \binom{N+1}{5}\beta_3\mathcal{O}(-2) \longrightarrow$$

$$\binom{N+1}{4}\beta_3\mathcal{O}(-1) \oplus \alpha\mathcal{O} \longrightarrow I_{V_{t,Z'}}(2) \longrightarrow 0$$

where $\alpha = h^0(\mathbb{P}^N, I_{V_{t,Z'}}(2))$ and $\beta_i = h^3(\mathbb{P}^N, I_{V_{t,Z'}}(2-i))$, $3 \le i \le N$.

Proof. It is enough to apply Proposition 1.1 to the locally free resolution of $I_{V_{t,Z'}}(2)$ given in Proposition 1.3 and use the free resolution of Ω^p given in Lemma 1.2.

Corollary 1.5. *The homogeneous ideal $I(V_{t,Z'})$ of $V_{t,Z'}$ is generated by quadrics and cubics and the minimal generators of the p-th module of syzygies of $I(V_{t,Z'})$ occur in degrees $\le 3+p$.*

Proof. It follows from Theorem 1.4 that $I_{t,Z'}$ is 3-regular. Therefore, for every $p \ge 0$ the minimal generators of the p-th module of syzygies of $I(V_{t,Z'})$ occur in degrees $\le 3+p$.

Remark 1.5.1. Examples 1-4 of [4] show that the result of Corollary 1.5 is the best we can hope for. In fact,

Example 1: $I(V_{3,Z})$, $Z' = Z = \{P_1, P_2, P_3\}$, has one 3-syzygy of degree 6.

Example 2: $I(V_{4,Z})$, $Z' = Z = \{P_1, \cdots, P_6\}$ has: 24 third syzygies of degree 6; 12 fourth syzygies of degree 7; 3 fifth syzygies of degree 8.

Example 3: $I(V_{5,Z})$, $Z' = Z = \{P_1, \cdots, P_{10}\}$ has: 50 third syzygies of degree 6; 120 fourth syzygies of degree 7; 105 fifth syzygies of degree 8; 40 sixth syzygies of degree 9; 6 seventh syzygies of degree 10.

Example 4: $I(V_{6,Z})$, $Z' = Z = \{P_1, \cdots, P_{15}\}$ has: 175 third syzygies of degree 6; 630 fourth syzygies of degree 7; 945 fifth syzygies of degree 8; 760 sixth syzygies of degree 9; 351 seventh syzygies of degree 10; 90 eighth syzygies of degree 11; 10 ninth syzygies of degree 12.

Remark 1.6. One cannot expect that cancelling all the "ghost terms" appearing in the free resolution given in Theorem 1.4 will give, in general, a minimal free resolution. A simple counterexample occurs for $V_{5,Z}$ with $Z' = Z = \{P_1, \cdots, P_{10}\}$ (see [4, Example 3]). In view of Corollary 1.5, however, it follows that the homogeneous ideal $I(V_{t,Z'})$ of $V_{t,Z'} \subset \mathbb{P}^N$ has a minimal free resolution of the following type:

$$0 \longrightarrow \alpha_{N-3}^{N}S(-N)\oplus\alpha_{N-3}^{N-1}S(-N+1) \longrightarrow \alpha_{N-4}^{N-1}S(-N+1)\oplus\alpha_{N-4}^{N-2}S(-N+2) \longrightarrow$$

$$\longrightarrow \alpha_{N-5}^{N-2}S(-N+2) \oplus \alpha_{N-5}^{N-3}S(-N+3) \longrightarrow \dots \longrightarrow \alpha_2^5 S(-5) \oplus \alpha_2^4 S(-4) \longrightarrow$$

$$\longrightarrow \alpha_1^4 S(-4) \oplus \alpha_1^3 S(-3) \longrightarrow \alpha_0^3 S(-3) \oplus \alpha_0^2 S(-2) \longrightarrow I(V_{t,Z'}) \longrightarrow 0$$

where $S = k[X_0, \cdots, X_N]$ is the homogeneous coordinate ring of $\mathbb{P}^N$ and α_p^i is the number of minimal generators of degree i of the p-th module of syzygies of $I(V_{t,Z})$, $p \geq 0$.

Our next goal is to bound α_p^i. To this end we will use the following lemma:

Lemma 1.7. *Let $S \subset \mathbb{P}^N$ be a smooth a.C.M. surface. Let $\mathcal{I}_S$ (respectively $I(S)$) be its ideal sheaf (respectively homogeneous ideal). Then, the sequence*

$$0 \longrightarrow \oplus_n \alpha_{N-3}^n \mathcal{O}(-n) \longrightarrow \oplus_n \alpha_{N-4}^n \mathcal{O}(-n) \longrightarrow \dots \longrightarrow \oplus_n \alpha_0^n \mathcal{O}(-n) \longrightarrow \mathcal{I}_S \longrightarrow 0$$

is exact if and only if the sequence

$$0 \longrightarrow \oplus_n \alpha_{N-3}^n S(-n) \longrightarrow \oplus_n \alpha_{N-4}^n S(-n) \longrightarrow \dots \longrightarrow \oplus_n \alpha_0^n S(-n) \longrightarrow I(S) \longrightarrow 0$$

is exact.

Proof. We leave this simple proof to the reader.

Finally using Theorem 1.4, Remark 1.6, Lemma 1.7 and the fact that any free resolution of $I(V_{t,Z'})$ is the direct sum of a minimal free resolution of $I(V_{t,Z'})$ and a free resolution of 0, we get:

Proposition 1.8. *Let α_p^i be the number of minimal generators of degree i of the p-th module of syzygies of $I(V_{t,Z'})$, $p \geq 0$. Set $\beta_i = h^3(\mathbb{P}^N, I_{V_{t,Z'}}(2-i))$, $3 \leq i \leq N$. We have:*

Case (a) *(for $t \geq \sigma + 1$):*

(1) *For all $p \geq 0$, $\alpha_p^i = 0$ if $p \neq p+2, p+3$.*

(2) *$\alpha_0^2 = h^0(\mathbb{P}^N, I_{V_{t,Z'}}(2))$ and $\alpha_0^3 = 0$.*

(3) *For all $p > 0$, $\max(0, \sum_{i=p+4}^{N+1}(-1)^{i-p}\binom{N+1}{i}\beta_{i-p-1}) \leq \alpha_p^{p+3} \leq \binom{N+1}{p+4}\beta_3$. In particular, $\alpha_{N-3}^N = \beta_3$.*

(4) *$\alpha_1^3 = \sum_{i=4}^{N+1}(-1)^{i-1}\binom{N+1}{i}\beta_{i-1}$ and, for all $p > 1$, $0 \leq \alpha_p^{p+2} \leq \sum_{i=p+4}^{N+1}(-1)^{i-p}\binom{N+1}{i}\beta_{i-p}$.*

Case (b) *(for $t = \sigma$)*

(1) *For all $p \geq 0$, $\alpha_p^i = 0$ if $p \neq p+2, p+3$.*

(2) *$\alpha_0^2 = h^0(\mathbb{P}^N, I_{V_{t,Z'}}(2))$ and $0 \leq \alpha_0^3 \leq \binom{N+1}{N-3}\beta_3$.*

(3) *For all $p > 0$, $\max(0, \sum_{i=p+4}^{N+1}(-1)^{i-p}\binom{N+1}{i}\beta_{i-p-1}) \leq \alpha_p^{p+3} \leq \binom{N+1}{p+4}\beta_3$. In particular, $\alpha_{N-3}^N = \beta_3$.*

(4) *For all* $p > 0$, $0 \leq \alpha_p^{p+2} \leq \sum_{i=p+4}^{N+1}(-1)^{i-p}\binom{N+1}{i}\beta_{i-p}$.

2. Examples

In this last section, we collect some examples.

Example 2.1. Let $Z = \{P_1, P_2, P_3\}$ be a set of points in $\mathbb{P}^2$ not lying on a line. Let $\mathbb{P}^2(Z)$ be the surface obtained from $\mathbb{P}^2$ by blowing up the points of Z. The divisor $H = 3E_0 - E_1 - E_2 - E_3$ is very ample and defines a morphism $\varphi : \mathbb{P}^2(Z) \longrightarrow \mathbb{P}^6$ whose image, $V_{3,Z}$, is a Del Pezzo sestic in $\mathbb{P}^6$. It is known that its ideal sheaf has the following minimal free resolution :

$$0 \longrightarrow \mathcal{O}(-6) \longrightarrow 9\mathcal{O}(-4) \longrightarrow 16\mathcal{O}(-3) \longrightarrow 9\mathcal{O}(-2) \longrightarrow I_{V_{3,Z}} \longrightarrow 0$$

Hence: $\alpha_3^6 = 1$; $\alpha_2^4 = 9$; $\alpha_1^3 = 16$; $\alpha_0^2 = 9$ and $\alpha_i^j = 0$ otherwise.

On the other hand, applying the Riemann-Roch Theorem we get: $h^0(\mathbb{P}^6, I_{V_{3,Z}}(2) = 9$; $\beta_3 = 1$; $\beta_4 = 7$; $\beta_5 = 19$ and $\beta_6 = 37$ ($\beta_i = h^3(\mathbb{P}^6, I_{V_{3,Z}}(2-i))$. Hence, for all $3 \geq p > 0$, $\max(0, \sum_{i=p+4}^{7}(-1)^{i-p}\binom{7}{i}\beta_{i-p-1}) = \alpha_p^{p+3}$ and $\alpha_1^3 = \sum_{i=4}^{7}(-1)^{i-1}\binom{7}{i}\beta_{i-1}$.

Example 2.2. Let $Z = \{P_1, \cdots, P_6\}$ be a set of points in $\mathbb{P}^2$ not lying on a conic and with no 3 of the points of Z on a line. Let $\mathbb{P}^2(Z)$ be the surface obtained from $\mathbb{P}^2$ by blowing up the points of Z. The divisor $H = 4E_0 - E_1 - \cdots - E_6$ is very ample and defines a morphism $\varphi : \mathbb{P}^2(Z) \longrightarrow \mathbb{P}^8$ whose image, $V_{4,Z}$, is a smooth arithmetically Cohen-Macaulay surface in $\mathbb{P}^8$. This degree 10 surface in $\mathbb{P}^8$ is known as Room's surface. From [4], the ideal sheaf of this surface has a minimal free resolution of the following type:

$$0 \longrightarrow 3\mathcal{O}(-8) \longrightarrow 12\mathcal{O}(-7) \longrightarrow 10\mathcal{O}(-6) \oplus 24\mathcal{O}(-5) \longrightarrow 60\mathcal{O}(-4) \longrightarrow$$

$$\longrightarrow 52\mathcal{O}(-3) \longrightarrow 18\mathcal{O}(-2) \longrightarrow I_{V_{4,Z}} \longrightarrow 0$$

Hence: $\alpha_5^8 = 3$; $\alpha_4^7 = 12$; $\alpha_3^6 = 10$; $\alpha_3^5 = 24$; $\alpha_2^4 = 60$; $\alpha_1^3 = 52$; $\alpha_0^2 = 18$ and $\alpha_i^j = 0$ otherwise.

On the other hand, applying the Riemann-Roch Theorem we get: $h^0(\mathbb{P}^8, I_{V_{4,Z}}(2) = 18$; $\beta_3 = 3$; $\beta_4 = 15$; $\beta_5 = 37$; $\beta_6 = 69$; $\beta_7 = 111$ and $\beta_8 = 163$. Hence, for all $5 \geq p > 0$, $\max(0, \sum_{i=p+4}^{9}(-1)^{i-p}\binom{9}{i}\beta_{i-p-1}) = \alpha_p^{p+3}$ and $\alpha_1^3 = \sum_{i=4}^{9}(-1)^{i-1}\binom{9}{i}\beta_{i-1}$.

Example 2.3. Let $Z = \{P_1, \cdots, P_{10}\}$ be a set of distinct points in $\mathbb{P}^2$ not lying on a cubic and with no four points of Z on a line. Let $\mathbb{P}^2(Z)$ be the surface obtained from $\mathbb{P}^2$ by blowing up the points of Z. The divisor $H = 5E_0 - E_1 - \cdots - E_{10}$ is very ample and defines a morphism $\varphi : \mathbb{P}^2(Z) \longrightarrow \mathbb{P}^{10}$ whose image, $V_{5,Z}$, is a smooth arithmetically Cohen-Macaulay surface of degree 15 in $\mathbb{P}^{10}$. The ideal sheaf of this embedded surface has a minimal free resolution of the following type:

$$0 \longrightarrow 6\mathcal{O}(-10) \longrightarrow 40\mathcal{O}(-9) \longrightarrow 105\mathcal{O}(-8) \longrightarrow 120\mathcal{O}(-7) \oplus 50\mathcal{O}(-6) \longrightarrow$$

$$\longrightarrow 50\mathcal{O}(-6) \oplus 168\mathcal{O}(-5) \longrightarrow 210\mathcal{O}(-4) \longrightarrow 120\mathcal{O}(-3) \longrightarrow$$

$$\longrightarrow 30\mathcal{O}(-2) \longrightarrow I_{V_{5,Z}} \longrightarrow 0$$

Hence: $\alpha_7^{10} = 6$; $\alpha_6^9 = 40$; $\alpha_5^8 = 105$; $\alpha_4^7 = 120$; $\alpha_4^6 = 50$; $\alpha_3^6 = 50$; $\alpha_3^5 = 168$; $\alpha_2^4 = 210$; $\alpha_1^3 = 120$; $\alpha_0^2 = 30$ and $\alpha_i^j = 0$ otherwise.

On the other hand, applying the Riemann-Roch Theorem we get:
$h^0(\mathbb{P}^{10}, I_{V_{5,Z}}(2) = 30$; $\beta_3 = 6$; $\beta_4 = 26$; $\beta_5 = 61$; $\beta_6 = 111$; $\beta_7 = 176$; $\beta_8 = 256$; $\beta_9 = 351$ and $\beta_{10} = 481$. Hence, for all $7 \geq p > 0$, $p \neq 3$,
$\max(0, \sum_{i=p+4}^{11}(-1)^{i-p}\binom{11}{i}\beta_{i-p-1}) = \alpha_p^{p+3}$ and $\alpha_1^3 = \sum_{i=4}^{11}(-1)^{i-1}\binom{11}{i}\beta_{i-1}$.

Example 2.4. Let $Z = \{P_1, \cdots, P_{15}\}$ be a set of distinct points in $\mathbb{P}^2$ not lying on a quartic with no 5 points of Z on a line. Let $\mathbb{P}^2(Z)$ be the surface obtained from $\mathbb{P}^2$ by blowing up the points of Z. The divisor $H = 5E_0 - E_1 - \cdots - E_{15}$ is very ample and defines a morphism $\varphi : \mathbb{P}^2(Z) \longrightarrow \mathbb{P}^{12}$ whose image $V_{5,Z}$ is a smooth arithmetically Cohen-Macaulay surface of degree 15 in $\mathbb{P}^{12}$. The ideal sheaf of this embedded surfaces has a minimal free resolution of the following type:

$$0 \longrightarrow 10\mathcal{O}(-12) \longrightarrow 90\mathcal{O}(-11) \longrightarrow 351\mathcal{O}(-10) \longrightarrow 760\mathcal{O}(-9) \longrightarrow$$

$$\longrightarrow 945\mathcal{O}(-8) \oplus 90\mathcal{O}(-7) \longrightarrow 630\mathcal{O}(-7) \oplus 385\mathcal{O}(-6) \longrightarrow 175\mathcal{O}(-6) \oplus 648\mathcal{O}(-5)$$

$$\longrightarrow 540\mathcal{O}(-4) \longrightarrow 230\mathcal{O}(-3) \longrightarrow 45\mathcal{O}(-2) \longrightarrow I_{V_{5,Z}} \longrightarrow 0$$

Hence: $\alpha_9^{12} = 10$; $\alpha_8^{11} = 90$; $\alpha_7^{10} = 351$; $\alpha_6^9 = 760$; $\alpha_5^8 = 945$; $\alpha_5^7 = 90$; $\alpha_4^7 = 630$; $\alpha_4^6 = 385$; $\alpha_3^6 = 175$; $\alpha_3^5 = 648$; $\alpha_2^4 = 540$; $\alpha_1^3 = 230$; $\alpha_0^2 = 45$ and $\alpha_i^j = 0$ otherwise.

On the other hand, applying the Riemann-Roch Theorem we get:
$h^0(\mathbb{P}^{12}, I_{V_{6,Z}}(2) = 45$; $\beta_3 = 10$; $\beta_4 = 40$; $\beta_5 = 91$; $\beta_6 = 163$; $\beta_7 = 256$; $\beta_8 = 370$; $\beta_9 = 505$; $\beta_{10} = 661$; $\beta_{11} = 838$ and $\beta_{12} = 1243$. Hence, for all $9 \geq p > 0$, $p \neq 3, 4$,
$\max(0, \sum_{i=p+4}^{13}(-1)^{i-p}\binom{13}{i}\beta_{i-p-1}) = \alpha_p^{p+3}$ and $\alpha_1^3 = \sum_{i=4}^{13}(-1)^{i-1}\binom{13}{i}\beta_{i-1}$.

Example 2.5. Our last example will be a surface $V_{t,Z'}$ contained in a rational normal scroll of dimension 3.

Let $Z = \{P_1, \cdots, P_s\}$ be a set of s generic enough points in $\mathbb{P}^2$ and let $J = \wp_1^{t-2} \cap \wp_2 \cap \cdots \cap \wp_s$ with $s = 2t - k$, $k \geq 0$. The divisor $D = tE_0 - (t-2)E_1 - E_2 - \cdots - E_s$ on $\mathbb{P}^2(Z)$ is very ample and defines a morphism $\varphi_{t,Z'} : \mathbb{P}^2(Z) \longrightarrow \mathbb{P}^N$ where $N = 3t - s = t + k$. We will denote by $V = V_{t,Z'}$ the image of $\varphi_{t,Z'}$. Note that the divisors $L_j \in |E_0 - E_1|$ form a pencil of curves of degree $L_j \cdot D = 2$, so the image of L_j is a conic in $\mathbb{P}^N$ (perhaps reducible) and hence V is a conic bundle.

Let Π_j be the plane spanned by the image of L_j. Then the 3-fold $X = \cup_{j\in\mathbb{P}^1}(\Pi_j)$ is a rational normal scroll (See eg. [9; §5]) of degree $N-2 = t+k-2$ which contains V. Let H be a hyperplane section of X and let R be one of the planes of the ruling of X. Then, Pic(X) is generated by R and H, and $H^3 = N-2$, $H^2 \cdot R = 1$, $R^2 = 0$.

Let $V \equiv aH - bR$ in $\mathrm{Pic}(X)$. Then:

(1) $V \cdot R \cdot H = aR \cdot H^2 = a$, but $V \cdot R$ is a conic so $V \cdot R \cdot H = a = 2$;

(2) $V \cdot H^2 = aH^3 + b = 2N - 4 - b$. But $V \cdot H^2 = \deg(V) = D^2 = 4t - s - 3 = N + t - 3$. Hence $b = N - t - 1 = k - 1$.

Now, we consider the exact sequence

$$0 \to \mathcal{O}_X(-2H + bR) \to \mathcal{O}_X \to \mathcal{O}_V \to 0.$$

Twisting by $\mathcal{O}_X(2)$, we get

$$0 \to \mathcal{O}_X(bR) \to \mathcal{O}_X(2) \to \mathcal{O}_V(2) \to 0. \qquad (**)$$

When $b \geq -1$ (i.e. when $k = 2, 1, 0$), the complex $\mathcal{C}^b$ defined in [9] is a resolution of $\mathcal{O}_X(bR)$. Since the resolution of $\mathcal{O}_X$ is well known, a mapping cone over (**) will gives us a resolution of $\mathcal{O}_V$.

Let us note that the case $k = 0$ ($b = -1$) is studied in [5]. This is a case when V is a Bordiga-White surface and there is a complete description of the ideal sheaf as a determinantal ideal whose resolution may be computed using the Eagon-Northcott complex.

In case $k = 1$ the surface V is cut on X by a quadric hypersurface ($V \equiv 2H$), so the computations are quite easy. In this case the (minimal) locally free resolution of V turns out to be :

$$0 \longrightarrow 3\binom{N-2}{4}\mathcal{O}(-6) \longrightarrow 2\binom{N-2}{3}\mathcal{O}(-5) \oplus 3\binom{N-2}{4}\mathcal{O}(-4) \longrightarrow$$

$$\longrightarrow \binom{N-2}{2}\mathcal{O}(-4) \oplus 2\binom{N-2}{3}\mathcal{O}(-3) \longrightarrow \left(\binom{N-2}{2} + 1\right)\mathcal{O}(-2) \longrightarrow I_V \longrightarrow 0$$

The case $k = 2$ is thus the first new case. The complex $\mathcal{C}^1$ is given, in this case, by the sheaves:

$$\mathcal{C}^1_j = \begin{cases} 2\mathcal{O} & \text{if } j = 0 \\ (N-2)\mathcal{O}(-1) & \text{if } j = 1 \\ \binom{N}{j+1}(j+1)\mathcal{O}(-j-1) & \text{if } j \geq 2. \end{cases}$$

The resolution of $\mathcal{O}_X$ is:

$$0 \to 3\binom{N-2}{4}\mathcal{O}(-4) \to 2\binom{N-2}{3}\mathcal{O}(-3) \to \binom{N-2}{2}\mathcal{O}(-2) \to \mathcal{O} \to \mathcal{O}_X \to 0$$

so the mapping cone gives:

$$0 \to (N-4)\mathcal{O}(-N) \to \cdots \to 2\binom{N-2}{4}\mathcal{O}(-6)$$

$$\to \binom{N-2}{3}\mathcal{O}(-5) \oplus 3\binom{N-2}{4}\mathcal{O}(-4) \longrightarrow \left(\binom{N-2}{3} + N - 2\right)\mathcal{O}(-3)$$

$$\longrightarrow (\binom{N-2}{2} + 2)\mathcal{O}(-2) \longrightarrow \mathcal{O}_X \longrightarrow \mathcal{O}_V \longrightarrow 0$$

Note that there is only one place in this resolution where a double shift appears. It follows that the resolution is minimal. Thus, for this V, we obtain:

$$\alpha_p^{p+2} = 0 \text{ for } p \geq 3;\ \alpha_0^2 = \binom{N-2}{2} + 2;\ \alpha_1^3 = \binom{N-2}{3} + N - 2;$$
$$\alpha_p^{p+3} = 0 \text{ for } p \leq 1;\ \alpha_p^{p+3} = (p-1)\binom{N-2}{p+1},\ 2 \leq p \leq N-3.$$

Now let us compute the numbers $\beta_i = h^3(I_V(2-i))$. We have $\beta_3 = g(D) = \binom{t-1}{2} - \binom{t-1}{2} = t - 2 = N - 4$, and since $\beta_i = \beta_{i-1} + g - 1 + (i-3)\deg(V)$, $\beta_i = (i-2)(g-1) + 1 + \deg(V)\binom{i-2}{2} = (N-5)\binom{i-1}{2} + N\binom{i-2}{2} + 1$.

In the specific case when $t = 4$ we get that $N = 6$ and V is a surface of degree 7 in $\mathbb{P}^6$. In this situation we have:

$$\beta_3 = 2,\ \beta_4 = 10,\ \beta_5 = 25,\ \beta_6 = 47$$

while the bounds of Proposition 1.8 give:

$$0 \leq \alpha_1^4 \leq 42,\ 4 \leq \alpha_2^5 \leq 7,\ \alpha_3^6 = 2$$

The lower bounds are sharp, since the resolution of $\mathcal{I}_V$ is:

$$0 \longrightarrow 2\mathcal{O}(-6) \longrightarrow 4\mathcal{O}(-5) \oplus 3\mathcal{O}(-4) \longrightarrow 12\mathcal{O}(-3) \longrightarrow 8\mathcal{O}(-2) \longrightarrow I_V \longrightarrow 0.$$

References

[1] E.D. Davis, A.V. Geramita and P. Maroscia, Perfect homogeneous ideals: Dubreil's theorems revisited, Bull. Sci. Math. 108 (1984), 143–185.

[2] E.D. Davis and A.V. Geramita, Birational morphism to $\mathbb{P}^2$: An ideal-theoretic perspective, Math. Ann. 279 (1988), 435–448.

[3] A.V. Geramita and P. Maroscia, The ideal of forms vanishing at a finite set of points in $\mathbb{P}^n$, J. Algebra 90 (1984), 528–555.

[4] A.V. Geramita and A. Gimigliano, Generators for the defining ideal of certain rational surfaces, Duke Math. J. 62 (1991), 61–83.

[5] A.V. Geramita , A. Gimigliano, Y. Pitteloud, (preprint, 1993).

[6] A.V. Geramita, A. Gimigliano, and B. Harbourne, Intrinsic and Extrinsic Properties of Special Blowings up of $\mathbb{P}^2$, J. Algebra (to appear).

[7] A. Gimigliano, On Veronesean surfaces, Proc. Konin. Ned. Akads. van Wetenschappen, Ser. A, 92 (1989), 71–85.

[8] C. Okonek, M. Scheneider and H. Spindler, Vector Bundles and Complex Projective Spaces, Progress in Math., vol. 3.

[9] F.Schreyer, Syzygies of canonical curves and special linear series, Math. Ann. 275 (1986), 105–137.

Resolutions of 0-dimensional subschemes of a smooth quadric

*S. Giuffrida, R. Maggioni, A. Ragusa**

Abstract. For a 0-dimensional subscheme X of a smooth quadric $\mathbf{Q} \cong \mathbf{P}^1 \times \mathbf{P}^1$ we investigate the relationship between the graded Betti numbers of a minimal free resolution of X as an $\mathcal{O}_{\mathbf{Q}}$-module and as an $\mathcal{O}_{\mathbf{P}}$-module. Some particular cases are considered, as the arithmetically Cohen-Macaulay 0-dimensional subschemes and those which have a length two resolution and balanced syzygies. For points X in generic position, we relate the generators' degrees with the postulation of X.

AMS 1991 subject classification: 14C20, 14C25, 13D02, 13D40.

Introduction

To find resolutions for 0-dimensional schemes of $\mathbf{P}^3$ is the object of study in several papers during the last few years. In particular the graded Betti numbers of a minimal free resolution of the ideal sheaf of the scheme are studied in terms of the geometric properties of the scheme. In this direction, since we deal with non-degenerate subschemes of $\mathbf{P}^3$, the next step is to study resolutions for points lying on a quadric. When a scheme X lying on a smooth quadric of $\mathbf{P}^3$ is given, one can consider it both as a subscheme of $\mathbf{Q}$ and of $\mathbf{P}^3$, so the idea is to find relationships between the minimal free resolution of the ideal sheaf $\overline{\mathcal{I}}_X$ of X on $\mathbf{Q}$ and the minimal free resolution of the ideal sheaf $\mathcal{I}_X$ of X on $\mathbf{P}^3$, and then to find minimal free resolutions of $\overline{\mathcal{I}}_X$ according to the postulation of X in $\mathbf{Q}$. We can obtain the graded Betti numbers of $\mathcal{I}_X$ on $\mathbf{P}^3$ when we know the graded Betti numbers of $\overline{\mathcal{I}}_X$ on $\mathbf{Q}$ and on which elements each syzygy acts. So the minimal free resolution of $\mathcal{I}_X$ is completely determined when X is an arithmethically Cohen-Macaulay (ACM for short) subscheme of $\mathbf{Q}$ and for some X which is complete intersection of two curves of $\mathbf{Q}$; this is the argument of sections 2 and 3. In the last section we deal with points which are in generic position (as subscheme of $\mathbf{Q}$), i.e. points with maximal Hilbert matrix. The results that we found on these subschemes give information on the degrees of any minimal set of generators in terms of the postulation of X. Our feeling is that, at least for points in uniform

* Work done with financial support of M.U.R.S.T., while the authors were members of C.N.R.

position, one can read from the Hilbert matrix all the degrees of a minimal set of generators: the results of section 4 are all in this direction. For all general facts not explicitly mentioned we refer to Hartshorne's book [H].

1. Notation and preliminaries

Let $\mathbf{P}^1 = \mathbf{P}^1_k$ (k an algebraically closed field), and let $\mathbf{Q} \cong \mathbf{P}^1 \times \mathbf{P}^1$ be a quadric and $\mathcal{O}_{\mathbf{Q}}$ its structure sheaf. Since $\operatorname{Pic}\mathbf{Q} \cong \mathbf{Z} \times \mathbf{Z}$, as usual we can assume the classes of the two rulings as basis of $\operatorname{Pic}\mathbf{Q}$. If $D \subset \mathbf{Q}$ is a divisor, the class of D is singled out by a couple of integers (a, b), the *type* of D. If $D \subset \mathbf{Q}$ is a divisor of type (a, b) we denote by $\mathcal{O}_{\mathbf{Q}}(a, b)$ the corresponding sheaf, and for any sheaf $\mathcal{F}$ on $\mathbf{Q}$, we set $\mathcal{F}(a, b) = \mathcal{F} \otimes \mathcal{O}_{\mathbf{Q}}(a, b)$. We also use the notation

$$H^i(a, b) = H^i(\mathbf{Q}, \mathcal{O}_{\mathbf{Q}}(a, b)), \quad h^i(a, b) = \dim_k H^i(a, b)$$
$$H^i(\mathcal{F}(a, b)) = H^i(\mathbf{Q}, \mathcal{F}(a, b)), \quad h^i(\mathcal{F}(a, b)) = \dim_k H^i(\mathcal{F}(a, b))$$

for $i = 0, 1, 2$. The dimensions $h^i(a, b), i = 0, 1, 2$, are easily computed ([GMR], §1). For any 0-dimensional subscheme $X \subset \mathbf{Q}$ one can consider the Hilbert matrix of X , defined as the function

$$M_X : \mathbf{Z} \times \mathbf{Z} \to \mathbf{N}$$
$$M_X(i, j) = (m_{ij}) = h^0(i, j) - h^0(\overline{\mathcal{I}}_X(i, j))$$

where $\overline{\mathcal{I}}_X$ is the ideal sheaf of X on $\mathbf{Q}$. In the sequel we will use the first difference matrix $\Delta M_X = (c_{ij})$ with $c_{ij} = a_{ij} + a_{i-1j-1} - a_{ij-1} - a_{i-1j}$, and the second difference matrix $\Delta^2 M_X = (d_{ij})$ with $d_{ij} = c_{ij} + c_{i-1j-1} - c_{ij-1} - c_{i-1j}$. For properties of these matrices see [GMR], §2. Moreover we will use the following ring (called the *global ring* of $\mathbf{Q}$ in [GM1]) :

$$G(\mathbf{Q}) = S = \bigoplus_{m,n \geq 0} H^0(m, n)$$

which is generated by $H^0(1, 0)$ and $H^0(0, 1)$: denoting by u, u' a basis of $H^0(1, 0)$, and by v, v' a basis of $H^0(0, 1)$ we obtain the bigraded ring

$$S \cong k[u, u'; v, v'].$$

The scheme X is represented in S by a bigraded homogeneous saturated ideal $\overline{I}(X)$, which has a minimal free resolution as a bigraded module of the type:

$$\begin{aligned} 0 \to \bigoplus_{i=1}^{p} S(-a_{2i}, -a'_{2i}) \to \bigoplus_{i=1}^{n} S(-a_{1i}, -a'_{1i}) \to \\ \to \bigoplus_{i=1}^{m} S(-a_{0i}, -a'_{0i}) \to \overline{I}(X) \to 0 \end{aligned} \tag{1}$$

where the morphisms are of bidegree $(0,0)$. From this, taking sheaves, one gets an $\mathcal{O}_{\mathbf{Q}}$-free resolution of the ideal sheaf $\overline{\mathcal{I}}_X$.

Take now any $\mathcal{O}_{\mathbf{Q}}$-free minimal resolution of $\overline{\mathcal{I}}_X$

$$0 \to \mathcal{F}_2 \to \mathcal{F}_1 \xrightarrow{\varphi} \mathcal{F}_0 \to \overline{\mathcal{I}}_X \to 0$$

such that

$$\begin{cases} H^0(\mathcal{F}_0(r,s)) \to H^0(\overline{\mathcal{I}}_X(r,s)) & \text{is surjective for every (r,s)} \\ H^0(\mathcal{F}_1(r,s)) \to H^0(\mathcal{E}(r,s)) & \text{is surjective for every (r,s), with } \mathcal{E} = Im\varphi. \end{cases} \qquad (*)$$

With this choice, for every (r,s) we obtain the exact sequence

$$0 \to H^0(\mathcal{F}_2(r,s)) \to H^0(\mathcal{F}_1(r,s)) \to H^0(\mathcal{F}_0(r,s)) \to H^0(\overline{\mathcal{I}}_X(r,s)) \to 0$$

and since $H^0_*(\overline{\mathcal{I}}_X) = \bigoplus_{r,s\geq 0} H^0(\overline{\mathcal{I}}_X(r,s)) \cong \overline{I}(X)$, taking sums on (r,s) we obtain a resolution which is isomorphic to (1). Thus the resolution

$$\begin{aligned} 0 \to \bigoplus_{i=1}^{p} \mathcal{O}_{\mathbf{Q}}(-a_{2i}, -a'_{2i}) \to \bigoplus_{i=1}^{n} \mathcal{O}_{\mathbf{Q}}(-a_{1i}, -a'_{1i}) \to \\ \to \bigoplus_{i=1}^{m} \mathcal{O}_{\mathbf{Q}}(-a_{0i}, -a'_{0i}) \to \overline{\mathcal{I}}_X \to 0 \end{aligned} \qquad (2)$$

obtained by taking sheaves in (1), satisfies conditions $(*)$. We will refer to (2) as *the minimal free resolution* of $\overline{\mathcal{I}}_X$ without further specification. For more facts about the ring G(**Q**), the Hilbert matrix and properties of 0-dimensional subschemes of **Q** we refer to [GMR]. In the sequel, when no confusion can arise, we shall omit the prefix "bi" in the words bidegree, bihomogeneous, ...

2. Relationships between resolutions

Given a 0-dimensional subscheme $X \subset \mathbf{Q}$ we can also consider X as a subscheme of $\mathbf{P}^3$ via the usual embedding $\mathbf{Q} \hookrightarrow \mathbf{P}^3$: our purpose in this section is to determine a resolution of the ideal sheaf $\mathcal{I}_X \subset \mathcal{O}_{\mathbf{P}^3}$ starting from a resolution of $\overline{\mathcal{I}}_X$. We begin with some general facts. Let $T \subset \mathbf{Q}$ be any subscheme, let $\mathcal{I}_T \subset \mathcal{O}_{\mathbf{P}^3}, \overline{\mathcal{I}}_T \subset \mathcal{O}_{\mathbf{Q}}$ be its ideal sheaves and let $I(T) \subset k[x_0,x_1,x_2,x_3]$ be its homogeneous ideal as a subscheme of $\mathbf{P}^3$. Suppose that the quadric **Q** is a minimal generator of $I(T)$. Denoting by $\mathcal{I}_{\mathbf{Q}} \subset \mathcal{O}_{\mathbf{P}^3}$ the ideal sheaf of **Q**, we have the exact sequence of $\mathcal{O}_{\mathbf{P}^3}$-modules

$$0 \to \mathcal{I}_{\mathbf{Q}} \to \mathcal{I}_T \to \overline{\mathcal{I}}_T \to 0$$

from which we see that any resolution of $\mathcal{I}_T$ can be obtained from a resolution of $\overline{\mathcal{I}}_T$ (as $\mathcal{O}_{\mathbf{P}^3}$ -module) just adding $\mathcal{O}_{\mathbf{P}^3}(-2) \cong \mathcal{I}_{\mathbf{Q}}$ to the sheaf of generators. Using this fact for a curve $C \subset \mathbf{Q}$ of type (a,b) (we assume $b \geq a \geq 0$), we see that

$\overline{\mathcal{I}}_C \cong \mathcal{O}_{\mathbf{Q}}(-a,-b)$ has a well known resolution as $\mathcal{O}_{\mathbf{P}^3}$-module (see [GM], Theorem 4.8):

if $a < b$ $\quad 0 \to \mathcal{O}_{\mathbf{P}^3}(-b-2)^{b-a-1} \to \mathcal{O}_{\mathbf{P}^3}(-b-1)^{2(b-a)} \to$
$$\to \mathcal{O}_{\mathbf{P}^3}(-b)^{b-a+1} \to \overline{\mathcal{I}}_C \to 0;$$

if $a = b$ $\quad 0 \to \mathcal{O}_{\mathbf{P}^3}(-b-2) \to \mathcal{O}_{\mathbf{P}^3}(-b) \to \overline{\mathcal{I}}_C \to 0.$

Now, let $X \subset \mathbf{Q}$ be a 0-dimensional subscheme, and let be given the minimal free resolution of its ideal sheaf as $\mathcal{O}_{\mathbf{Q}}$-module

$$\begin{array}{ccccccccc} 0 \to \mathcal{F}_2 \to & & \mathcal{F}_1 & \to & \mathcal{F}_0 & \to \overline{\mathcal{I}}_X \to 0 \\ & & & \searrow & \nearrow & & \\ & & & & \mathcal{E} & & \\ & & & \nearrow & \searrow & & \\ & & 0 & & & 0 & \end{array} \tag{3}$$

where $\mathcal{F}_i = \bigoplus_{j=1}^{n_i} \mathcal{O}_{\mathbf{Q}}(-a_{ij}, -a'_{ij})$ are decomposed sheaves, and $\mathcal{E}$ is locally free. For each $\mathcal{F}_i$ $(i = 0,1,2)$ we can write its minimal free resolution $\mathcal{L}.^{(i)} \to \mathcal{F}_i \to 0$ as $\mathcal{O}_{\mathbf{P}^3}$-module just taking the direct sum of the resolutions of all components. Our first step is to take the mapping cone of the diagram

$$\begin{array}{ccccccccc} & & 0 & & 0 & & & & \\ & & \uparrow & & \uparrow & & & & \\ 0 & \to & \mathcal{F}_2 & \to & \mathcal{F}_1 & \to & \mathcal{E} & \to & 0. \\ & & \uparrow & & \uparrow & & & & \\ & & \mathcal{L}.^{(2)} & & \mathcal{L}.^{(1)} & & & & \end{array}$$

So we have a resolution of $\mathcal{E} : \mathcal{L}. \to \mathcal{E} \to 0$. Notice that the maps

$$H^0(\mathbf{P}^3, \mathcal{L}_0(n)) \to H^0(\mathbf{P}^3, \mathcal{E}(n))$$

are surjective for any n because $H^0(\mathbf{P}^3, \mathcal{L}_0^{(1)}(n)) \to H^0(\mathbf{P}^3, \mathcal{F}_1(n))$ trivially are, and also the maps $H^0(\mathbf{P}^3, \mathcal{F}_1(n)) \to H^0(\mathbf{P}^3, \mathcal{E}(n))$ are surjective by the condition $(*)$. Hence, $H^i(\operatorname{coker}(\mathcal{L}_3 \to \mathcal{L}_2)) = 0$ for $i = 1,2$, which means, by Horrocks' criterion, that $\mathcal{L}_3 \to \mathcal{L}_2$ splits. Thus, $\mathcal{L}.$ can be shortened and precisely $\mathcal{L}_3 = \mathcal{L}_2^{(2)}$ is isomorphic to a free component of $\mathcal{L}_2 = \mathcal{L}_2^{(1)} \oplus \mathcal{L}_1^{(2)}$ (more precisely, $\mathcal{L}_3$ is isomorphic to a free component of $\mathcal{L}_2^{(1)}$). So we obtain a length three free resolution

$$\mathcal{L}'. \to \mathcal{E} \to 0.$$

Our second step is to take the mapping cone of the diagram

$$\begin{array}{ccccccccc} & & 0 & & 0 & & & & \\ & & \uparrow & & \uparrow & & & & \\ 0 & \to & \mathcal{E} & \to & \mathcal{F}_0 & \to & \overline{\mathcal{I}}_X & \to & 0. \\ & & \uparrow & & \uparrow & & & & \\ & & \mathcal{L}'. & & \mathcal{L}.^{(0)} & & & & \end{array}$$

Again, we get a length four resolution of $\overline{\mathcal{I}}_X$ as $\mathcal{O}_{\mathbf{P}^3}$-module: $\mathcal{R}. \to \overline{\mathcal{I}}_X \to 0$. By the same argument as in step one, this resolution can be shortened to a length three one: $\mathcal{R}'_. \to \overline{\mathcal{I}}_X \to 0$. Hence we have for $\overline{\mathcal{I}}_X$ the resolution

$$0 \to \mathcal{R}'_2 \to \mathcal{R}'_1 \to \mathcal{R}'_0 \to \overline{\mathcal{I}}_X \to 0. \tag{4}$$

In general this resolution is not minimal with respect to the Betti numbers, but if one concretely knows the maps in (3), then one can obtain from (4) the minimal resolution by deleting isomorphic components. Indeed, a little less is enough: to simplify (4) one can apply the following lemma every time it is possible.

Lemma 2.1. *Let*

$$0 \to \mathcal{F} \oplus \mathcal{O}_{\mathbf{Q}}(-a,-b) \xrightarrow{\varphi} \mathcal{D} \oplus \mathcal{O}_{\mathbf{Q}}(-a',-b) \to \operatorname{Coker}\varphi \to 0$$

be an exact sequence of $\mathcal{O}_{\mathbf{Q}}$-modules, with $\mathcal{F}, \mathcal{D}$ decomposable sheaves, $b \geq a > a' \geq 0$. Suppose that the restriction of φ to $\mathcal{O}_{\mathbf{Q}}(-a,-b)$ is not zero on $\mathcal{O}_{\mathbf{Q}}(-a',-b)$ (i.e. the image of the restriction is not contained in $\mathcal{D}$). In case $b > a$, taking the resolutions (as $\mathcal{O}_{\mathbf{P}^3}$-modules)

$$\mathcal{L}_. \quad 0 \to \mathcal{L}_2 \oplus \mathcal{O}_{\mathbf{P}^3}(-b-2)^{b-a-1} \to \mathcal{L}_1 \oplus \mathcal{O}_{\mathbf{P}^3}(-b-1)^{2(b-a)} \to$$
$$\to \mathcal{L}_0 \oplus \mathcal{O}_{\mathbf{P}^3}(-b)^{b-a+1} \to \mathcal{F} \oplus \mathcal{O}_{\mathbf{Q}}(-a,-b) \to 0$$

$$\mathcal{L}'_. \quad 0 \to \mathcal{L}'_2 \oplus \mathcal{O}_{\mathbf{P}^3}(-b-2)^{b-a'-1} \to \mathcal{L}'_1 \oplus \mathcal{O}_{\mathbf{P}^3}(-b-1)^{2(b-a')} \to$$
$$\to \mathcal{L}'_0 \oplus \mathcal{O}_{\mathbf{P}^3}(-b)^{b-a'+1} \to \mathcal{D} \oplus \mathcal{O}_{\mathbf{Q}}(-a',-b) \to 0$$

then we have for Coker φ *(as $\mathcal{O}_{\mathbf{P}^3}$-module) the following free resolution*

$$0 \to \mathcal{L}_2 \to \mathcal{L}'_2 \oplus \mathcal{L}_1 \oplus \mathcal{O}_{\mathbf{P}^3}(-b-2)^{a-a'} \to \mathcal{L}'_1 \oplus \mathcal{L}_0 \oplus \mathcal{O}_{\mathbf{P}^3}(-b-1)^{2(a-a')} \to$$
$$\to \mathcal{L}'_0 \oplus \mathcal{O}_{\mathbf{P}^3}(-b)^{a-a'} \to \operatorname{Coker}\varphi \to 0.$$

The same result is obtained when $b = a$.

Proof. The final resolution is obtained by two steps. First we take the mapping cone of the diagram $\mathcal{L}_. \xrightarrow{\varphi} \mathcal{L}'_.$. The second step consists in a simplification of that mapping cone: the liftings φ_i of φ give, by the hypothesis on φ, injections (we call φ_i also the restrictions) $\varphi_i : \mathcal{O}_{\mathbf{P}^3}(-b-i)^r \to \mathcal{O}_{\mathbf{P}^3}(-b-i)^s$ with $r < s$. □

Remark 2.2. Up to now, what we have seen can be summarized saying: we can obtain a minimal free resolution of $\mathcal{I}_X \subset \mathcal{O}_{\mathbf{P}^3}$ just knowing in (3) the graded Betti numbers, and on which elements each syzygy acts, i.e. which entries of the matrices associated to the maps of the resolution (3) are zero. This happens, for instance, for ACM 0-dimensional subschemes of $\mathbf{Q}$ (see [GMR], §4). For reader's convenience we recall some basic properties of the resolutions of these subschemes. Let X be a 0-dimensional ACM subscheme of $\mathbf{Q}$ and $\overline{I}(X)$ its homogeneous saturated ideal in $G(\mathbf{Q})$. Then every minimal set of generators of $\overline{I}(X)$ contains elements which have degrees $(a_i, b_{m-i+1}), i = 1, 2, \ldots, m$ such that no two of them are comparable with respect to the partial ordering induced on $\mathbf{Z}^2$ by the usual one. Moreover, setting $a_1 > a_2 > \cdots > a_m = 0$ and $b_1 > b_2 > \cdots > b_m = 0$, one sees that the syzygies' degrees are (a_i, b_{m-i}) with $i = 1, 2, \ldots, m-1$, and that the syzygy of degree (a_i, b_{m-i}) acts just on the generators of degree $(a_i, b_{m-i+1}), (a_{i+1}, b_{m-i})$. In the next proposition we completely determine the graded Betti numbers of the resolution of the ideal sheaf $\overline{\mathcal{I}}_X \subset \mathcal{O}_{\mathbf{P}^3}$ for any ACM 0-dimensional subscheme $X \subset \mathbf{Q}$. For any such subscheme X, we use the following notation (recall that $\Delta M_X = (c_{ij})$):

$$\begin{aligned} h =& \min\{r \mid c_{rr} = 0\}, \\ \alpha_i =& \min\{r \mid c_{ir} = 0,\} \text{ for any } i \geq 0 , \\ \beta_i =& \min\{s \mid c_{si} = 0,\} \text{ for any } i \geq 0 , \\ k_i =& \alpha_i + \beta_i, \text{ for } i \geq h, \\ t+1 =& \min\{r \mid k_i = 0\}. \end{aligned}$$

Proposition 2.3. *Let $X \subset \mathbf{Q}$ be an ACM 0-dimensional subscheme. Then the minimal free resolution of the ideal sheaf $\mathcal{I}_X \subset \mathcal{O}_{\mathbf{P}^3}$ is of the following type:*

$$0 \to \bigoplus_{i=h+1}^{t+1} \mathcal{O}_{\mathbf{P}^3}(-i-2)^{k_{i-1}-k_i} \oplus \mathcal{O}_{\mathbf{P}^3}(-h-2)^{2h-1-k_h} \to$$

$$\to \bigoplus_{i=h+1}^{t+1} \mathcal{O}_{\mathbf{P}^3}(-i-1)^{2(k_{i-1}-k_i)} \oplus \mathcal{O}_{\mathbf{P}^3}(-h-1)^{4h-2k_h} \to$$

$$\to \bigoplus_{i=h+1}^{t+1} \mathcal{O}_{\mathbf{P}^3}(-i)^{k_{i-1}-k_i} \oplus \mathcal{O}_{\mathbf{P}^3}(-h)^{2h+1-k_h} \oplus \mathcal{O}_{\mathbf{P}^3}(-2) \to \mathcal{I}_X \to 0$$

Proof. Using [GMR] Theorem 4.1 we see that $\overline{\mathcal{I}}_X$ has a free resolution of this kind:

$$\begin{aligned} 0 \to \mathcal{O}_{\mathbf{Q}}(-h,-h) \oplus \bigoplus_{i=h+1}^{t+1} (\mathcal{O}_{\mathbf{Q}}(-i,-\alpha_{i-1}) \oplus \mathcal{O}_{\mathbf{Q}}(-\beta_{i-1},-i)) \xrightarrow{\delta} \\ \to \bigoplus_{i=h}^{t+1} (\mathcal{O}_{\mathbf{Q}}(-i,-\alpha_i) \oplus \mathcal{O}_{\mathbf{Q}}(-\beta_i,-i)) \to \overline{\mathcal{I}}_X \to 0 \end{aligned} \tag{†}$$

where the restriction maps

$$\delta_{|\mathcal{O}_{\mathbf{Q}}(-h,-h)}, \quad \delta_{|\mathcal{O}_{\mathbf{Q}}(-i,-\alpha_{i-1})}, \quad \delta_{|\mathcal{O}_{\mathbf{Q}}(-\beta_{i-1},-i)}$$

are isomorphisms whenever (respectively)

$$\begin{gathered} \alpha_h = h \quad \text{and/or} \quad \beta_h = h \\ \alpha_{i-1} = \alpha_i \\ \beta_{i-1} = \beta_i. \end{gathered}$$

Note that in the minimal free resolution of $\overline{\mathcal{I}}_X$ obtained in [GMR], theorem 4.1, for every $i \geq h$ there is a generator of degree (i, α_i) (resp. (β_i, i)) if and only if $\alpha_i < \alpha_{i-1}$ (resp. $\beta_i < \beta_{i-1}$). So, the resolution (†) is obtained from the minimal one by adding, for every i such that $\alpha_i = \alpha_{i-1}$ or $\beta_i = \beta_{i-1}$, the identity isomorphisms

$$\mathcal{O}_{\mathbf{Q}}(-i, -\alpha_{i-1}) \xrightarrow{\sim} \mathcal{O}_{\mathbf{Q}}(-i, -\alpha_i)$$
$$\mathcal{O}_{\mathbf{Q}}(-\beta_{i-1}, -i) \xrightarrow{\sim} \mathcal{O}_{\mathbf{Q}}(-\beta_i, -i).$$

Now, applying Lemma 2.1 to the resolution (†) we get the result. □

3. Determinantal subschemes

We have seen how to obtain the resolution of $\mathcal{I}_X \subset \mathcal{O}_{\mathbf{P}^3}$ starting from the minimal free resolution of $\overline{\mathcal{I}}_X \subset \mathcal{O}_{\mathbf{Q}}$. Sometimes one has not all the information to write such a resolution, even if one can have enough data to define the subscheme. For instance, a surjection $\mathcal{F} \to \overline{\mathcal{I}}_X$ with $\mathcal{F}$ decomposed sheaf on $\mathcal{O}_{\mathbf{Q}}$ is enough to define X as a subscheme of $\mathbf{Q}$, but not in general to know the minimal free resolution. Indeed, many subschemes of $\mathbf{Q}$ can be defined by a "short" free resolution: the first example, of course, are the ACM subschemes, for which the minimal free resolution has length two. But other examples can be done: the simplest ones are the subschemes which are ideally complete intersection. A subscheme $X \subset \mathbf{Q}$ having a length 2 free resolution

$$0 \to \mathcal{F}_1 \xrightarrow{\varphi} \mathcal{F}_0 \to \overline{\mathcal{I}}_X \to 0 \tag{5}$$

will be called a *determinantal subscheme* since the maximal minors of the matrix associated to φ generate an ideal I of S such that the sheaf associated to I is $\overline{\mathcal{I}}_X$. In this section we deal with the relationships between the short resolution of a determinantal subscheme and its minimal free resolution as a subscheme of $\mathbf{P}^3$. As a first result we will show that a determinantal subscheme for which the shifts of the components of $\mathcal{F}_0$ have the same configuration as the generators of an ACM subscheme is indeed ACM. Hence the "structure" of the degrees of the generators characterizes the ACM subschemes of $\mathbf{Q}$.

Theorem 3.1. *Let $X \subset \mathbf{Q}$ be a 0-dimensional determinantal subscheme with a "short" resolution of the following type*

$$0 \to \bigoplus_{i=1}^{m-1} \mathcal{O}_{\mathbf{Q}}(-x_i, -y_i) \to \bigoplus_{i=1}^{m} \mathcal{O}_{\mathbf{Q}}(-a_i, -b_{m+1-i}) \to \overline{\mathcal{I}}_X \to 0.$$

Suppose that $0 = a_m < \cdots < a_2 < a_1$ and that $0 = b_m < \cdots < b_2 < b_1$. Then X is ACM.

Proof. Notice that the hypothesis on the generators' degrees says that we have a generator of degree $(a_1, 0)$ and a generator of degree $(0, b_1)$; moreover the degrees of the generators are not comparable. From the given resolution we have the equalities

$$\sum_{i=1}^{m-1} x_i = \sum_{i=1}^{m} a_i \tag{i}$$

$$\sum_{i=1}^{m-1} y_i = \sum_{i=1}^{m} b_i \tag{ii}$$

by the exactness of the resolution and by the fact that X has dimension zero. Let us order the x_i's so that $x_{m-1} \leq \cdots \leq x_2 \leq x_1$.

Claim: $x_i \geq a_i$ for $i = 1, 2, \cdots, m-1$.

From the claim and from equality (i) it follows that $x_i = a_i$ for $i = 1, 2, \cdots, m-1$; from this it follows that $y_i \geq b_{m-i}$ for $i = 1, 2, \cdots, m-1$. In fact, the syzygy of degree (x_i, y_i), since $x_i = a_i < a_{i-1} < \cdots < a_1$, can act only on generators of degree $(a_i, b_{m+1-i}), \ldots, (0, b_1)$, so $y_i \geq b_{m-i}$. Hence, again by equality (ii), we have $y_i = b_{m-i}$ for $i = 1, 2, \ldots, m-1$. In particular, the degrees of the syzygies are the expected ones for X to be ACM. From the resolution we have also (see [GMR], Proposition 3.3)

$$\deg X = \sum_{i=1}^{m-1} x_i y_i - \sum_{i=1}^{m} a_i b_{m+1-i}. \tag{iii}$$

Taking the Hilbert matrix M_X of X, we know ([GMR],Proposition 2.7) that the first differences matrix $\Delta M_X = (c_{ij})$ has the following property:

$$c_{rs} \leq 0 \text{ for } (r, s) \geq (a_i, b_{m+1-i})$$

and since $\deg X = \sum_{r,s \in \mathbf{Z}} c_{rs}$, we see that $\deg X \leq$ area of A, where A is the union of rectangles having vertices in (x_i, y_i) (see Figure 1, below).

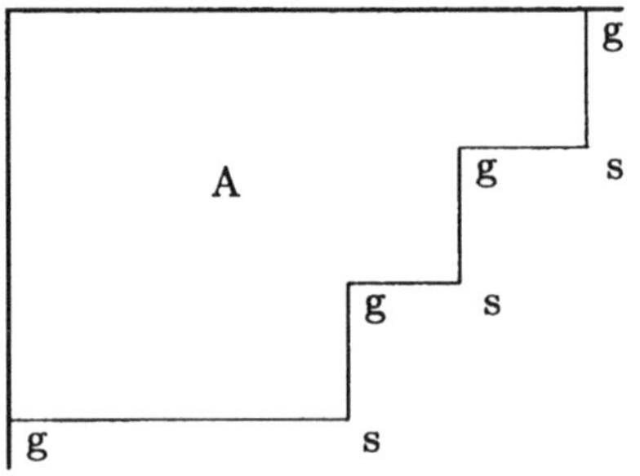

Figure 1

Now, since the area of A is $\sum_{i=1}^{m-1} x_i y_i - \sum_{i=1}^{m} a_i b_{m+1-i}$, which is equal to $\deg X$, it follows that the entries of the matrix ΔM_X are: $c_{rs} = 1$ if $(r, s) \in A$ and $c_{rs} = 0$ if $(r, s) \notin A$. This means that X is ACM (see [GMR],Theorem 4.1).

Proof of the claim. We use induction on m. $x_1 \geq a_1$ is clear; assuming $x_1 \geq a_1, x_2 \geq a_2, \ldots, x_{r-1} \geq a_{r-1}$ we have to prove that $x_r \geq a_r$. Assume the converse,

$x_i < a_r$ for $i = r, r+1, \dots, m-1$. In this case we cannot have any first syzygy of degree (x_i, y_i) linking generators of degree (a_j, b_j) with $j = 1, 2, \dots, r$; hence the matrix of φ looks like

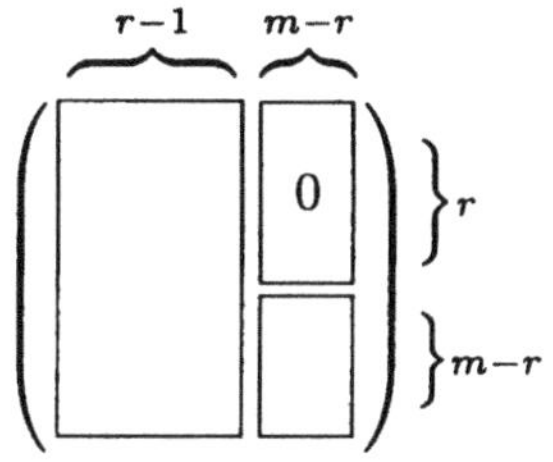

i.e. the submatrix obtained intersecting the first r rows with the last $m-r$ columns is zero. But in this case the $m-1 \times m-1$ minors obtained by deleting one of the last $m-r$ rows are zero (apply Laplace Theorem to the first r rows), and the minors obtained by deleting one of the first r rows have a common component, namely the determinant of the submatrix of the last $m-r$ rows and columns (again by Laplace Theorem). This is a contradiction since X is zero-dimensional. □

Now we deal with determinantal subschemes of $\mathbf{Q}$ which are not ACM. Let X be such a subscheme and (5) its resolution. Using the resolutions of $\mathcal{F}_1$ and $\mathcal{F}_0$ as $\mathcal{O}_{\mathbf{P}^3}$-modules and taking the mapping cone we obtain a free resolution of $\overline{\mathcal{I}}_X$ as $\mathcal{O}_{\mathbf{P}^3}$-module, hence a free resolution of $\mathcal{I}_X$

$$0 \to \mathcal{R}_3 \to \mathcal{R}_2 \to \mathcal{R}_1 \to \mathcal{R}_0 \oplus \mathcal{O}_{\mathbf{P}^3}(-2) \to \mathcal{I}_X \to 0 \tag{6}$$

which, generally, has not the correct length. If we want (6) to have the correct length we must require the map

$$H^0_*(\mathcal{R}_0 \oplus \mathcal{O}_{\mathbf{P}^3}(-2)) \to H^0_*(\mathcal{I}_X)$$

to be surjective: this is equivalent to have the map $H^0(\mathcal{F}_0(n,n)) \to H^0(\overline{\mathcal{I}}_X(n,n))$ surjective for all n, i.e. the map $H^1(\mathcal{F}_1(n,n)) \to H^1(\mathcal{F}_0(n,n))$ injective for all n. This happens, for instance, when $H^1(\mathcal{F}_1(n,n)) = 0$ for any $n > 0$, hence, setting $\mathcal{F}_1 = \oplus_{i=1}^r \mathcal{O}_{\mathbf{Q}}(-a_i, -b_i)$, we can require $|a_i - b_i| \le 1$ for $i = 1, 2, \dots, r$ (balanced syzygies or ACM syzygies). In fact, this condition is equivalent to $H^1(\mathcal{F}_1(n,n)) = 0$ for any $n > 0$. The next proposition summarizes the above argument.

Proposition 3.2. *Let $X \subset \mathbf{Q}$ be a 0-dimensional determinantal subscheme and let*

$$0 \to \mathcal{F}_1 \xrightarrow{\varphi} \mathcal{F}_0 \to \overline{\mathcal{I}}_X \to 0 \tag{5}$$

be a short free resolution of $\overline{\mathcal{I}}_X$ as $\mathcal{O}_{\mathbf{Q}}$-module. If $\mathcal{F}_1 = \oplus_{i=1}^r \mathcal{O}_{\mathbf{Q}}(-a_i, -b_i)$, and for $i = 1, 2, \dots, r$ we have $|a_i - b_i| \le 1$, then the mapping cone on the resolutions of $\mathcal{F}_1$ and $\mathcal{F}_0$ as $\mathcal{O}_{\mathbf{P}^3}$-modules gives a length three resolution of $\overline{\mathcal{I}}_X$.

Proof. Since $|a_i - b_i| \leq 1, \mathcal{F}_1$ has a length two minimal free resolution as $\mathcal{O}_{\mathbf{P}^3}$-module, so that the mapping cone gives a length three free resolution of $\overline{\mathcal{I}}_X$ as $\mathcal{O}_{\mathbf{P}^3}$-module. □

Notice that the resolution (5) does not give all the postulation of X with respect to the curves of $\mathbf{Q}$, i.e. it does not determine all the entries of the Hilbert matrix M_X of X, not even on its diagonal (which is the Hilbert function of X as a subscheme of $\mathbf{P}^3$). Nevertheless, in the hypothesis of Proposition 3.2 we have $H^1(\mathcal{F}_1(n,n)) = 0$ for any $n > 0$, so the diagonal of M_X is completely determined. Let $X \subset \mathbf{Q}$ be a 0-dimensional subscheme which is ideally generated by two elements, i.e. it is set theoretically a complete intersection. Denoting by (a,b) and (c,d) the degrees of the two curves C_1, C_2 such that $X = C_1 \cap C_2$, we see that X is determinantal and has the following resolution as $\mathcal{O}_{\mathbf{Q}}$-module (Koszul)

$$0 \to \mathcal{O}_{\mathbf{Q}}(-a-c,-b-d) \xrightarrow{\varphi} \mathcal{O}_{\mathbf{Q}}(-a,-b) \oplus \mathcal{O}_{\mathbf{Q}}(-c,-d) \to \overline{\mathcal{I}}_X \to 0.$$

If $|a+c-b-d| \leq 1$, then Proposition 3.2 can be applied, so one obtains a length three free resolution of $\overline{\mathcal{I}}_X$ as $\mathcal{O}_{\mathbf{P}^3}$-module, hence a resolution of $\mathcal{I}_X$. Moreover, an explicit computation shows that we obtain a minimal free resolution. Supposing for instance $a \leq b$ and $c \leq d$ this resolution is

$$\begin{array}{c}
0\\
\uparrow\\
\mathcal{I}_X\\
\uparrow\\
\mathcal{O}_{\mathbf{P}^3}(-b)^{b-a+1} \oplus \mathcal{O}_{\mathbf{P}^3}(-d)^{d-c+1} \oplus \mathcal{O}_{\mathbf{P}^3}(-2)\\
\uparrow\\
\mathcal{O}_{\mathbf{P}^3}(-b-d)^{b+d-a-c+1} \oplus \mathcal{O}_{\mathbf{P}^3}(-b-1)^{2(b-a)} \oplus \mathcal{O}_{\mathbf{P}^3}(-d-1)^{2(d-c)}\\
\uparrow\\
\mathcal{O}_{\mathbf{P}^3}(-b-d-1)^{2(b+d-a-c)} \oplus \mathcal{O}_{\mathbf{P}^3}(-b-2)^{b-a-1} \oplus \mathcal{O}_{\mathbf{P}^3}(-d-2)^{d-c-1}\\
\uparrow\\
0
\end{array}$$

In the other cases one gets a similar resolution.

4. Points in generic position

In this section we deal with the graded Betti numbers of the minimal free resolution (as an $\mathcal{O}_{\mathbf{Q}}$-module) of the ideal sheaf $\overline{\mathcal{I}}_X$ of a 0-dimensional subscheme $X \subset \mathbf{Q}$ in generic position. We start with some definition.

Definition 4.1. A 0-dimensional subscheme $X \subset \mathbf{Q}$ is said to be in *generic position* if its Hilbert matrix M_X is maximal, i.e.

$$M_X(i,j) = \min\{\deg X, (i+1)(j+1)\}.$$

X is said to be in *uniform position* (as a subscheme of $\mathbf{Q}$) if every subscheme $X' \subset X$ is in generic position.

We remark that for a subscheme $X \subset \mathbf{Q}$ the above definition of uniformity is equivalent to say that X is in generic position and has the strong-UPP (see [PRR], Definition 1.8). For a subscheme $X \subset \mathbf{Q}$ in generic position it is easy to describe the second difference matrix $\Delta^2 M_X = (d_{ij})$: this description can be very useful because we know that $d_{ij} = -\alpha_{ij} + \beta_{ij} - \gamma_{ij}$ for $(i,j) > (0,0)$ ([GMR], Theorem 3.3, item vi)), where $\alpha_{ij}, \beta_{ij}, \gamma_{ij}$ are the numbers of generators, first syzygies and second syzygies respectively of degree (i,j) of the minimal free resolution of $\overline{\mathcal{I}}_X$.

Denote, for every $i > 0$, $\deg X = i \cdot q_i + r_i$ with $0 \leq r_i < i$, then

$$d_{i-1j} = \begin{cases} r_i - i & \text{for } j = q_i \\ -r_i & \text{for } j = q_i + 1 \\ 2(i-1-r_{i-1}) & \text{for } j = q_{i-1} \\ 2r_{i-1} & \text{for } j = q_{i-1} + 1 \\ r_{i-2} - (i-2) & \text{for } j = q_{i-2} \\ -r_{i-2} & \text{for } j = q_{i-2} + 1 \\ 0 & \text{otherwise} \end{cases}$$

where we use the convention $d_{00} = 1, d_{iq_t} = 0$ if $t \leq 0$, and whenever $q_i = q_{i+1}$ or $q_i = q_{i+1} + 1$ we just add the corresponding values to get d_{ij}. Note that if $q_i + 1 < q_{i-1}$ then there are at least $(i - r_i) + r_i = i$ minimal generators of bidegree $(i-1, -)$ and no more of them (see [GMR] Proposition 3.3 vi) and Remark 2.6). Of course, again by the same Remark, if $d_{ij} = -(i+1)$ (resp. $-(j+1)$) then in degree $(i, j+1)$ (resp. $(i+1, j)$) there are no new generators at all.

Lemma 4.2. *Let $X \subset \mathbf{Q}$ be a 0-dimensional subscheme in generic position. If $H^0(\overline{\mathcal{I}}_X(i,j)) \neq 0$ then in any minimal set of generators of the homogeneous saturated ideal of X, $\overline{I}(X) \subset S$, there is no element of degree (i,n) with $n \geq j+2$ and no element of degree (m,j) with $m \geq i+2$.*

Proof. Since M_X is maximal, in any degree (i,j) such that $(i+1)(j+1) \geq N$ one has $h^0(\overline{\mathcal{I}}_X(i,j)) = (i+1)(j+1) - N$. It is enough to show that the cup-product morphisms

$$\varphi : H^0(\overline{\mathcal{I}}_X(i,n-1)) \otimes H^0(0,1) \to H^0(\overline{\mathcal{I}}_X(i,n))$$
$$\psi : H^0(\overline{\mathcal{I}}_X(m-1,j)) \otimes H^0(1,0) \to H^0(\overline{\mathcal{I}}_X(m,j))$$

are surjective for $n \geq j+2$ and $m \geq i+2$. Applying Lemma 2.2 of [GMR] we have

$$\dim \operatorname{Im} \varphi = 2h^0(\overline{\mathcal{I}}_X(i,n-1)) - h^0(\overline{\mathcal{I}}_X(i,n-2)) =$$
$$2(i+1)n - 2N - (i+1)(n-1) + N = (i+1)(n+1) - N = h^0(\overline{\mathcal{I}}_X(i,n)).$$

The same argument shows that ψ is surjective too. □

As a consequence of the above lemma one can find minimal generators for $\overline{I}(X)$ only in "consecutive" degrees, i.e. (i,j) and $(i,j+1)$, or (i,j) and $(i+1,j)$.

Let $X \subset \mathbf{Q}$ be a 0-dimensional subscheme of degree N in generic position. We want to see what can be said about the degrees of the elements of a minimal system of generators of $\overline{I}(X) \subset G(\mathbf{Q})$, just looking at the Hilbert matrix M_X of X, or, more precisely, looking at the second difference matrix $\Delta^2 M_X$. Since $\Delta^2 M_X$ is symmetric, in the sequel we assume $i \leq j$. Forgetting $d_{00} = 1$, for each row i (resp. column j), the first non-zero entry d_{ij} is negative, then there are $-d_{ij}$ minimal generators of degree (i,j). Further generators can only be found, by Lemma 4.2, in degree $(i+1,j)$ or $(i,j+1)$. So, the degrees (i,j) in which we cannot deduce by Lemma 4.2, from the matrix $\Delta^2 M_X$, the number of minimal generators of $\overline{I}(X)$ can be characterized as follows:

$$\begin{gathered} d_{rj} = 0 \text{ for } r \leq i-2 \text{ and } j > 0\,;\, d_{is} = 0 \text{ for } s \leq j-2 \text{ and } i > 0 \\ d_{i-1j} < 0, \;\; d_{ij-1} < 0. \end{gathered} \tag{\#}$$

Recalling that $i \leq j$ we can reformulate the above conditions in terms of dimensions:

$$\begin{aligned} h^0(\overline{\mathcal{I}}_X(i,j-2)) &= 0 \text{ hence } (i+1)(j-1) \leq N \\ h^0(\overline{\mathcal{I}}_X(i-1,j)) &= -d_{i-1j} > 0 \text{ hence } ij + i > N. \end{aligned}$$

Now, setting $N = ij + x$, by the above inequalities we have $-1 \leq j-i-1 \leq x < i$, and one gets

$$\begin{aligned} d_{i-1j} &= -\, h^0(\overline{\mathcal{I}}_X(i-1,j)) = -i + x, \\ d_{ij-1} &= -\, h^0(\overline{\mathcal{I}}_X(i,j-1)) = -j + x, \end{aligned}$$

$$d_{i-1j-1} = \begin{cases} 0 & \text{if } x > -1 \\ -1 & \text{if } x = -1. \end{cases}$$

For a pair (i,j) satisfying (#) we consider the following diagram

$$\begin{array}{ccc} & & H^0(\overline{\mathcal{I}}_X(i-1,j)) \otimes H^0(1,0) \\ & & \downarrow \varphi \\ H^0(\overline{\mathcal{I}}_X(i,j-1)) \otimes H^0(0,1) & \xrightarrow{\psi} & H^0(\overline{\mathcal{I}}_X(i,j)) \end{array}$$

and we observe that ψ is surjective if and only if

$$\dim \operatorname{Im}\psi = 2(j-x) = h^0(\overline{\mathcal{I}}_X(i,j)) = i + j + 1 - x$$

i.e. $h^0(\overline{\mathcal{I}}_X(i,j-1)) = j - x = i+1$; so, ψ is surjective if and only if all the possible $i+1$ generators of the i-th row effectively occur in degree $(i,j-1)$. Notice that this happens, in particular, when $x = -1$ since in this case one has $i = j$.

Proposition 4.3. *Let $X \subset \mathbf{Q}$ be a 0-dimensional scheme. Suppose that in degree (i,j) the conditions (#) occur, and that $d_{i-1j-1} = 0, d_{i-1j} = d_{ij-1} = -1$; moreover suppose that either the generator F of degree $(i-1,j)$ or the generator F' of degree $(i,j-1)$ is irreducible. Then $d_{ij} \leq 0$ (indeed $d_{ij} < 0$ if $\deg X > 5$) and in any minimal system of generators of $\overline{I}(X)$ there are just $-d_{ij}$ elements of degree (i,j).*

Proof. By the hypotheses it follows $i = j, x = i-1$ and $\deg X > 1$. Moreover, since $d_{ij} = -3x + i + j - 1$, we have $d_{ij} = 2 - i \leq 0$, thus $d_{ij} = 0$ only when $i = 2$, i.e. $N = 5$. Notice that the hypotheses are satisfied only for schemes of degree $N = i^2 + i - 1$ with $i > 1$. In order to show that in degree (i, j) there are just $i - 2$ minimal generators it is equivalent to control that for such a degree we have no syzygies. Now, a syzygy in degree (i, j) should be of the type $F \cdot L + F' \cdot L' = 0$ with $L \in H^0(1, 0)$ and $L' \in H^0(0, 1)$, but this would imply that both F and F' are reducible, a contradiction. □

Remark 4.4. Notice that the irreducibility hypothesis is satisfied whenever X is in uniform position (see [PRR], Remark 3.2). If the irreducibility hypothesis of the Proposition is not fulfilled the assertion is no more true. This can be seen by considering 5 points, four of which lying on a conic and the last elsewhere: the resulting scheme is generic but in degree $(2, 2)$ there is a syzygy (hence a generator) despite the fact that $d_{22} = 0$.

Proposition 4.5. *Let $X \subset \mathbf{Q}$ be a 0-dimensional scheme in generic position. Suppose that conditions (#) are satisfied in degree (i, j) and that $N = \deg X = ij$. Then in any minimal system of generators of $\overline{I}(X)$ there are no elements of degree (i, j).*

Proof. The hypothesis implies $h^0(\overline{\mathcal{I}}_X(i-1, j)) = i, h^0(\overline{\mathcal{I}}_X(i, j-1)) = j$ and since $i \leq j \leq i+1$ one can have $j = i+1$ or $j = i$. In the first case ψ is surjective so we must look at $j = i$. Now $\dim \mathrm{Im}\varphi = \dim \mathrm{Im}\psi = 2i$ and $h^0(\overline{\mathcal{I}}_X(i-1, j)) = 2i+1$, so it is enough to prove that $\mathrm{Im}\varphi$ is not contained in $\mathrm{Im}\psi$. On the contrary suppose that $\mathrm{Im}\varphi \subseteq \mathrm{Im}\psi$. Choose a basis L_1, L_2 of $H^0(0, 1)$ such that $L_1 L_2$ vanishes on no points of X; then for any element $D \in H^0(\overline{\mathcal{I}}_X(i-1, j))$ one would have:

$$DL_1' = C'L_1 + C''L_2; DL_2' = \overline{C'}L_1 + \overline{C''}L_2. \tag{‡}$$

Since $h^0(\overline{\mathcal{I}}_X(i-1, j)) = i$, for $i-1$ generic points on L_1 there exists a unique element $\overline{D} \in H^0(\overline{\mathcal{I}}_X(i-1, j))$ vanishing on them; the genericity of the points imply that there is a unique section $C'' \in H^0(\overline{\mathcal{I}}_X(i, j-1))$ through them. Using the relations (‡) for $\overline{D}$ one gets $C'' = \overline{C''}$ and this implies, in particular, that $\overline{D}$ contains L_1, and consequently there is a $(i-1, j-1)$ curve through X, a contradiction. □

The same argument can be used when $i - 1 = 1$ and $j - x = 2$: but this implies $x \leq 1$, so the only new case is $x = 1$, i.e. $N = 7$. Note that the hypothesis of the above Proposition implies $d_{ij} > 0$, and indeed generators were not expected; for $N = 7$ the hypothesis does not apply but we have again $d_{23} = 1$.

Remark 4.6. The first case which is not included in the previous results occurs for $N = 10$. Nevertheless, we are able to handle this case with *ad hoc* arguments.

All the cases we have already solved state that for every scheme X of degree N there are a lot of places (i, j) in which any minimal set of generators of $\overline{I}(X)$ contains $-d_{ij}$ elements in that degree. So, the following question arises

Question 4.7. Does any minimal set of generators of a 0-dimensional subscheme $X \subset \mathbf{Q}$ of degree N in uniform position consist of $\max\{-d_{ij}, 0\}$ elements of degree (i,j) for every i and $j = q_i, j = q_i + 1$?

One can also ask something more general regarding all the Betti numbers of such subschemes

Question 4.8. Do the graded Betti numbers of a 0-dimensional subscheme $X \subset \mathbf{Q}$ of degree N in uniform position consist of $\max\{-d_{ij}, 0\}$ generators of degree (i,j) for every i and $j = q_i, j = q_i + 1$; $\max\{d_{ij}, 0\}$ first syzygies of degree (i,j) for every i and $j = q_i - 1, j = q_{i-1} + 1$; $\max\{-d_{ij}, 0\}$ second syzygies of degree (i,j) for every i and $j = q_{i-2}, j = q_{i-2} + 1$?

References

[GM] Giuffrida, S., Maggioni, R., On the resolution of a curve lying on a smooth cubic surface in $\mathbf{P}^3$. Trans. Amer. Math. Soc. 331 (1) (1992), 181–201.

[GM1] —, —, On the global ring associated with a smooth projective surface. Preprint.

[GMR] Giuffrida, S., Maggioni, R., Ragusa, A., On the postulation of 0-dimensional subschemes on a smooth quadric. Pac. J. Math. 155 (1992), 251–282.

[H] Hartshorne, R., Algebraic Geometry. GTM 52, Springer-Verlag, Berlin, 1977.

[PRR] Paxia, G., Raciti, G., Ragusa, A., Uniform position properties and Hilbert functions for points on a smooth quadric. J. Algebra 149 (1992), 102–121.

Points in Uniform Position and Maximum Distance Separable Codes

*Johan P. Hansen**

Abstract. We present a construction of linear error-correcting codes from families of points in a projective space. Among the codes constructed we obtained the geometric Goppa codes. The parameters of the codes are calculated by help of the Hilbert-functions of the family of points and certain subsets thereof. The results are used to devise methods for obtaining Maximum Distance Separable codes.

Keywords: Linear error-correcting codes, Maximum Distance Separable, Cayley-Bacharach schemes, uniform position

Introduction

This paper presents a general construction of linear error-correcting codes from families of points in projective space. The codewords are values of rational functions obtained from all homogenous polynomials of degree r in a set of points $\mathcal{P} = \{P_1, P_2, \ldots, P_n\} \subset \mathbf{P}^l$. Assume that $\mathcal{P} \cap Z(f_0) = \emptyset$ and let $\mathbf{F}_q[X_0, \ldots, X_l]_{(r)}$ be the homogenous polynomials of degree r. The code is obtained as the image of the map

$$\begin{aligned} \phi : \mathbf{F}_q[X_0, \ldots, X_l]_{(r)} &\rightarrow \mathbf{F}_q{}^n \\ f &\mapsto \left(\frac{f(P_1)}{f_0(P_1)}, \ldots, \frac{f(P_n)}{f_0(P_n)} \right) \end{aligned}$$

where $f_0 \in \mathbf{F}_q[X_0, \ldots, X_l]_{(r)}$.

In the first section the parameters of the resulting codes are calculated. The parameters depend very much on the relative position of the points and are calculated by help of the Hilbert-functions of the set of points and certain subsets thereof.

In the third section the results are used to determine conditions for our construction to yield Maximum Distance Separable (MDS) codes. If a linear code of length n has dimension k and minimum distance d, then the Singleton bound

* This research was supported by a grant from The Danish Natural Science Research Council.

asserts that

$$d \leq n - k + 1. \tag{1}$$

When equality holds in (1) the code is called Maximum Distance Separable (MDS), cf. [4]. A code is thus MDS if and only if there are no non-trivial codewords with k or more zero entries.

The MDS-conjecture states that for $1 < k < q$ the maximal length of an MDS code over $\mathbf{F}_q$ of dimension k is

$$\begin{cases} q+2, & \text{for } k = 3 \text{ or } k = q-1,\ q \text{ even;} \\ q+1, & \text{otherwise.} \end{cases}$$

The conjecture is known to be true for $q \leq 11$ or $k \leq 5$.

However if the points $\{P_1, P_2, \ldots, P_n\} \subset \mathbf{P}^l$ are in uniform position the construction yields several MDS codes. The MDS-conjecture and the above results thus makes it very interesting to construct more than $1 + q$ $\mathbf{F}_q$-points in uniform position in $\mathbf{P}^l$.

Construction and Parameters

Before defining the codes we fix our notation.

Notation 1. Let $R = \mathbf{F}_q[X_0, \ldots, X_l]$ be the polynomial ring with r-th graded piece $R_{(r)} = \mathbf{F}_q[X_0, \ldots, X_l]_{(r)}$ of homogenous polynomials of degree r. Let f_0 be a homogenous polynomial of degree r. Let $\mathcal{P} = \{P_1, P_2, \ldots, P_n\} \subset \mathbf{P}^l$ be a set of points such that $\mathcal{P} \cap Z(f_0) = \emptyset$. Let

$$I = \oplus_{i=\alpha}^{\infty} I_{(i)}$$

be the homogenous graded ideal of $\mathcal{P}$ with $I_{(\alpha)} \neq 0$, so that α is the lowest degree of a homogenous polynomial through $\mathcal{P}$. Let the Hilbert-function of $\mathcal{P}$ be the function given by

$$H_{\mathcal{P}}(r) = H_{R/I}(r) = \dim_{\mathbf{F}_q} R_{(r)} - \dim_{\mathbf{F}_q} I_{(r)}.$$

The codes in question are obtained by evaluating rational functions in points in a projective space.

Definition 2. The code is the image

$$C_{(r)} = \phi(\mathbf{F}_q[X_0, \ldots, X_l]_{(r)}) \subseteq \mathbf{F}_q{}^n$$

of the $\mathbf{F}_q$-linear map

$$\begin{aligned} \phi : \mathbf{F}_q[X_0, \ldots, X_l]_{(r)} &\rightarrow \mathbf{F}_q{}^n \\ f &\mapsto \left(\frac{f(P_1)}{f_0(P_1)}, \ldots, \frac{f(P_n)}{f_0(P_n)}\right) \end{aligned}$$

where $f_0 \in \mathbf{F}_q[X_o, \ldots, X_l]_{(r)}$.

In order to determine the parameters we recollect some basic facts about the Hilbert-function of a set of points in projective space.

Remark 3. There is a number $a = \operatorname{reg} R/I$ such that the Hilbert-function $H_{\mathcal{P}}(r) = H_{R/I}(r)$ satiesfies:

i) $H_{\mathcal{P}}(r) = \binom{r+l}{l}$ for $r < \alpha$
ii) $H_{\mathcal{P}}(r) < \binom{r+l}{l}$ for $r \geq \alpha$
iii) $H_{\mathcal{P}}(r) < H_{\mathcal{P}}(s)$ for $r < s < a$
iv) $H_{\mathcal{P}}(a) = H_{\mathcal{P}}(r)$ for $a \leq r$.

The functions $H : \mathbf{N} \to \mathbf{N}$ that can appear as Hilbert-functions are known [7], [8].

Proposition 4. *In the notation above we have that the code $C_{(r)}$ of length n has dimension*

$$k = H_{\mathcal{P}}(r)$$

where $H_{\mathcal{P}}(r)$ is the Hilbert-function of $\mathcal{P}$.

Proof. The claim follows immediately from the short exact sequence

$$0 \to I_{(r)} \to \mathbf{F}_q[X_o, \ldots, X_l]_{(r)} \to C_{(r)} \to 0.$$

□

Lemma 5. *Let $\mathcal{S} \subseteq \mathcal{P}$ be a set of points and let $C^{\mathcal{S}}_{(r)}$ be the codewords that have zeros in the positions corresponding to the points of $\mathcal{S}$, that is*

$$C^{\mathcal{S}}_{(r)} = \{\phi(f) \mid f(P) = 0 \quad for \quad P \in \mathcal{S}\}.$$

The dimension of $C^{\mathcal{S}}_{(r)}$ is $H_{\mathcal{P}}(r) - H_{\mathcal{S}}(r)$.

Proof. Let $J_{(r)}$ denote the homogenous polynomials of degree r through the points of $\mathcal{S}$. Then ϕ induces an isomorphism between $J_{(r)}/I_{(r)}$ and $C^{\mathcal{S}}_{(r)}$ such that

$$\dim\ C^{\mathcal{S}}_{(r)} = \dim\ J_{(r)} - \dim\ I_{(r)} = H_{\mathcal{P}}(r) - H_{\mathcal{S}}(r)$$

□

The following proposition makes it easy to estimate the minimum distance d of the code, that is the minimal number of non-zero entries in a non-zero codeword $C_{(r)}$.

Proposition 6. *There are no non-zero codewords in the code $C_{(r)}$ with at least s zero entries if and only if*

$$H_{\mathcal{P}}(r) = H_{\mathcal{S}}(r)$$

for all $\mathcal{S} \subseteq \mathcal{P}$ with $|\mathcal{S}| = s$.

Proof. Follows from Lemma 5. □

Geometric Goppa Codes

In [2] V. D. Goppa devised a method for obtaining error-correcting codes from points and divisors on algebraic curves.

Remark 7. Let X be an algebraic curve of genus g defined over the finite field $\mathbf{F}_q$. Let $P_1, P_2, \ldots, P_n$ be $\mathbf{F}_q$-rational points on X and let D be an $\mathbf{F}_q$-rational divisor on X. Assume that D is *very ample*, e.g. deg $D \geq 2g+1$, such that the complete linear system $|D|$ gives rise to a closed immersion of X into projective space.

$$|D| : X \to \mathbf{P}^{l(D)-1}$$
$$P \mapsto (b_0(P), \ldots, b_{l(D)}(P))$$

where $b_0, \ldots, b_{l(D)}$ is a basis for $L(D)$. Then the image of X in $\mathbf{P}^{l(D)-1}$ has degree $d = \deg D$. Denoting the images of P_i by P_i we see that the code of Definition 2

$$C_{(1)} = \phi(\mathbf{F}_q[X_0, \ldots, X_l]_{(1)}) \subseteq \mathbf{F}_q{}^n$$

obtained as the image of the $\mathbf{F}_q$-linear map

$$\phi : \mathbf{F}_q[X_0, \ldots, X_l]_{(1)} \to \mathbf{F}_q{}^n$$
$$f \mapsto \left(\frac{f(P_1)}{f_0(P_1)}, \ldots, \frac{f(P_n)}{f_0(P_n)}\right)$$

where f_0 is a linear form chosen such that $P_i \notin Z(f_0)$, is the Goppa code of [2] of length n, dimension $l(D)$ and minimum distance at least $n - \deg D$.

Maximum Distance Separable Codes

The relative position of the points in $\mathcal{P}$ as expressed by the Hilbert-functions of various subsets determines the parameters of the codes and in particular if it is MDS.

Theorem 8. *The code $C_{(r)}$ of length n and dimension $k = H_{\mathcal{P}}(r)$ is Maximum Distance Separable if and only if there are no non-trivial codewords with k or more zero entries. That is, if and only if*

$$H_{\mathcal{P}}(r) = H_{\mathcal{S}}(r)$$

for all $\mathcal{S} \subseteq \mathcal{P}$ with k elements.

Proof. Follows from Proposition 6. □

In [1] the concept of Cayley-Bacharach schemes is studied. Such configurations of points give some trivial MDS-codes by evaluating polynomials of large degree.

Definition 9. The points $\mathcal{P} = \{P_1, P_2, \dots, P_n\} \subset \mathbf{P}^l$ form a Cayley-Bacharach scheme if the Hilbert-functions

$$H_{\mathcal{P}-\{\mathcal{P}\}}$$

are independent of $P \in \mathcal{P}$.

Corollary 10. *Let $\mathcal{P} = \{P_1, P_2, \dots, P_n\} \subset \mathbf{P}^l$ be a Cayley-Bacharach scheme. Let a be the regularity of $\mathcal{P}$. Assume that $H_{\mathcal{P}}(a-1) = n-1$. Then $C_{(a-1)}$ is an* MDS *code of length n and dimension $n-1$.*

Proof. According to [1, 2.7] we have that

$$H_{\mathcal{P}-\{\mathcal{P}\}}(a-1) = H_{\mathcal{P}}(a-1)$$

and the claim then follows from Theorem 8. □

In [1] various methods for obtaining Cayley-Bacharach schemes are discussed. The points of affine space provide a concrete example of a Cayley-Bacharach scheme producing trivial MDS-codes.

Example 11 - A trivial MDS code. Let $\mathcal{P} = \{P_1, P_2, \dots, P_8\} = \mathbf{A}^3(\mathbf{F}_2) \subset \mathbf{P}^3(\mathbf{F}_2)$. We have that $\mathcal{P}$ is a Cayley-Bacharach scheme and that $C_{(2)}$ is an MDS code of length 8 and dimension 7. The generator-matrix is

$$\begin{pmatrix} 1 & 1 & 1 & 1 & 1 & 1 & 1 & 1 \\ 0 & 0 & 0 & 0 & 1 & 1 & 1 & 1 \\ 0 & 0 & 1 & 1 & 0 & 0 & 1 & 1 \\ 0 & 1 & 0 & 1 & 0 & 1 & 0 & 1 \\ 0 & 0 & 0 & 0 & 0 & 0 & 1 & 1 \\ 0 & 0 & 0 & 0 & 0 & 1 & 0 & 1 \\ 0 & 0 & 0 & 1 & 0 & 0 & 0 & 1 \end{pmatrix}$$

The Hilbert-function of $\mathcal{P}$ is

$$H_{\mathcal{P}}(r) = \begin{cases} 1, & \text{for } r = 0; \\ 4, & \text{for } r = 1; \\ 7, & \text{for } r = 2; \\ 8, & \text{for } r \geq 3; \end{cases}$$

such that the regularity of $\mathcal{P}$ is 3. Let $P \in \mathcal{P}$ be an arbitrary point. The Hilbert-function of $\mathcal{P} - \{\mathcal{P}\}$ is

$$H_{\mathcal{P}-\{\mathcal{P}\}}(r) = \begin{cases} 1, & \text{for } r = 0; \\ 4, & \text{for } r = 1; \\ 7, & \text{for } r \geq 2; \end{cases}$$

and the claims follow from Corollary 10.

As for calculating the generator-matrix, we use $f_0 = x_0^2$ and let subscripts indicate affine coordinates of the points in $\mathcal{P}$:

	P_{1000}	P_{1001}	P_{1010}	P_{1011}	P_{1100}	P_{1101}	P_{1110}	P_{1111}
$\frac{x_0^2}{x_0^2}$	1	1	1	1	1	1	1	1
$\frac{x_0x_1}{x_0^2}$	0	0	0	0	1	1	1	1
$\frac{x_0x_2}{x_0^2}$	0	0	1	1	0	0	1	1
$\frac{x_0x_3}{x_0^2}$	0	1	0	1	0	1	0	1
$\frac{x_1x_2}{x_0^2}$	0	0	0	0	0	0	1	1
$\frac{x_1x_3}{x_0^2}$	0	0	0	0	0	1	0	1
$\frac{x_2x_3}{x_0^2}$	0	0	0	1	0	0	0	1

The configuration of $\mathbf{F}_q$-rational points on a normal rational curve gives Generalized Reed-Solomon codes by evaluating polynomials of degree 1.

Example 12 - Generalized Reed-Solomon codes. Let

$$\psi : \mathbf{P}^1 \to \mathbf{P}^d$$
$$(x : y) \mapsto (x^d y^0 : \ldots : x^i y^{d-i} : \ldots : x^0 y^d)$$

be the d-uple embedding of $\mathbf{P}^1$ in $\mathbf{P}^d$. The image $\psi(\mathbf{P}^1)$ is not contained in any hyperplane of $\mathbf{P}^d$ and has degree d. Consider the $\mathbf{F}_q$-rational points of $\psi(\mathbf{P}^1)$, that is the subset $\mathcal{P} \subset \psi(\mathbf{P}^1) \subset \mathbf{P}^d$, $d \leq q$, where

$$\mathcal{P} = \{(x^d y^o : \ldots : x^i y^{d-i} : \ldots : x^o y^d) | \quad (x : y) \in \mathbf{P}^1(\mathbf{F}_q)\}.$$

The $1 + q$ points of $\mathcal{P}$ are not contained in any hyperplane of $\mathbf{P}^d$ as $\psi(\mathbf{P}^1)$ is of degree $d < 1 + q$, hence

$$H_{\mathcal{P}}(1) = 1 + d.$$

Let $\mathcal{S} \subseteq \mathcal{P}$ with $|\mathcal{S}| = 1 + d$. Again as $\psi(\mathbf{P}^1)$ has degree d in $\mathbf{P}^d$ the set $\mathcal{S}$ is not contained in a hyperplane ($\mathcal{P}$ is an $(1 + q)$-arc, see [6]) so that

$$H_{\mathcal{S}}(1) = 1 + d.$$

According to Theorem 8, the code $C_{(1)}$ of length $1 + q$ and dimension $1 + d$ is Maximum Distance Separable. This is the Generalized Reed-Solomon code.

Following [3] the points in $\mathcal{P}$ are said to be in *uniform position* if subsets of points of the same size have the same Hilbert-function. Uniform point configurations give MDS codes by evaluating polynomials of all degrees.

Definition 13. The points $\mathcal{P} = \{P_1, P_2, \ldots, P_n\} \subset \mathbf{P}^l$are in *uniform position* if subsets of the same size have the same Hilbert-functions

$$|S_1| = |S_2| \Rightarrow H_{S_1} = H_{S_2}.$$

Corollary 14. *Let $\mathcal{P} = \{P_1, P_2, \ldots, P_n\} \subset \mathbf{P}^l$ be points in uniform position. Let a be the regularity of $\mathcal{P}$ and let $r < a$. Then $C_{(r)}$ is an MDS code of length n and dimension $H_{\mathcal{P}}(r)$.*

Proof. Let S be a subset of size $H_{\mathcal{P}}(r)$. We have that

$$H_S(r) = H_{\mathcal{P}}(r)$$

and the claim then follows from Theorem 8. □

Rathmann [5] etablished that the points of generic hypersurface intersections with a curve are in uniform position over algebraically closed fields of characteristic p. Further work is needed for explicit construction of $\mathbf{F}_q$-rational points in uniform position.

References

[1] A.V. Geramita, M. Kreuzer and L. Robbiano., Cayley-Bacharach Schemes and their Canonical Modules. Preprint N. 159, Univ. di Genova (1991).

[2] V.D. Goppa, Codes on algebraic curves. Soviet Math. Dokl. 24 (1981), 170–172.

[3] J. Harris, The genus of space curves. Math. Ann. 249 (1980), 191–204.

[4] F.J. MacWilliams and N.J.A. Sloane, The Theory of Error-Correcting Codes. North-Holland Publishing Company, Amsterdam, New York, Oxford 1977.

[5] J. Rathmann, The uniform position principle for curves in characteristic p. Math. Ann. 276 (1987) 565–579.

[6] B. Segre, Lectures on Modern Geometry. Edizioni Cremonese, Roma 1961.

[7] R. Stanley, Hilbert functions of graded algebras. Adv. in Math. 28 (1978), 57-83.

[8] R. Stanley, On the Hilbert function of a graded Cohen-Macaulay domain. Preprint, Massachusetts Inst. of Technology, Cambridge 1990

Points in Good Position in $\mathbf{P}^2$

*Brian Harbourne**

Abstract. Motivated by the problem of determining Hilbert functions of 0-cycles of generic fat points in $\mathbf{P}^2$, the notions of similarity and of points being in good position are defined; it is shown that similar points in good position have the same Hilbert function. It is conjectured that generic sets of points are in good position; the question is raised of what structure (if any) the space of all points in good position may exhibit. Partial answers are obtained for sets of 8 or fewer points; in particular, it is shown that any set of 8 or fewer distinct points are in good position. Throughout this paper the ground field is assumed to be an algebraically closed field of arbitrary characteristic.

1991 Mathematics Subject Classification: 14J26, 14F17, 14D05, 14C20.

Keywords and phrases: Hilbert functions, fat points, blow up, similarity, good position, effectivity, numerical effectivity.

1. Introduction

Let $p_1, \ldots, p_n$ be points of $\mathbf{P}^2$, possibly infinitely near. To be more precise, suppose $p_1 \in \mathbf{P}^2 = P_0$, $p_2 \in P_1$, $p_3 \in P_2$, ..., $p_n \in P_{n-1}$, $P_n = P$, where P_i is the surface obtained by blowing up p_i on P_{i-1}; then we will say that the points $p_1, \ldots, p_n$ are *essentially distinct.* Thus the notion of points being essentially distinct subsumes the usual notion of a set of distinct points, but also allows for some of the points to be infinitely near. Moreover, the space of all sets of n essentially distinct points of $\mathbf{P}^2$, as discussed in more detail in section 3, is projective and universal [**H4**], and so provides a natural compactification of the space of n distinct points.

A 0-*cycle* of essentially distinct points $p_1, \ldots, p_n$ of $\mathbf{P}^2$ is a formal sum $m_1p_1 + \cdots + m_np_n$ with integer coefficients $m_1, \ldots, m_n$. It is called, respectively, *positive, nonnegative, etc.,* if the coefficients m_i are. For later use we also now define the *support* of a 0-cycle: for essentially distinct points $p_1, \ldots, p_n$, the support of the

* We wish to thank Tony Geramita for inviting us to write [**H3**], a forerunner to the present paper, the National Science Foundation for partially supporting our work on this paper, and the organizers of the June 1992 Ravello conference on 0-dimensional schemes for the opportunity to present these results.

0-cycle $m_1p_1+\cdots+m_np_n$ is the collection of points p_i, $1 \leq i \leq n$, for which either $m_i \neq 0$, or, for some $j > i$, $m_j \neq 0$ and p_j is infinitely near to p_i.

This paper, which extends and subsumes **[H3]**, is motivated by trying to understand conditions which determine Hilbert functions of 0-cycles, and when these conditions hold. Recall that the *Hilbert function* h_Z of a nonnegative 0-cycle $Z = m_1p_1 + \cdots + m_np_n$ is defined for positive integers d so that $\delta - h(d)$ is the dimension of the vector space of forms (in the homogeneous coordinate ring $k[x_1, x_2, x_3]$ of $\mathbf{P}^2$) of degree d vanishing at each point p_i with multiplicity at least m_i, $1 \leq i \leq n$, where $\delta = (d^2 + 3d + 2)/2$ is the dimension of the vector space of all forms of degree d.

Let $P \to \mathbf{P}^2$ be the morphism obtained by blowing up the points $p_1, \ldots, p_n$ consecutively. Let $e_1, \ldots, e_n$ be the divisor classes of the scheme theoretic inverse images of the points $p_1, \ldots, p_n$, and let e_0 be the class of the total transform of a line in $\mathbf{P}^2$. Then we recall that $h_Z(d) = (d^2 + 3d + 2)/2 - h^0(P, F_d)$, where $F_d = de_0 - m_1e_1 - \cdots - m_ne_n$. I.e., computing Hilbert functions of 0-cycles of essentially distinct points of $\mathbf{P}^2$ is equivalent to computing dimensions of the spaces of sections of divisor classes on the surface P obtained by blowing up the points. Thus studying geometrical properties of P can be of value in understanding the behavior of the Hilbert function h_Z.

We begin in section 2 of this paper by describing geometrical conditions (similarity and good position) on P that determine the Hilbert function h_Z. In section 3 we adduce conjectures (equivalent to conjectures put forward previously in **[H2]**, **[Hi]** and **[Gi]**) that would allow the determination of the Hilbert function of any nonnegative 0-cycle with support at generic points of $\mathbf{P}^2$. One of these conjectures, that any n generic points of $\mathbf{P}^2$ should be in good position, suggests studying the space of all sets of n essentially distinct points in good position. We do this in section 4, obtaining, among other results, that any 8 distinct points are in good position, and any 6 essentially distinct points are in good position. We conjecture that any 8 essentially distinct points of $\mathbf{P}^2$ are also in good position.

2. Good Position and Similarity

Let $p_1, \ldots, p_n$ be essentially distinct points of $\mathbf{P}^2$. By blowing up the points in order, as described above, we obtain a smooth, projective rational surface P. We say that $p_1, \ldots, p_n$ are in *good position* if $h^1(P, \mathcal{O}_P(D)) = 0$ for every effective, numerically effective divisor D on P. (Recall that D is said to be *effective* if it is a nonnegative integer linear combination of prime divisors; a divisor D is said to be *numerically effective* if $D \cdot C \geq 0$ whenever C is an effective divisor. A divisor class will be said to be *effective, numerically effective, etc.*, if it has a representative which is.)

Clearly, $p_1, \ldots, p_n$ being in good position is really a property of the surface P. In particular, let $p_1, \ldots, p_n$ and $q_1, \ldots, q_n$ be sets of essentially distinct points which, when blown up, give isomorphic surfaces P and Q; then $p_1, \ldots, p_n$ are in good position if and only if $q_1, \ldots, q_n$ are. The next property we wish to define is not, however, purely intrinsic to P and thus requires us to recall some standard facts.

If e_i is, as above, the class in Pic(P) of the total transform of the point p_i under the blowing up $P \to P_{i-1}$ of the points p_j, $i \leq j \leq n$, and if e_0 is the pullback to P of the class of a line in $\mathbf{P}^2$, then $e_0, \ldots, e_n$ give a basis for Pic(P), orthogonal with respect to the intersection product on Pic(P), such that $-e_0 \cdot e_0 = e_1 \cdot e_1 = \cdots = e_n \cdot e_n = -1$. With respect to this basis the canonical class K_P of P is $-3e_0 + e_1 + \cdots + e_n$. If h is the Hilbert function of a nonnegative 0-cycle $m_1p_1 + \cdots + m_np_n$, then, setting $F_d = de_0 - m_1e_1 - \cdots - m_ne_n$, we have $h(d) = (d^2+3d+2)/2 - h^0(P, F_d) = \sum_{i=1}^n (m_i^2 + m_i)/2 - h^1(P, F_d)$.

Say P is obtained, as above, by blowing up essentially distinct points $p_1, \ldots, p_n$; let $e_0, \ldots, e_n$ be the corresponding basis of Pic(P). Likewise, say Q is obtained by blowing up essentially distinct points $q_1, \ldots, q_n$, with $e'_0, \ldots, e'_n$ being the corresponding basis of Pic(Q). Then the $\mathbf{Z}$-linear homomorphism taking e_i to e'_i defines a $\mathbf{Z}$-module isomorphism π : Pic(P) $\to$ Pic(Q) taking K_P to K_Q and preserving the intersection product. We will say that $p_1, \ldots, p_n$ and $q_1, \ldots, q_n$ are *similar* given that for each $D \in$ Pic(P), D is an effective divisor class if and only if $\pi(D)$ is. (Note that this implies that D is numerically effective if and only if $\pi(D)$ is.) We now show that the Hilbert function of a nonnegative 0-cycle supported at points $p_1, \ldots, p_n$ in good position is determined by the coefficients and the similarity class of the points:

Theorem 1.1. *If $p_1, \ldots, p_n$ and $q_1, \ldots, q_n$ are similar sets of essentially distinct points of $\mathbf{P}^2$ in good position, then the Hilbert function h of any nonnegative 0-cycle $m_1p_1 + \cdots + m_np_n$ is equal to the Hilbert function h' of $m_1q_1 + \cdots + m_nq_n$.*

Proof. As above, define F_d to be $F_d = de_0 - m_1e_1 - \cdots - m_ne_n$ and define $F'_d = de'_0 - m_1e'_1 - \cdots - m_ne'_n = \pi(F_d)$. If F_d is not the class of a divisor which is effective, then by similarity neither is F'_d, hence $h(d) = (d^2+3d+2)/2 = h'(d)$. So assume that F_d and hence F'_d are the classes of effective divisors. If F_d and hence F'_d are also numerically effective, then by being in good position we have $h(d) = \sum_{i=1}^n (m_i^2 + m_i)/2 - h^1(P, F_d) = \sum_{i=1}^n (m_i^2 + m_i)/2 = \sum_{i=1}^n (m_i^2 + m_i)/2 - h^1(P, F'_d) = h'(d)$. Thus we are left with the case that F_d is the class of an effective but not numerically effective divisor. Thus there exist classes $C_1, \ldots, C_r$ of effective prime divisors of negative self-intersection such that $F(i) \cdot C_{i+1} < 0$ for $0 \leq i < r$ and such that $F(r)$ is a numerically effective and effective divisor class, where $F_d = F(0)$ and, for $i > 0$, $F(i) = F_d - C_1 - \cdots - C_i$. Thus $h^0(P, F_d) = h^0(P, F(r))$. Note $\pi(C_i) = C'_i$ is the class of a prime divisor for each i. (If not, there exist nonzero effective classes A' and B' such that $C'_i = A' + B'$. But then $C_i = A + B$, where $A = \pi^{-1}(A')$ and $B = \pi^{-1}(B')$ are nonzero effective classes. Since $0 > C_i \cdot C_i = C_i \cdot (A+B)$, either $A \cdot C_i < 0$ or $B \cdot C_i < 0$. If, say, $A \cdot C_i < 0$, then primality of C_i implies $A - C_i$ is effective. But this means $-B = (A - C_i)$ and B are both effective so $B = 0$.) Therefore, taking $F'(i) = \pi(F(i))$, we have $F'(i) \cdot C'_{i+1} < 0$ for $0 \leq i < r$ such that $F'(r)$ is numerically effective and effective. Thus $h^0(Q, F'(r)) = h^0(Q, F'_d)$.

But the hypothesis that $p_1, \ldots, p_n$ and $q_1, \ldots, q_n$ are in good position implies $h^1(P, F(r)) = 0 = h^1(Q, F'(r))$, so $h^0(P, F(r)) = (F(r)^2 - K \cdot F(r))/2 + 1 = (F'(r)^2 - K \cdot F'(r))/2 + 1 = h^0(Q, F'(r))$ follows by Riemann-Roch (and the fact that $h^2(S, H) = h^0(S, K_S - H) \leq h^0(S, K_S) = 0$ for any effective class H on any smooth complete rational surface S), whence $h(d) = h'(d)$. □

The proof of the preceding theorem makes clear that to actually compute the Hilbert function of a 0-cycle supported at points in good position, it suffices to be able to determine the prime divisors of negative self-intersection on the blowing up P of the points. Thus it is of interest to find criteria for points to be in good position, and to be able to determine the set of prime divisors of negative self-intersection on the blowing up of the points. Regarding these points we now raise some conjectures.

3. Conjectures

We begin with the following combined pair of conjectures:

Conjecture 3.1. *Any n sufficiently general points $p_1, \ldots, p_n$ in $\mathbf{P}^2$ are in good position, and the only reduced irreducible curves of negative self-intersection on the surface P obtained by blowing up $p_1, \ldots, p_n$ are the exceptional curves (i.e., smooth rational curves of self-intersection -1).*

We note that 3.1 is a coordinate free reformulation of the conjectures stated in Remark I.6.2 of [**H2**]. Although these two sets of conjectures are equivalent, the current formulation is better suited to our present purposes of a conceptual discussion, while the formulation given in [**H2**] is more useful for actually implementing an algorithm for computing the resulting conjectural Hilbert function of an arbitrary nonnegative 0-cycle supported at generic points of $\mathbf{P}^2$.

We will reformulate 3.1, using Kleiman's [**K**] inductively defined parameter spaces X_n of all sequences of n essentially distinct points, constructed as follows. We set X_0 equal to a point and X_1 equal to $\mathbf{P}^2$. Then we have the obvious morphism $\phi_1 : X_1 \to X_0$. Now assume that we have defined X_m and $\phi_m : X_m \to X_{m-1}$ for $1 \leq m \leq n$. Then we take $X_{n+1} = \mathrm{Bl}_\Delta(X_n \times_{X_{n-1}} X_n)$, where $X_n \times_{X_{n-1}} X_n$ is the fiber product of ϕ_n with itself, where Δ is the diagonal of the product (isomorphic, therefore, to X_n), and where $\mathrm{Bl}_\Delta(X_n \times_{X_{n-1}} X_n)$ is the blowing up of $X_n \times_{X_{n-1}} X_n$ along Δ. We obtain ϕ_{n+1} by composing the blowing up $X_{n+1} \to X_n \times_{X_{n-1}} X_n$ with the projection $X_n \times_{X_{n-1}} X_n \to X_n$ to the first factor.

If $x \in X_n$, and if $y = \phi_n(x)$, then y represents a sequence of $n-1$ essentially distinct points of $\mathbf{P}^2$, and $\phi_n^{-1}(y)$ is the surface obtained by blowing up these $n-1$ points. But $x \in \phi_n^{-1}(y)$, so x represents a choice of a point on $\phi_n^{-1}(y)$, thereby giving an n-th essentially distinct point. Thus X_n parametrizes sequences of n essentially distinct points of $\mathbf{P}^2$, and it is easy to see that the parametrization is bijective. In fact, this parametrization is universal and hence gives a fine moduli space [**H4**].

Let $\mathcal{G}_n$ denote the locus in X_n representing sequences of n points in good position. Likewise let $\mathcal{S}_n$ denote the locus in X_n of points x for which the corresponding surface $\phi_{n+1}^{-1}(x)$ has no prime divisors of negative self-intersection but exceptional curves. A more precise version of 3.1 can now be stated:

Conjecture 3.2. *The set $\mathcal{G}_n \cap \mathcal{S}_n$ is Zariski-dense in X_n; i.e., a sufficiently general set of n points of $\mathbf{P}^2$ will be in good position and the surface obtained by blowing these points up will have no curves of negative self-intersection but exceptional curves.*

To see that this bears on the problem of computing Hilbert functions, or, equivalently, dimensions of complete linear systems, let P be a blowing up of n points of $\mathbf{P}^2$ corresponding to a point z of $\mathcal{G}_n \cap \mathcal{S}_n$, and let $e_0, \ldots, e_n$ be the corresponding basis of $\operatorname{Pic}(P)$. Let $F = a_0e_0 + \cdots + a_ne_n$ be a divisor class on P; the problem is to compute $h^0(P, F)$. If $a_0 < 0$, then F is not effective so $h^0(P, F) = 0$ and we are done. Otherwise, using an algorithm (cf. [**H1,** §2], [**H2,** §1.6]) based on [**N**] and [**L**], one can (under the hypothesis that $z \in \mathcal{S}_n$) explicitly write F as $M + N_1 + \cdots + N_r$, where, for $1 \le i \le r$, N_i is an exceptional curve such that $N_i \cdot (M + N_1 + \cdots + N_i) < 0$, and where either M is not effective or M is effective and numerically effective. In either case $h^0(P, F) = h^0(P, M)$, with both 0 in the former case, and with the hypothesis $z \in \mathcal{G}_n$ implying that $h^1(P, M)$ vanishes in the latter case, and hence that $h^0(P, F) = h^0(P, M) = (M^2 - M \cdot K_P)/2 + 1$.

For comparison, Hirschowitz has conjectured ([**Hi**]; see [**Gi**] for a discussion of other equivalent variants) as follows. Let P be a blowing up of $\mathbf{P}^2$ at a set z of n sufficiently general points and let $D = a_0e_0 - \cdots - a_ne_n$ be a divisor class on P such that $a_0 \ge a_1 \ge \cdots \ge a_n \ge 0$. If $D \cdot E \ge -1$ for every exceptional divisor E with $E \cdot e_0 \le a_0/2$, then the conjecture is that either $h^0(P, D)$ or $h^1(P, D)$ vanishes.

Hirschowitz's conjecture is equivalent to Conjecture 3.1. Indeed, Hirschowitz points out in [**Hi**] that his conjecture implies that the only integral curves of negative self-intersection on a blowing up of $\mathbf{P}^2$ at a set z of sufficiently general points are exceptional curves; i.e., $z \in \mathcal{S}_n$. To see his conjecture also implies $z \in \mathcal{G}_n$, we recall the notion of an *exceptional configuration*. Let S be any rational surface. A basis $b_0, \ldots, b_n$ (over the integers) of $\operatorname{Pic}(S)$ is called an exceptional configuration if there exists an isomorphism $f : S \to Q$ such that $f^*(e'_i) = b_i$, $i \ge 0$, where Q is any surface obtained by blowing up n essentially distinct points $q_1, \ldots, q_n$ of $\mathbf{P}^2$, $e'_0, \ldots, e'_n$ being the corresponding basis of $\operatorname{Pic}(Q)$. (In particular, and for example, $e'_0, \ldots, e'_n$ is an exceptional configuration of Q.) The pertinent fact, which can be obtained by applying the algorithm mentioned above, is that for any numerically effective divisor class D on a blowing up P of $\mathbf{P}^2$ at sufficiently general points, there is always some exceptional configuration $e'_0, \ldots, e'_n$ such that $a'_0 \ge a'_1 \ge \cdots \ge a'_n \ge 0$, where $a'_0, \ldots, a'_n$ are the uniquely determined integers for which $D = a'_0e'_0 - a'_1e'_1 - \cdots - a'_ne'_n$. Thus the hypothesis of Hirschowitz's conjecture applies, with respect to some exceptional configuration $e'_0, \ldots, e'_n$, to any numerically effective divisor D. But numerical effectivity gives $D \cdot E > -1$

for *every* exceptional divisor E, and so Hirschowitz's conjecture implies $h^1(P,D)$ vanishes for any effective numerically effective class D. In particular, it follows that $z \in \mathcal{G}_n$.

Conversely, suppose that P is a blowing up of $\mathbf{P}^2$ at a set z of n essentially distinct points with $z \in \mathcal{G}_n \cap \mathcal{S}_n$, and that D is an effective class satisfying the hypotheses of Hirschowitz's conjecture. If E is an exceptional divisor with $D \cdot E < -1$, then E is a fixed component of D of multiplicity at least 2, and so $a_0 = D \cdot e_0 \geq 2E \cdot e_0$, contradicting a hypothesis of Hirschowitz's conjecture. I.e., Hirschowitz's hypotheses imply that $D \cdot E \geq -1$ for every exceptional divisor E. Since $z \in \mathcal{S}_n$ by Conjecture 3.1, we may write $D = M + m_1 N_1 + \cdots + m_r N_r$, where N_i, $1 \leq i \leq r$ are distinct exceptional divisors occurring as fixed components of D of multiplicities m_i, $1 \leq i \leq r$, and where M is numerically effective and effective, none of its fixed components (if any) having negative self-intersection. If $i \neq j$ but N_i and N_j meet, it is easy to see that $h^0(P, N_i + N_j) > 1$; thus $N_1, \ldots, N_r$ must all be disjoint. If we now write $D = M' + N_1' + \cdots + N_s'$, where $N_1', \ldots, N_s'$ are the exceptional divisors meeting D negatively (and thus comprise a subset of $\{N_1, \ldots, N_r\}$), we see M' must be numerically effective. Since $z \in \mathcal{G}_n$, it follows from Conjecture 3.1 that $h^1(P, M')$ vanishes. Clearly, $h^0(P, M') = h^0(P, D)$, while direct calculation shows $M'^2 - K \cdot M' = D^2 - K \cdot D$. By Riemann-Roch $h^0(P, M') - h^1(P, M') = (M'^2 - K \cdot M')/2 + 1$ and $h^0(P, D) - h^1(P, D) = (D^2 - K \cdot D)/2 + 1$, which are thus equal, so

$$h^0(P,D) = h^0(P,M') = h^0(P,M') - h^1(P,M') = h^0(P,D) - h^1(P,D),$$

which implies the conclusion $h^1(P,D) = 0$ of Hirschowitz's conjecture.

This establishes the equivalence of Hirschowitz's conjecture with Conjecture 3.1 (and thereby with those made in [**H2**]). However, Conjectures 3.1 and 3.2 suggest (and recall) other questions of interest. One, which we offer for its own interest but which we will not pursue here, is: what can one say about the curves of negative self-intersection on a surface? (This is of interest not just for rational surfaces. Indeed, given any smooth projective surface S, not necessarily a minimal model but for which no positive multiple of $-K_S$ is effective, an old folklore question is whether the set of integers occurring as self-intersections of reduced irreducible curves on S is bounded below.)

A second, more pertinent one, is: what structure (if any) can be ascribed to the space of sets of essentially distinct points in good position? That is, rather than (directly) try to show that sufficiently general points are in good position, one could simply study the space of all points in good position. It is this which will interest us for the remainder of this paper. Some limited results are known, but nothing of a global nature. For example, it follows from [**H1**] that sufficiently general sets of essentially distinct points on an irreducible plane cubic are in good position, and indeed the results of [**H1**] characterize the essentially distinct sets of points on an irreducible plane cubic which are in good position. These results also allow one to conclude that any $n \leq 9$ sufficiently general points of $\mathbf{P}^2$ are in good position. But precisely which sets of n essentially distinct points are in good position is not known. We wish to start here by considering this problem for blowings-up of $\mathbf{P}^2$

at $n \leq 8$ points. For $n \leq 8$ one obtains three technical simplifications: first, if P is the surface obtained by blowing up the n points, then $-K_P$ is effective (since there is always a plane cubic through any 8 or fewer points in the plane); second, the subspace $K^{\perp} \subset \mathrm{Pic}(P)$ (of divisor classes whose intersection with K_P is 0) is negative definite; third, numerically effective classes on P are necessarily effective. Since a smooth cubic passes through any 8 sufficiently general points, effectivity of numerically effective classes is a corollary of the following fact:

Proposition 3.3. *Suppose that P is obtained by blowing up n essentially distinct points of $\mathbf{P}^2$ which are a specialization (in the universal parametrization X_n) of n essentially distinct points of a smooth plane cubic. Then any numerically effective divisor class on P is effective.*

Proof. For points on a smooth cubic, it follows from **[H1]** that every numerically effective class is effective. But the set of effective classes enlarges under specialization, so the set of numerically effective classes contracts under specialization. Hence numerically effective classes remain effective under specialization. □

Our results, which we present in section 4 and which use these three technical simplifications heavily, suggest the following global conjecture regarding $n \leq 8$ points in good position:

Conjecture 3.4. *Any* 8 *or fewer essentially distinct points of* $\mathbf{P}^2$ *are in good position.*

This conjecture implies, for a nonnegative 0-cycle Z with support at 8 or fewer essentially distinct points, that the extent to which the conditions imposed by Z on homogeneous forms of degree d fail to be independent is completely accounted for by fixed components in the linear system of forms of degree d vanishing on Z.

For example, let $Z = 2p_1 + 2p_2$. The subspace of conics having a given double point has codimension 3 among the 6 dimensional space of all plane conics, so $2p_1$ and $2p_2$ both impose 3 conditions on homogeneous forms of degree 2. Since there is indeed a conic passing through two points twice, we see that the total of 3+3=6 conditions imposed by Z cannot be independent. This lack of independence is explained by looking at the divisor class $F = 2e_0 - 2e_1 - 2e_2$ on the blowing up P of $\mathbf{P}^2$ at p_1 and p_2. We see that the proper transform E to P of the line through the two points is a prime divisor of negative self-intersection which occurs as a fixed component in $|F|$ with multiplicity 2; removing these fixed components gives $F - 2E = 0$, which corresponds to considering forms of degree 0 vanishing on the zero 0-cycle. Of course, forms of degree 0 are constants, of which there is 1 dimension, and the zero 0-cycle imposes 0 conditions, from which we obtain the correct number, $6 - (1 - 0) = 5$, for the number of conditions imposed by Z on forms of degree 2.

Thus the conjecture says that as long as Z has support at no more than 8 points, nothing more subtle than this can occur, and thus, to compute the Hilbert function of such a Z, one merely needs to know the set of prime divisors of negative self-intersection on the surface obtained by blowing up the points at which Z has support. This is no longer true for $n = 9$; for example, the Hilbert function for $Z = p_1 + \cdots + p_9$ differs depending on whether the points $p_1, \ldots, p_9$ comprise the base points of a pencil of cubics.

In the next section we will prove two special cases of Conjecture 3.4; namely, that any 8 or fewer *distinct* points of $\mathbf{P}^2$ are in good position, and that any 6 or fewer *essentially distinct* points of $\mathbf{P}^2$ are in good position.

4. Points in Good Position

Our consideration of Conjecture 3.4 relies on a number of results, for which we use the notation $K_P^\perp$ (or simply $K^\perp$ when P is understood) for the space of divisor classes orthogonal (with respect to the intersection form on Pic(P)) to the canonical class K_P of the surface P.

Lemma 4.1. *Let P be a smooth projective rational surface such that $-K_P$ is effective and $K^\perp$ is negative definite. If C is a reduced irreducible divisor on P with $C^2 \geq 0$, then $h^1(P, \mathcal{O}_P(rC)) = 0$ for every integer $r \geq 0$.*

Proof. First we note that C is numerically effective; in particular, $-K_P \cdot C \geq 0$. If $-K_P \cdot C = 0$, then $C \in K^\perp$, and since $C^2 \geq 0$ on the one hand while $K^\perp$ is negative definite on the other, we see that we then must have $C = 0$. But P is rational, hence $0 = h^1(P, \mathcal{O}_P) = h^1(P, \mathcal{O}_P(C))$, as desired.

So suppose that $-K_P \cdot C > 0$. Then $rC \cdot C = (r-1)C^2 + C^2 \geq \deg(K_C) - K_P \cdot C > \deg(K_C)$, so we see that $h^1(C, \mathcal{O}_C(rC)) = 0$ for $r > 0$. Now, from $h^1(P, \mathcal{O}_P) = 0$ and

$$0 \to \mathcal{O}_P((r-1)C) \to \mathcal{O}_P(rC) \to \mathcal{O}_C(rC) \to 0$$

we see by taking cohomology and by induction on r that $h^1(P, \mathcal{O}_P(rC)) = 0$. □

Theorem 4.2. *Let P be a smooth projective rational surface such that $-K_P$ is effective and $K^\perp$ is negative definite. If F is an effective divisor class on P such that the complete linear system $|F|$ has no fixed components, then $h^1(P, F) = 0$.*

Proof. Since F has no fixed components, we can write $F = r_1C_1 + \cdots + r_tC_t$ for some positive integers r_i and some distinct reduced irreducible classes C_i with $C_i^2 \geq 0$. (This follows, for example, from an analysis of the proof of Bertini's Theorem [**Z**, Theorem I.6.3].) By 4.1, each summand r_iC_i is regular (i.e., has $h^1(P, \mathcal{O}_P(r_iC_i)) = 0$). Thus it would suffice to show that a sum of effective regular divisors with no common components is regular. This is the conclusion of the next lemma. □

Lemma 4.3. *Let C and D be effective regular divisors on a smooth rational surface P and suppose that C and D have no common components. Then $C + D$ is regular.*

Proof. We have the following commutative diagram with exact rows and columns:

$$\begin{array}{ccccccccc}
 & & 0 & & 0 & & 0 & & \\
 & & \downarrow & & \downarrow & & \downarrow & & \\
0 & \rightarrow & \mathcal{O}_P & \rightarrow & \mathcal{O}_P(C) & \rightarrow & \mathcal{O}_C(C) & \rightarrow & 0 \\
 & & \downarrow & & \downarrow & & \downarrow & & \\
0 & \rightarrow & \mathcal{O}_P(D) & \rightarrow & \mathcal{O}_P(C+D) & \rightarrow & \mathcal{O}_C(C+D) & \rightarrow & 0 \\
 & & \downarrow & & \downarrow & & \downarrow & & \\
0 & \rightarrow & \mathcal{O}_D(D) & \rightarrow & \mathcal{O}_D(C+D) & \rightarrow & \mathcal{O}_{C\cdot D} & \rightarrow & 0 \\
 & & \downarrow & & \downarrow & & \downarrow & & \\
 & & 0 & & 0 & & 0 & &
\end{array} \qquad (*)$$

where $\mathcal{O}_{C\cdot D}$ denotes the structure sheaf of the 0-dimensional scheme $C \cdot D$ which is the complete intersection of C and D.

Now $h^2(P, \mathcal{O}_P) = 0$ since P is rational, so by taking cohomology of the left column of (*) and using regularity of D, we conclude $h^1(D, \mathcal{O}_D(D)) = 0$. Thus $\mathcal{O}_D(D)$ is regular, so, since $C \cdot D$ has dimension zero, taking cohomology of the bottom row we find that $\mathcal{O}_D(C+D)$ is regular. Now taking cohomology of the middle column, using regularity of C and $\mathcal{O}_D(C+D)$, we conclude that $C+D$ is regular. □

We now make a definition. Let F and G be effective divisor classes, and let $N_1, \dots, N_r$ be the classes of the prime divisors which occur as fixed components of either F or G. Thus we can write $F = C + n_1N_1 + \cdots + n_rN_r$ and $G = D + m_1N_1 + \cdots + m_rN_r$, where $n_1, \dots, n_r$ and $m_1, \dots, m_r$ are nonnegative integers and C and D are fixed component free and move in linear systems of the same dimensions as F and G respectively. We say that F *dominates* G if $n_i^2 \geq n_im_i$ for every i; i.e., if whenever a prime divisor is a fixed component of $|G|$, of multiplicity t, then either it is not a fixed component of $|F|$ or it is a fixed component of $|F|$ of multiplicity at least t. Also, for each i, let $\mu_i = m_i - n_i$ if $m_i > n_i > 0$, and let $\mu_i = 0$ otherwise; now set $\mu(G/F) = \sum_{i=1}^r \mu_iN_i$. Then we say that F *nearly dominates* G if $2n_i^2 \geq n_im_i$ for every i, and if $\mu(G/F)^2 + K_P \cdot \mu(G/F) \leq 0$. (Note that $\mu(G/F) = 0$ if and only if F dominates G. In particular, if F dominates G, then F nearly dominates G.) Using these definitions we now state the key lemma:

Lemma 4.4. *Let F be a numerically effective and effective divisor class on a smooth projective rational surface P such that $h^0(P, -K_P) > 1$ and $K^\perp$ is negative definite. Then F is regular and $\mu(-K_P/F)^2 + K_P \cdot \mu(-K_P/F) = 0$ if F nearly dominates $-K_P$. Moreover, F is also fixed component free if F dominates $-K_P$.*

Proof. Write $F = C + N$, where N is the sum, counted with multiplicities, of the fixed components of F, and so C is effective and fixed component free. By Theorem 4.2, C is regular and so we have

$$1 + (F^2 - F \cdot K_P)/2 \leq h^0(P, F) = h^0(P, C) = 1 + (C^2 - C \cdot K_P)/2. \qquad (**)$$

Replacing F by $C+N$ and simplifying, we obtain $0 \geq (C+N) \cdot N + (C - K_P) \cdot N$. Since F is numerically effective, we see $0 \leq F \cdot N = (C+N) \cdot N$. Thus $0 \geq$

$(C - K_P) \cdot N$. Therefore, showing $(C - K_P) \cdot N \geq 0$ shows that (**) is an equality and hence that F is regular. From this we would also obtain $F \cdot N = 0$. Assuming F nearly dominates $-K_P$, we now show that $(C - K_P) \cdot N \geq 0$.

Let $\{N_1, \cdots, N_r\}$ be the set of classes of prime divisors occurring as fixed components of either $-K_P$ or F. Thus there are nonnegative integers m_i and n_i, $i = 1, \cdots, r$, such that $-K_P = D + \sum_{i=1}^r m_i N_i$ and $N = \sum_{i=1}^r n_i N_i$, where $h^0(P, -K_P) = h^0(P, D)$. Let $A = D + \sum_{n_i=0} m_i N_i$ and $N' = \sum_{n_i>0} m_i N_i$; then: $-K_P = A + N'$, A and F have no fixed components in common, and every irreducible component of N' is a fixed component of F. We can also write $N' = M' + T'$, where $M' = \sum_{n_i \geq m_i} m_i N_i$ and $T' = \sum_{m_i > n_i > 0} m_i N_i$. Similarly, we write $N = M + T$, where $M = \sum_{n_i \geq m_i} n_i N_i$ and $T = \sum_{m_i > n_i} n_i N_i$. Finally, let $T'' = T' - T$; i.e., $T'' = \mu(-K_P/F)$. Note that the hypothesis that F nearly dominates $-K_P$ means that $K_P \cdot T'' + (T'')^2 \leq 0$ and that $T - T''$ is effective. Substitution into $(C - K_P) \cdot N$ gives

$$(C - K_P) \cdot N = (C + A + M' + T') \cdot (M + T) =$$

$$A \cdot M + F \cdot (M' + T) + C \cdot (M - M') + A \cdot T + T'' \cdot (M + T).$$

Now $A \cdot M \geq 0$ because A and F have no fixed components in common, and $F \cdot (M' + T) \geq 0$ and $C \cdot (M - M') \geq 0$ since F and C are numerically effective. But $A \cdot T + T'' \cdot (M + T)$ is just $A \cdot (T - T'') + T'' \cdot (M - M') - (K_P \cdot T'' + (T'')^2)$. Since each of these terms is nonnegative, we see $0 \leq (C - K_P) \cdot N$. Thus, as observed above, F is regular and $F \cdot N = 0$. It also follows that $A \cdot M = 0$, $C \cdot (M - M') = 0$ and $K_P \cdot T'' + (T'')^2 = 0$.

Now suppose that F actually dominates $-K_P$. Then $T = T' = 0$, so $N = M$, $N' = M'$ and $-K_P = A + M'$. From $A \cdot M = 0$, from A and F having no components in common, and from $M - M'$ being effective, we see that $A \cdot M' = 0$. But the facts $A \cdot M' = 0$, $-K_P = A + M'$, and A and M' having no components in common imply either that $-K_P$ is not connected or $M' = 0$ or $A = 0$. Since $h^0(P, -K_P) > 1$, we see $A \neq 0$. Moreover, every section of $-K_P$ is connected. (Indeed, let L be a section of $-K_P$. Then $0 = h^1(P, \mathcal{O}_P) = h^1(P, K_P)$ by regularity of P and duality, so $1 = h^0(P, \mathcal{O}_P) = h^0(L, \mathcal{O}_L)$ follows from $0 \to \mathcal{O}_P(K_P) \to \mathcal{O}_P \to \mathcal{O}_L \to 0$.) Hence $M' = 0$, so $-K_P \cdot N = A \cdot M = 0$; i.e., N lies in $K^\perp$. But $0 = C \cdot (M - M') = C \cdot N$ and $(C + N) \cdot N = F \cdot N = 0$ imply $N^2 = 0$, so negative definiteness of $K^\perp$ gives $N = 0$. □

Lemma 4.5. *Let P be a blowing up of $\mathbf{P}^2$ at some $n \leq 8$ essentially distinct points $p_1, \ldots, p_n$ such that $-K_P$ has no fixed components of multiplicity 3 or more. Then any effective class on P nearly dominates $-K_P$, and any numerically effective class on P dominates $-K_P$.*

Proof. Recall that numerically effective classes are also effective, by Proposition 3.3. First suppose F is effective. Let $T = \mu(-K_P/F)$, defined above. Since we assume all components of $-K_P$ have multiplicity 2 or less, it follows from the definitions that T is reduced and that to show F nearly dominates $-K_P$, it suffices to show either $T = 0$ or $T^2 + T \cdot K_P < 0$. Moreover, in case F is also numerically effective, $T^2 + T \cdot K_P < 0$ contradicts Lemma 4.4, so we see $T = 0$, whence F dominates $-K_P$. Thus the result is established if we show that $T \neq 0$ implies $T^2 + T \cdot K_P < 0$.

Suppose that $T \neq 0$ but $e_0 \cdot T = 0$; i.e., the components of T are all components of the exceptional divisors $e_1, \ldots, e_n$. Then each connected component of T has no loops and the irreducible components of T meet transversely, no three at a point (because $e_1, \ldots, e_n$ have these properties).

In this situation we claim that $T^2 + T \cdot K_P = -2c$, where c is the number of connected components of T. To see this, let C be a connected component of T; we must show that $C^2 + C \cdot K_P = -2$. If C is irreducible, this follows since C is smooth and rational. Otherwise, since there are no loops and no three components meet at a point, there is a component of C which meets precisely one other component of C; write $C = C' + E$, where E is a component of C meeting only one other component. By induction on the number of components we may assume that $(C')^2 + C' \cdot K_P = -2$ and so we have

$$C^2 + C \cdot K_P = ((C')^2 + C' \cdot K_P) + (E^2 + E \cdot K_P) + 2E \cdot C' = (-2) + (-2) + 2.$$

This proves the claim in case $e_0 \cdot T = 0$.

Now suppose that $e_0 \cdot T \neq 0$; since e_0 is numerically effective we have $e_0 \cdot T > 0$. Since $-K_P - 2T$ is effective, we have $e_0 \cdot (-K_P - 2T) = 3 - 2(e_0 \cdot T) \geq 0$, so $e_0 \cdot T = 1$. Thus T is the union of the proper transform L of a line, possibly together with some components of $e_1, \ldots, e_n$. But again each connected component of T has no cycles and the irreducible components of T meet transversely, no three at a point. As above we have $T^2 + T \cdot K_P = -2c$. □

From 4.5 and 4.4 we obtain:

Corollary 4.6. *Let P be a blowing up of $\mathbf{P}^2$ at $n \leq 8$ essentially distinct points $p_1, \ldots, p_n$. If $-K_P$ has no fixed components of multiplicity 3 or more, then the points $p_1, \ldots, p_n$ are in good position and every numerically effective divisor on P is free of fixed components.*

Using Corollary 4.6, we prove Conjecture 3.4 in the case of distinct points.

Theorem 4.7. *Any $n \leq 8$ distinct points of $\mathbf{P}^2$ are in good position.*

Proof. To establish this result it will suffice by Corollary 4.6 to show that $-K_P$ has no fixed components of multiplicity three or more, where P is the blowing up of $\mathbf{P}^2$ at the n points. (What is more, it will then also follow that any numerically effective class on P is fixed component free.)

Suppose that D is a fixed component of $-K_P$ of multiplicity 3 or more. Since e_0 is numerically effective we have $D \cdot e_0 \geq 0$, and since $-K_P - 3D$ is effective by hypothesis and moves in a linear system (because $n \leq 8$ implies $h^0(P, -K_P) \geq 2$), we have $0 < (-K_P - 3D) \cdot e_0 = 3 - 3(D \cdot e_0)$. Thus $D \cdot e_0 = 0$. But since distinct points were blown up to obtain P, the only irreducible divisors not meeting e_0 are $e_1, \ldots, e_n$. Since $-K_P = 3e_0 - e_1 - \cdots - e_n$, the only way for a section of $-K_P$ to have, for some $i > 0$, e_i as a component of multiplicity r is if e_i is the blowing up of a point of multiplicity $r + 1$ on a plane cubic. Since plane cubics have multiple points of multiplicity at most 3, we see that no section of $-K_P$ has e_i occurring as a component of multiplicity more than 2. □

The proof of Theorem 4.7 depends on showing that on the blowing up P of any $n \leq 8$ distinct points of $\mathbf{P}^2$, $-K_P$ has a section with no components having multiplicity more than 2. In fact, with somewhat more effort, one can show that $-K_P$ has reduced sections in this situation. Moreover, the proof of 4.7 (via 4.6) shows that numerically effective classes on P are fixed component free, but again with somewhat more work more can be said:

Proposition 4.8. *Let P be a blowing up of $n \leq 7$ distinct points $p_1, \ldots, p_n$ of $\mathbf{P}^2$. Then every numerically effective class is base point free.*

Proof. By 4.7 and its proof, any numerically effective class on P is regular and fixed component free. Thus 4.8 is a consequence of the following lemma. □

Lemma 4.9. *Let P be a blowing up of $\mathbf{P}^2$ at $n \leq 7$ essentially distinct points. Then any effective and fixed component free class is base point free.*

Proof. Let F be effective and fixed component free; then F is numerically effective and dominates $-K_P$, hence is regular by 4.4. Let $p \in P$, let r be the multiplicity of a general section of F at p, and let P' be the blowing up of P at p, with e the class of the blowing up of p. Then $F - re$ is fixed component free, hence regular by 4.4. But the spaces of global sections of F and $F - re$ have the same dimension and F is regular, so $F - re$ being regular implies $r = 0$. I.e., p is not a base point of F. □

To obtain further results we will consider sets of essentially distinct points up to an equivalence determined by the fixed components of the anticanonical classes of the surfaces obtained by blowing up the points. (More precisely, let P and P' be the blowings up of $\mathbf{P}^2$ at essentially distinct sets of points $p_1, \ldots, p_n$ and $p'_1, \ldots, p'_n$. Let $\pi : \mathrm{Pic}(P) \to \mathrm{Pic}(P')$ be the canonical homomorphism defined by sending e_i to e'_i for $i = 0, \ldots, n$. Suppose D is a fixed component of $-K_P$ if and only if $\pi(D)$ is a fixed component of $-K_{P'}$; then we say $p_1, \ldots, p_n$ and $p'_1, \ldots, p'_n$ are *fixed component equivalent*.)

Since $n \leq 8$, a fixed component of $-K_P$ is either the proper transform of a line (and thus has the form $e_0 - e_{i_1} - \cdots - e_{i_r}$ for some $0 < i_1 < \cdots < i_r \leq n$), or a conic (and thus has the form $2e_0 - e_{j_1} - \cdots - e_{j_s}$ for some $0 < j_1 < \cdots < j_s \leq n$), or is a component of one of the exceptional classes $e_1, \ldots, e_n$ (and thus has the form $e_{l_1} - e_{l_2} - \cdots - e_{l_t}$ for some $0 < l_1 < \cdots < l_t \leq n$). In particular, for each $n \leq 8$ there are only finitely many fixed component equivalence classes. Moreover, the equivalence class determines the multiplicities of the fixed components that occur.

To see this, suppose that $N_1, \ldots, N_r$ are the classes of the fixed components of $-K_P$. Then for some s we can write $-K_P = M + N_{i_1} + \cdots + N_{i_s}$, where $(M + N_{i_1} + \cdots + N_{i_j}) \cdot N_{i_j} < 0$ for $1 \leq j \leq s$, and where $M \cdot N_i \geq 0$ for all i (and hence M is numerically effective, and of course satisfies $M \cdot e_0 \leq 3$, since $-K_P$ does). By our next result, this shows that M is fixed component free, and thus that the multiplicities of the fixed components of $-K_P$ are determined by the summands of $N_{i_1} + \cdots + N_{i_s}$.

Lemma 4.10. *Let F be an effective and numerically effective class on a blowing up P at n essentially distinct points $p_1, \ldots, p_n$ of $\mathbf{P}^2$, and let $e_0, \ldots, e_n$ be the corresponding basis of* Pic(P).

(a) *If $F \cdot e_0$ is at most 2, then F is regular and base point free.*
(b) *If $F \cdot e_0 = 3$ and n is at most 8, then F is regular and fixed-component free. Moreover, F is base point free unless $F = -K_P$ and $n = 8$.*

Proof. Since F is numerically effective, we must have $F \cdot e_i \geq 0$ for all i. Since F is also effective, we must have $F^2 \geq 0$.

(a) If $F \cdot e_0 = 0$, then $F^2 < 0$ unless $F = 0$. Hence $F = 0$ and F is base point free and regular.

If $F \cdot e_0 = 1$, then either $F = e_0$ or $F = e_0 - e_i$, for some $i > 0$. But e_0 is, of course, regular and base point free. If $F = e_0 - e_i$, then numerical effectivity implies $e_j - e_i$ is not effective for $0 < j < i$. Thus after reindexing, we may assume that $i = 1$. By regarding $F = e_0 - e_1$ as a class on a blowing up of $\mathbf{P}^2$ at a single point, it is easy to see that F is regular and base point free, and it remains so on any further blowings up.

Now say $F \cdot e_0 = 2$. Arguing as above, we reduce to having either $F = 2(e_0 - e_1)$ or $F = 2e_0 - e_1 - \cdots - e_i$ for $i \leq 4$. But $F = 2(e_0 - e_1)$ is clearly base point free, and regular by Lemma 4.3. Suppose that $F = 2e_0 - e_1 - \cdots - e_i$, for some $i \leq 4$. Then F can be regarded as a divisor on the blowing up P_i of i essentially distinct points of $\mathbf{P}^2$. Let K_i be the canonical class of P_i. Then $-K_i = F + e_0$, so $-K_i$ is numerically effective. Since $-K_i$ clearly dominates itself, Lemma 4.4 implies that $-K_i$ is free of fixed components, so every effective divisor dominates $-K_i$. Thus Lemma 4.4 shows that F is regular (and fixed component free) on P_i. This of course shows that F is regular on any blowing up of P_i (and so, in particular, on P). To see that F is base point free on P, it suffices to check that F is base point free on P_i, which follows from 4.9.

(b) We have $F \cdot e_0 = 3$. Arguing as above, either $F = 3(e_0 - e_1)$, $F = (3e_0 - e_1 - \cdots - e_j) - e_1$ for some $1 \leq j \leq 6$, or $F = 3e_0 - e_1 - \cdots - e_j$ for some $j \leq 8$.

Now $F = 3(e_0 - e_1)$ is regular and base point free, so consider $F = (3e_0 - e_1 - \cdots - e_j) - e_1$. If $1 \leq j \leq 2$, defining P_j and K_j as in (a), there are up to isomorphism only three possibilities for P_j, and it is easy to check $-K_j$ has a smooth section for each of them. Thus F dominates $-K_j$, hence F is regular and, on P_j, fixed component free. By 4.9, F is base point free on P_j, and hence on P.

So say $2 < j \leq 6$. First suppose $-K_j$ is not numerically effective. Then, since F is numerically effective and $-K_j = F + e_1$, we see there is a component $D = \sum_{0<i\leq j} a_i e_i$ of e_1, reduced and irreducible with $-K_j \cdot D < 0$. Hence $-a_1 = D \cdot e_1 =$

$D\cdot(-K_j-F)<0$. Thus D has support on (and so is equal to) the proper transform of the section of e_1 on P_1; i.e., $D = e_1 - e_{t_1} - \cdots - e_{t_r}$, where $e_{t_1}, \ldots, e_{t_r}$ are blowings up infinitely near e_1. Now $0 \le F\cdot D = 2-r$, so $r \le 2$. On the other hand, $1-r = -K_j \cdot D < 0$ gives $r > 1$ so in fact r is exactly 2. Thus up to reindexing we have $D = e_1 - e_2 - e_3$.

Suppose $j=3$; there are only two 3 point blowings up of $\mathbf{P}^2$ with $e_1-e_2-e_3$ reduced and irreducible (one in which p_3 is not infinitely near p_2, the other in which it is). For each of them one checks directly that $-K_j = -K_3$ has a reduced section. But this means that F (and indeed any effective divisor) dominates $-K_3$ on P_3, so, by Lemma 4.4, F is regular, and, on P_3, fixed component free. By 4.9, as above, F is base point free.

Suppose that $j \ge 4$. If p_4 is not infinitely near p_3, then e_3-e_4 is not effective; i.e., $\{p_1,p_2,p_4\}$ are essentially distinct points of $\mathbf{P}^2$. Thus primality of $D=e_1-e_2-e_3$ implies that $e_1-e_2-e_4$ is not effective. Nor can $e_0-e_1-e_2-e_4$ be effective, since that would contradict numerical effectivity of F. Thus, applying the quadratic Cremona transformation based at p_1, p_2 and p_4, we can regard P_j as a blowing up of (a different) $\mathbf{P}^2$ at j essentially distinct points $p'_1, \ldots, p'_j$, giving a new basis $e'_0, \ldots, e'_j$ for $\mathrm{Pic}(P_j)$, where $e'_0 = 2e_0 - e_1 - e_2 - e_4$, $e'_1 = e_0 - e_2 - e_4$, $e'_2 = e_0 - e_1 - e_4$, $e'_3 = e_0 - e_1 - e_2$, $e'_4 = e_3$, and, if j is 5 or 6, $e'_5 = e_5$ and $e'_6 = e_6$. But with respect to this new basis, we have $F\cdot e'_0 = 2$, which is a case considered above. Again we see that F is regular and base point free on P.

Now suppose that $j\ge 4$, but that p_4 is infinitely near p_3. If p_3 is not infinitely near p_2, we may switch the labels on p_3 and p_2, so that p_4 is now not infinitely near p_3. Arguing as in the last paragraph again shows F is regular and base point free on P. Thus we are left with considering the case that p_4 is infinitely near p_3 which is infinitely near p_2. If $j=5$ (or 6), either we must also have p_5 (or p_5 and p_6) infinitely near p_3, or relabeling again reduces us to the situation of the previous paragraph. Thus we may assume the former, which with the numerical effectivity of F implies that the irreducible components of e_1 are $e_1-e_2-e_3$, e_2-e_3, and e_3-e_4 (and e_4-e_5, or e_4-e_5 and e_5-e_6, if $j=5$ or 6, respectively).

Since $-K_j\cdot(e_1-e_2-e_3)<0$, $(e_3-e_4)\cdot(-K_j-(e_1-e_2-e_3))<0$ and $(e_2-e_3)\cdot(-K_j-(e_1-e_2-e_3)-(e_3-e_4))<0$, we see $e_1-e_2-e_3$, e_2-e_3, and e_3-e_4 are all fixed components of $-K_j$, which when subtracted off gives $3e_0-2e_1-e_2$, when $j=4$. But $3e_0-2e_1-e_2$ is base point free on any two point blowing up of $\mathbf{P}^2$, so $3e_0-2e_1-e_2$ is fixed component free on P_j. Thus, when $j=4$, $-K_j$ has a reduced section, so as above F is regular and base point free.

Next suppose that $j=5$; arguing as above, the fixed components of $-K_j$ contain the sum $(e_1-e_2-e_3)+(e_2-e_3)+2(e_3-e_4)+(e_4-e_5)$, which when subtracted from $-K_j$ leaves $3e_0-2e_1-e_2-e_3$. But $3e_0-2e_1-e_2-e_3 = 3(e_0-e_1)+D$ meets D nonnegatively and so is numerically effective and therefore corresponds to the $j=3$ case treated above. Thus $3e_0-2e_1-e_2-e_3$ is base point free, so we see that $-K_j$ has no fixed components of multiplicity 3 or more, whence F is regular and base point free.

This leaves the case that $j=6$; this time the fixed components of $-K_j$ contain the sum $(e_1-e_2-e_3)+(e_2-e_3)+2(e_3-e_4)+2(e_4-e_5)+(e_5-e_6)$, which when subtracted from $-K_j$ leaves $3e_0 - 2e_1 - e_2 - e_3 - e_4$. Since we can write $3e_0 - 2e_1 - e_2 - e_3 - e_4$ as a sum $2(e_0 - e_1) + (e_0 - e_1 - e_2) + D + (e_2 - e_3) + (e_3 - e_4)$ of irreducible classes, each of which it meets nonnegatively, we see it is numerically effective and so, as above, we see $-K_j$ has no fixed components of multiplicity 3 or more and hence F is fixed component free and regular. Since $F^2 = 0$ in this case, the fact that F is base point free is now obvious.

Now consider the case that $-K_j$ is numerically effective. Since $-K_j$ dominates itself, 4.4 asserts that $-K_j$ has no fixed components, so every effective divisor (in particular, F) dominates $-K_j$. Thus F is regular and base point free.

Finally, suppose $F = -K_j$ for some $0 \leq j \leq 8$. Since $-K_j$ dominates itself, we see F is regular, and, on P_j, fixed component free, and, for $j < 8$, base point free. □

We now have:

Theorem 4.11. *Any* 6 *or fewer essentially distinct points* $p_1, \ldots, p_n$ *of* $\mathbf{P}^2$ *are in good position and any numerically effective divisor on the surface obtained by blowing up* 6 *or fewer essentially distinct points is base point free.*

Proof. We observed above that there are for each $n \leq 8$ only finitely many fixed component equivalence classes of n essentially distinct points, and by 4.10 we can for each equivalence class explicitly determine what the multiplicities are for each fixed component of the anticanonical class.

I.e., it is a finite computation to determine the equivalence classes of essentially distinct points giving rise to an anticanonical fixed component of multiplicity more than 2. We carried out this computation, partly with the aid of a computer, to find there are for $n < 6$ no such cases, and, up to the order of the points blown up, only two cases for $n = 6$. Therefore, apart from these two anomolous cases, 4.6 and 4.9 show that the points blown up are in good position and that the numerically effective divisor classes on the blowing up are base point free.

The two anomolous cases are

$$-K_P = (2e_0 - e_1 - e_2) + (e_0 - e_1 - e_2 - e_3) + (e_1 - e_2) + 2(e_2 - e_3 - e_4) +$$

$$2(e_3 - e_4) + 3(e_4 - e_5) + 2(e_5 - e_6) + e_6$$

and

$$-K_P = (2e_0 - e_1 - e_2) + (e_0 - e_1 - e_2 - e_3) + (e_1 - e_2) + 2(e_2 - e_3 - e_4) +$$

$$2(e_3 - e_4 - e_5) + 3e_4 + (e_5 - e_6),$$

where in each case the first summand is the numerically effective fixed component free part of the anticanonical class, and the remaining summands are the fixed components, with their multiplicities.

For the first anomolous case, let S denote the semigroup of divisor classes which meet each of the fixed components of $-K_P$ nonnegatively. Clearly, S contains the cone of numerically effective classes, so if we show that S is generated by numerically effective classes, then S is itself the cone of numerically effective classes. But S is generated by the classes e_0, $e_0 - e_1$, $2e_0 - e_1 - e_2$, $3e_0 - e_1 - e_2 - e_3$, $5e_0-2e_1-2e_2-e_3-e_4$, $5e_0-2e_1-2e_2-e_3-e_4-e_5$ and $5e_0-2e_1-2e_2-e_3-e_4-e_5-e_6$ and each of these is in fact numerically effective: e_0, $e_0 - e_1$ and $2e_0 - e_1 - e_2$ are since they are numerically effective on any 2 point blowing up of $\mathbf{P}^2$, and total transforms of numerically effective divisors remain numerically effective. Next, $3e_0-e_1-e_2-e_3$ is numerically effective since $3e_0-e_1-e_2-e_3 = -K_P+e_4+e_5+e_6$ and $3e_0 - e_1 - e_2 - e_3$ meets the components of each of these summands nonnegatively. Likewise, $5e_0 - 2e_1 - 2e_2 - e_3 - e_4 - e_5 - e_6 = -K_P + (2e_0 - e_1 - e_2)$, $5e_0 - 2e_1 - 2e_2 - e_3 - e_4 - e_5 = (5e_0 - 2e_1 - 2e_2 - e_3 - e_4 - e_5 - e_6) + e_6$ and $5e_0 - 2e_1 - 2e_2 - e_3 - e_4 = (5e_0 - 2e_1 - 2e_2 - e_3 - e_4 - e_5) + e_5$ are also numerically effective.

To establish our result, we now merely need to show that the generators we have found for S are regular and free of base points; arbitrary numerically effective classes then clearly lack base points, and are regular by 4.3. As observed above, the anticanonical class of a surface has no fixed components of multiplicity more than 2 when the surface is obtained by blowing up fewer than 6 essentially distinct points. Since e_0, $e_0 - e_1$, $2e_0 - e_1 - e_2$, $3e_0 - e_1 - e_2 - e_3$, $5e_0 - 2e_1 - 2e_2 - e_3 - e_4$ and $5e_0 - 2e_1 - 2e_2 - e_3 - e_4 - e_5$ can all be regarded as classes on a blowing up of $\mathbf{P}^2$ at 5 or fewer points, 4.6 and 4.9 show that these classes are regular and base point free. Now consider $5e_0 - 2e_1 - 2e_2 - e_3 - e_4 - e_5 - e_6$. If any of $e_0-e_1-e_2-e_3$, e_1-e_2, $e_2-e_3-e_4$, e_3-e_4, e_4-e_5 or e_5-e_6 are fixed components of $5e_0 - 2e_1 - 2e_2 - e_3 - e_4 - e_5 - e_6$, then intersection considerations show that each of them is. (For example, if $e_5 - e_6$ is a fixed component, then $((5e_0 - 2e_1 - 2e_2 - e_3 - e_4 - e_5 - e_6) - (e_5 - e_6)) \cdot (e_4 - e_5)$ is negative so $e_4 - e_5$ is also a fixed component, etc.) Subtracting them off (with multiplicity) leaves $4e_0-2e_1-e_2-e_3 = (e_0-e_1)+(3e_0-e_1-e_2-e_3)$, which is regular by 4.3 since both summands are. But then $10 = h^0(P, 4e_0 - 2e_1 - e_2 - e_3) = h^0(P, 5e_0 - 2e_1 - 2e_2 - e_3 - e_4 - e_5 - e_6)$, while $h^0(P, 5e_0 - 2e_1 - 2e_2 - e_3 - e_4 - e_5 - e_6) \geq 11$ follows from Riemann-Roch. Thus the only fixed component $5e_0 - 2e_1 - 2e_2 - e_3 - e_4 - e_5 - e_6$ and $-K_P$ could have in common is e_6, which has multiplicity 1 in the latter; i.e., $5e_0 - 2e_1 - 2e_2 - e_3 - e_4 - e_5 - e_6$ dominates $-K_P$ and so is regular and fixed component free, hence base point free by 4.9.

For the other anomolous case, let S be the semigroup of divisor classes which meet e_6 and each of the fixed components of $-K_P$ nonnegatively. Then S is generated by the classes e_0, e_0-e_1, $2e_0-e_1-e_2$, $3e_0-e_1-e_2-e_3$, $3e_0-e_1-e_2-e_3-e_5$, $3e_0 - e_1 - e_2 - e_3 - e_5 - e_6$, and $5e_0 - 2e_1 - 2e_2 - e_3 - e_4$ and, by arguments of the type used in the previous case, each of these is numerically effective, so S is in fact the numerically effective cone.

Now we need to show that these generators are regular and free of base points. But each can be regarded as a class on a blowing up of $\mathbf{P}^2$ at 5 or fewer points, so the result follows as above. □

All but two configurations of $n \leq 6$ essentially distinct points fall within the scope of Corollary 4.6; i.e., the anticanonical linear system of the surface obtained by blowing up the points has no fixed components of multiplicity more than 2. To prove that any $n \leq 6$ essentially distinct points are in good position, the proof of Theorem 4.11 handles the two anomolous configurations (for which multiplicities of 3 occur) by directly computing the cone of numerically effective divisors, finding generators for the cone, and proving each generator is regular and fixed component free. Then 4.11 follows by 4.3 and 4.9. When $7 \leq n \leq 8$, many more anomolous cases arise. Indeed, whereas, if $n < 7$, the anticanonical system never has fixed components of multiplicity more than 3, for $n = 7$ multiplicities can range at least to 5, and, for $n = 8$, an exhaustive computer search has shown that the maximum multiplicity which occurs is 9. If for $7 \leq n \leq 8$ things continue to work as they have for $n < 7$, a proof of Conjecture 3.4 is, by the approach developed in this paper, just a finite computation. But the large number of anomolous cases which occur, and the awkwardness required to handle them, makes it desirable to find a more efficient analysis for $7 \leq n \leq 8$. Moreover, such a method may also be useful in analyzing the interesting case of $n = 9$, for which even distinct points can fail to be in good position. One would also very much like to understand the cases $n \geq 10$, but these are beyond the power of the methods used here.

References

[Gi] Gimigliano, A., Our thin knowledge of fat points. Queen's papers in Pure and Applied Mathematics, no. 83, The Curves Seminar at Queen's, vol. VI (1989).

[H1] Harbourne, B., Complete linear systems on rational surfaces. Trans. Amer. Math. Soc. 289 (1985), 213–226.

[H2] —, The geometry of rational surfaces and Hilbert functions of points in the plane. Can. Math. Soc. Conf. Proc. 6 (1986), 95–111.

[H3] —, Hilbert functions of points in good position in $\mathbf{P}^2$. Queen's papers in Pure and Applied Mathematics, no. 83, The Curves Seminar at Queen's, vol. VI (1989).

[H4] —, Iterated blow-ups and moduli for rational surfaces; in Algebraic Geometry, Sundance, 1986. Lecture Notes in Mathematics 1311 (1988), 101–117.

[Hi] Hirschowitcz , A., Une conjecture pour la cohomologie des diviseurs sur les surfaces rationelles generiques. J. Reine Angew. Math. 397 (1989), 208–213.

[K] Kleiman, S., Multiple point formulas I: iteration. Acta Math. 147 (1981), 13–49.

[L] Looijenga, E., Rational surfaces with an anti-canonical cycle. Ann. of Math. 114 (1981), 267–322.

[N] Nagata, M., On rational surfaces II. Mem. Coll. Sci. Kyoto (A) 33 (1960), 271–293.

[Z] Zariski, O., Introduction to the Problem of Minimal Models in the Theory of Algebraic Surfaces. Publ. Math. Soc. of Japan, 1958.

The desingularization of $\mathrm{Hilb}^4\mathbf{P}^3$ and its Betti numbers

*Sheldon Katz**

Abstract. The Hilbert scheme $\mathrm{Hilb}^4\mathbf{P}^3$ of length 4 subschemes of $\mathbf{P}^3$ is shown to be singular along the locus of "fat points". The singularity is locally a cone over the Grassmannian $G(2,6)$ with a three dimensional vertex. The singularity is minimally resolved by a single blowup. The ranks of the Chow groups of $\mathrm{Hilb}^4\mathbf{P}^3$ as well as those of its minimal resolution are computed.

1991 Mathematics Subject Classification: 14C05, 14C15

Introduction

Let $\mathbf{P}^r$ denote complex projective r space, and let $\mathrm{Hilb}^d\mathbf{P}^r$ be the Hilbert scheme parametrizing length d subschemes of $\mathbf{P}^r$. $\mathrm{Hilb}^d\mathbf{P}^r$ has a component of dimension dr containing the d-tuples of distinct points, but may be reducible in general [11].

The Hilbert schemes $\mathrm{Hilb}^1\mathbf{P}^r$, $\mathrm{Hilb}^2\mathbf{P}^r$, $\mathrm{Hilb}^3\mathbf{P}^r$, and $\mathrm{Hilb}^4\mathbf{P}^2$ have already been explicitly described [1,5,6,7,13]. They are all smooth and irreducible. The Chow rings of the above spaces have also been computed (although not explicitly for $\mathrm{Hilb}^3\mathbf{P}^r$ with $r > 3$).

The first open case is $\mathrm{Hilb}^4\mathbf{P}^3$, which is not smooth. Let m_p be the maximal ideal of a closed point of $p \in \mathbf{P}^3$. Then m_p^2 is the ideal of a length 4 subscheme $F_p \subset \mathbf{P}^3$. Such a subscheme is called a "fat point". The fat points are parametrized by a subscheme $F \subset \mathrm{Hilb}^4\mathbf{P}^3$, which is naturally isomorphic to $\mathbf{P}^3$. The fat points are singular points of $\mathrm{Hilb}^4\mathbf{P}^3$ [8]. It can also be shown that $\mathrm{Hilb}^4\mathbf{P}^3$ is smooth outside F. Locally near any point of F, it is shown that $\mathrm{Hilb}^4\mathbf{P}^3$ is a cone over the Plücker embedding of the Grassmannian $G(2,6)$ with vertex equal to a neighborhood of $p \in \mathbf{P}^3$.

It is shown that $\mathrm{Hilb}^4\mathbf{P}^3$ can be resolved by a single blowup of F. The dimensions of the Chow groups of $\mathrm{Hilb}^4\mathbf{P}^3$ as well as those of its minimal resolution are computed, using results of Bialynicki-Birula [3,4].

The author wishes to thank J. Chistophersen, G. Ellingsrud, J. Stevens, and G. Trautmann for helpful conversations, and also S. A. Strømme for encouraging this investigation. The author especially wants to thank the organizers of the

* Supported in part by NSF grant DMS-9311386.

Ravello conference for providing an inspiring setting for the discussions that took place there.

1. The structure of $\mathrm{Hilb}^4\mathbf{P}^3$

Let $Z \subset \mathbf{P}^r$ be a length d subscheme. If Z is supported on the set $\{p_1, \ldots, p_k\}$, then Z is the union of length d_i subschemes Z_i supported at p_i, with $d = \sum_i d_i$. So to understand what finite length subschemes of $\mathbf{P}^r$ look like, it suffices to understand those that are concentrated at one point. This situation can be investigated locally, and for this, attention may be restricted to schemes concentrated at the origin of $\mathbf{C}^r$.

The schemes of length at most 3 concentrated at a single point are well known. The following table gives a list of all possibilities for the local form of the ideal of a length d subscheme of $\mathbf{C}^3$ concentrated at the origin, up to change of coordinates.

$$\begin{array}{cc} d & \text{ideal} \\ 1 & (x,y,z) \\ 2 & (x,y,z^2) \\ 3 & (x,y,z^3) \\ 3 & (x,y^2,yz,z^2) \\ 3 & (x,y+z^2,y^2,yz) \end{array} \tag{1.1}$$

Now assume Z is a length 4 subscheme of $\mathbf{C}^3$ concentrated at the origin. Let m denote the maximal ideal (x, y, z) of the origin. Choose a length 3 subscheme $Y \subset Z$. Without loss of generality, assume that Y is defined by one of the three ideals of the $d = 3$ case in (1.1). Letting J denote its ideal, we see that the ideal I of Z must satisfy

$$mJ \subset I \subset J \text{ and } \dim_{\mathbf{C}} J/I = 1. \tag{1.2}$$

Since the form of J is known explicitly, the various possibilities for I can be written down. In fact, the iterative method based on (1.2) can be used to construct (1.1). It turns out that there are infinitely many possibilities for I, even after coordinate changes.

For the rest of this paper, let $H = \mathrm{Hilb}^4\mathbf{P}^3$. Since $\mathrm{Hilb}^d\mathbf{P}^3$ is smooth for $d \leq 3$, it follows that H can only be singular at length 4 subschemes which are concentrated at a single point.

Let m_p be the ideal sheaf of the closed point $p \in \mathbf{P}^3$. Then the "fat point" m_p^2 defines a length 4 subscheme of $\mathbf{P}^3$, whose zero scheme $Z(m_p^2)$ is a point of H. The fat points are parametrized by a a subscheme $F \simeq \mathbf{P}^3$ of H. It is well known that H is singular along F [8].

H is known to be smooth outside F. This was pointed out to the author by G. Ellingsrud and by G. Trautmann (see also [12]). For completeness, a different proof will be sketched below, as there does not appear to be an explicit proof in the literature.

Given $p \in \mathbf{P}^3$, we can choose homogeneous coordinates (w, x, y, z) for $\mathbf{P}^3$ such that $p = (1,0,0,0)$ in this coordinate system, so that $m_p = (x, y, z)$. Consider the corresponding fat point with ideal sheaf $\mathcal{I} = m_p^2$. The tangent space to H at $Z(\mathcal{I})$ is $\mathrm{Hom}_{\mathcal{O}_{\mathbf{P}^3}}(\mathcal{I}, \mathcal{O}_{\mathbf{P}^3}/\mathcal{I})$ [10], which is an 18 dimensional vector space. Direct calculation of this Hom in fact gives that each of the 6 quadratic generators of $\mathcal{I}$ can map independently to any of the 3 generators for the degree 2 elements of $m_p \cdot w$ (mod $\mathcal{I}$). Thus, the first order neighborhood of $Z(\mathcal{I})$ in H is the subscheme of H determined by the family parametrized by Spec $\left(\mathbf{C}[a_1, \ldots, a_{18}]/(a_1, \ldots, a_{18})^2\right)$ with ideal

$$\begin{aligned}&(x^2 + a_1xw + a_2yw + a_3zw,\ xy + a_4xw + \ldots,\ xz + \ldots,\ y^2 + \ldots,\\ &\quad yz + \ldots, z^2 + a_{16}xw + a_{17}yw + a_{18}zw) \qquad (a_1, \ldots, a_{18})^2 = 0.\end{aligned} \tag{1.3}$$

H can be described locally near $Z(\mathcal{I})$ by finding the higher order deformations of (1.3) to bigger flat families. This is done by calculating the syzygies, and finding deformed ideals for which all of the syzygies extend.

The calculation of the higher order obstructions is simplified by the following fact.

Proposition 1.4 (Schlessinger-Stasheff, [14]). *The obstructions to extending this family are second order; there are no higher order obstructions.*

Thus, H is locally given by the vanishing of certain quadratic equations in $a_1, \ldots, a_{18}$. They can be calculated explicitly as follows.

The 6 quadrics $(x^2, \ldots, z^2)$ are related by eight independent linear syzygies, whose coefficients appear in the last eight rows of (1.5) below. The coefficients are to be multiplied by the monomials in the top row.

$$\begin{array}{cccccc} x^2 & xy & xz & y^2 & yz & z^2 \\ -y & x & 0 & 0 & 0 & 0 \\ -z & 0 & x & 0 & 0 & 0 \\ 0 & -z & y & 0 & 0 & 0 \\ 0 & -y & 0 & x & 0 & 0 \\ 0 & -z & 0 & 0 & x & 0 \\ 0 & 0 & 0 & -z & y & 0 \\ 0 & 0 & -z & 0 & 0 & x \\ 0 & 0 & 0 & 0 & -z & y \end{array} \tag{1.5}$$

One deforms the ideal $\mathcal{I}$ to first order as in (1.3), and chooses a lift of the syzygies in (1.5) to the deformed ideal, modulo $(a_1, \ldots, a_{18})^2$ (the deformation has been chosen to ensure that this can be done). In trying to extend them to second order, i.e. modulo $(a_1, \ldots, a_{18})^3$, we are free to add terms to the generators of the deformed ideal which are quadratic in $(a_1, \ldots, a_{18})$, and to make different choices of the extension of the syzygies of (1.5) to the first order ideal (1.3). By adding on general terms to the ideal and to the syzygies with variable coefficients, one arrives at an easily solved linear system. This system is seen to have non-trivial solutions if and only if certain quadratic equations in the a_i hold identically. In fact, the resulting quadratic equations span a 15 dimensional vector space. The same obstruction has been found independently by J. Stevens[15] and by D. Dias-Villet [16] (and probably by others as well). In fact, it can be seen via a coordinate change that these equations define an affine cone with 3 dimensional vertex over $G(2,6) \subset \mathbf{P}^{14}$, the embedding being the Plücker embedding. This observation had been independently made earlier by J. Christophersen after a suggestion by K. Ranestad (see also [2]). Rather than pursue this route, a different explanation will be given, which relies on somewhat less calculation. The 15 quadratic equations are displayed in Appendix A.

Let $V = H^0(\mathcal{O}_{\mathbf{P}^3}(1))$, so that $\mathbf{P}^3$ is canonically isomorphic to

$$\mathbf{P}(V) = \mathrm{Proj}\,(\mathrm{Symm}(V)).$$

We may replace $\mathbf{P}^3$ by F. There is the exact sequence of sheaves on F

$$0 \longrightarrow \mathcal{K} \longrightarrow V_F \xrightarrow{\pi} \mathcal{O}_F(1) \longrightarrow 0,$$

where V_F denotes the trivial bundle on F associated to V, π is the canonical surjection, and $\mathcal{K}$ is the kernel of π. Locally near any point of F, we can choose a basis $\{w, x, y, z\}$ of sections of V_F such that $\mathcal{K}$ is generated x, y, and z, while $\mathcal{O}_F(1)$ is generated by $\pi(w)$ (which will just be written as w in the sequel). The symbols w, x, y, z will be used to describe either an appropriate basis of V or an appropriate basis of local sections of V_F as above; no confusion should result from this abuse of notation.

Theorem 1.6.

1. *H is irreducible.*
2. *$H - F$ is smooth.*
3. *Near F, the scheme H is locally a cone over the Plücker embedding of $G(2,6) \subset \mathbf{P}^{14}$, with vertex F.*
4. *The normal sheaf of F in H is the rank 15 bundle $\wedge^2(S^2\mathcal{K}^*)$.*
5. *The projectivized normal cone of F in H is the Grassmann bundle $G_2(S^2\mathcal{K}^*)$ parametrizing rank 2 subbundles of $S^2\mathcal{K}^*$ and its pullbacks.*
6. *The projectived normal cone embeds in the projectivized normal bundle via the Plücker embedding.*
7. *The blowup $\tilde{H}$ of H along F is smooth and irreducible of dimension 12.*

Proof. The description of the tangent space to H at a fat point together with the local bases for $\mathcal{K}$, V_F, and $\mathcal{O}_F(1)$ described above show that the restriction to F of the tangent sheaf of H is

$$S^2\mathcal{K}^* \otimes \mathcal{K} \otimes \mathcal{O}_F(1). \tag{1.7}$$

The tangent sheaf of F is just $\mathcal{K}^*(1)$. Its embedding in (1.7) is induced by the natural map

$$\mathcal{K}^* \xrightarrow{\theta} S^2\mathcal{K}^* \otimes \mathcal{K}$$

defined by

$$\theta(s)(q) = s \cdot q.$$

Here, $s \in \mathcal{K}^*$ and $q \in S^2\mathcal{K}$, while the dot in $s \cdot q$ denotes the contraction map $\mathcal{K}^* \otimes S^2\mathcal{K} \to \mathcal{K}$. The sheaf $S^2\mathcal{K}^* \otimes \mathcal{K}$ has been identified with $\underline{\mathrm{Hom}}_{\mathcal{O}_F}(S^2\mathcal{K}, \mathcal{K})$ in the usual way. The normal sheaf is just the quotient of $S^2\mathcal{K}^* \otimes \mathcal{K} \otimes \mathcal{O}_F(1)$ by $\mathcal{K}^*(1)$.

Lemma 1.8. *There is an isomorphism*

$$\wedge^2(S^2\mathcal{K}^*) \simeq \left(S^2\mathcal{K}^* \otimes \mathcal{K} \otimes \mathcal{O}_F(1)\right)/\mathcal{K}^*(1).$$

Proof. Note that $\wedge^3\mathcal{K} \simeq \mathcal{O}_F(-1)$, so that

$$\wedge^2(S^2\mathcal{K}^*) \simeq \wedge^2(S^2\mathcal{K}^*) \otimes \wedge^3\mathcal{K} \otimes \mathcal{O}_F(1). \tag{1.9}$$

The isomorphism of the lemma is induced by a choice of the isomorphism (1.9) and the twist by $\mathcal{O}_F(1)$ of the map

$$\wedge^2(S^2\mathcal{K}^*) \otimes \wedge^3\mathcal{K} \xrightarrow{\Psi} S^2\mathcal{K}^* \otimes \mathcal{K} \tag{1.10}$$

defined by

$$\Psi\left((\omega_1 \wedge \omega_2) \otimes (k_1 \wedge k_2 \wedge k_3)\right)(d_1 d_2) = \sum_{\sigma \in S_3} (-1)^{\mathrm{sign}(\sigma)} k_{\sigma(3)} \omega_1(d_1 k_{\sigma(1)}) \omega_2(d_2 k_{\sigma(2)})$$

where the ω_i are in $S^2\mathcal{K}^*$, and the k_i and d_i are in $\mathcal{K}$. This proves the lemma, and with it, part (4) of Theorem 1.6. □

Continuing with the proof of the theorem, let $t^2 = 0$, and consider the first order deformation

$$(x^2 + tzw, xy, xz + txw, y^2 + tzw, yz - tyw, z^2) \tag{1.11}$$

of m_p^2. Extend the syzygies (1.5) to (1.11) as follows, modulo t^2:

$$\begin{matrix} x^2 + tzw & xy & xz + txw & y^2 + tzw & yz - tyw & z^2 \\ -y & x & 0 & 0 & tw & 0 \\ -z - tw & 0 & x & 0 & 0 & tw \\ 0 & -z - tw & y & 0 & 0 & 0 \\ 0 & -y & -tw & x & 0 & 0 \\ 0 & -z + tw & 0 & 0 & x & 0 \\ 0 & 0 & 0 & -z + tw & y & tw \\ 0 & 0 & -z + tw & 0 & 0 & x \\ 0 & 0 & 0 & 0 & -z - tw & y \end{matrix} \tag{1.12}$$

The first order deformation (1.11) can be extended to a second order deformation merely by deforming the ideal to

$$(x^2 + tzw - t^2w^2, xy, xz + txw, y^2 + tzw + t^2w^2, yz - tyw, z^2 - t^2w^2) \tag{1.13}$$

and leaving the syzygies as is; in fact, this is a deformation to all orders in t, as the syzygies hold identically in t without truncation. This is consistent with Proposition 1.4.

The first order deformation (1.11) is a "general" unobstructed deformation; most other choices would work just as well. We describe $S^2\mathcal{K}^*$ locally in terms of the dual basis to the basis $(x^2, \ldots, z^2)$ for $S^2\mathcal{K}$; so for a quadratic monomial m in the sections x, y, and z, the expression $(m)^*$ will denote the element of $S^2\mathcal{K}^*$ which evaluates to 1 on m and to 0 on the other monomials. Choose the isomorphism (1.9) induced by the nowhere vanishing section $(x \wedge y \wedge z) \otimes w$ of $\wedge^3\mathcal{K} \otimes \mathcal{O}_F(1)$. Then with this choice, the first order deformation (1.11) corresponds to the decomposable vector

$$\left(\frac{1}{2}(x^2)^* - \frac{1}{2}(y^2)^* - (z^2)^*\right) \wedge (xy)^* \tag{1.14}$$

under the isomorphism of Lemma 1.8. Taken modulo scalars, (1.14) may be thought of as describing a pencil of conics in $(\mathbf{P}^2)^*$. It is easy to see that the standard action of PGL(3) on the space of pencils of conics has a dense orbit, and that the particular pencil we have chosen is in that orbit. But the entire PGL(3) orbit must be unobstructed, as any coordinate change applied to (1.11) can also be applied to (1.13), showing that the transformed first order deformation is unobstructed. Thus our quadratic obstruction equations must vanish on *all* of the decomposable vectors. However, since there are $\binom{6}{2} = 15$ quadratic Plücker equations on $G(2,6)$, and 15 quadratic equations have been found, the 15 quadratic equations are *precisely* the Plücker equations in the coordinates of the normal space. This proves (3), (5), and (6).

Let $Z \in H - F$. It has already been mentioned that H is smooth at Z if Z is supported on more than one point. To investigate the schemes supported at a point, it suffices to look locally near the origin of $\mathbf{C}^3$. Earlier in this section, the method of classification of such schemes was outlined. Even though there are infinitely many possibilities for the form of such schemes up to change of coordinates, the omitted calculation shows that there are only finitely many possibilities which have the property that the closure of their PGL(3) orbits do not meet F, the PGL(3) action being change of coordinates. For each of these finitely many possibilities, one can directly calculate that the tangent space to H has dimension 12, and that Z can be deformed to a 4 tuple of distinct points. Note that the main component of H containing the 4 tuples of distinct points has dimension 12 as well. We see that every point of H is either a smooth point of the main component of H, or else its PGL(4) orbit closure meets F. Combining this with the local description of H near F, we get (1) and (2). Then (7) follows immediately. □

The exceptional divisor of $\tilde{H}$ will be denoted by E. By Theorem 1.6 (5), E is a $G(2,6)$ bundle over F.

2. Betti numbers

The method [5] of Ellingsrud and Strømme for $\mathrm{Hilb}^d\mathbf{P}^2$ can be adapted to the current situation. Many details will be omitted; but see [5] for details in a similar situation.

The goal of this section is the prove the following.

Theorem 2.1.

1. *The homology of $\tilde{H}$ is generated by the cells in a cell decomposition of $\tilde{H}$, and there is no odd dimensional homology.*
2. *The cycle map $A_*(\tilde{H}) \to H_*(\tilde{H})$ is an isomorphism.*
3. *The sequence of Betti numbers* $\dim H_{2i}(\tilde{H})$ *for* $0 \leq i \leq 12$ *is*

$$\{1, 3, 9, 19, 34, 44, 51, 44, 34, 19, 9, 3, 1\}.$$

Parts (1) and (2) will follow from results of Bialynicki-Birula [3,4] as applied by Ellingsrud and Strømme [5]. For the convenience of the reader, this circle of ideas will be reviewed here.

Let $D \subset \mathrm{SL}(4, \mathbf{C})$ be the rank 3 maximal torus of diagonal matrices. D acts on $\mathbf{P}^3$ with the 4 fixed points

$$p_1 = (1,0,0,0),\ p_2 = (0,1,0,0),\ p_3 = (0,0,1,0),\ p_4 = (0,0,0,1).$$

This action induces an action of D on H; since this action preserves F, it follows that D acts on $\tilde{H}$ as well.

The action of D on $\tilde{H}$ has finitely many fixed points. The fixed points of the action of D on H correspond to schemes whose support is contained in $\{p_1, p_2, p_3, p_4\}$; write $Z = \cup_{i=1}^4 Z_i$ as in §1. For such a Z to be a fixed point, it is necessary and sufficient that the ideal of Z_i be generated by monomials for all i. Since $\tilde{H} - E \simeq H - F$, this describes the fixed points of D outside of E. The fixed points of D in E correspond (via Theorem 1.6 and Lemma 1.8) to fat points concentrated at one of the p_i, together with a first order direction associated to a decomposable vector of the form $m_1^* \wedge m_2^*$, where the m_j^* are *monomials* in the 3 coordinates whose vanishing defines p_i.

It is shown in [5] that when the action of D is restricted to a general 1 parameter subgroup, this restricted action has the same fixed point set as does D. So a general 1 parameter subgroup of D has the same fixed point set in $\tilde{H}$ as does D. Such a choice of a $\mathbf{G}_m$ action will be assumed in the sequel.

Let X be a variety with a $\mathbf{G}_m$ action, and let x be a fixed point. Then $\mathbf{G}_m$ acts on the tangent space $T_{X,x}$. The invariant subspace of $T_{X,x}$ for which $\mathbf{G}_m$ has positive weights is denoted by $(T_{X,x})^+$.

For the convenience of the reader, the Bialynicki-Birula theorem [3,4] as stated in [5] is reproduced here.

Theorem 2.2 (Bialynicki-Birula). *Let X be a smooth projective variety with an action of $\mathbf{G}_m$. Suppose that the fixed point set $\{x_1, \ldots, x_n\}$ is finite and let $X_i = \{x \in X \mid \lim_{t\to 0} tx = x_i\}$. Then*
(i) *X has a cellular decomposition with cells X_i.*
(ii) $T_{X_i, x_i} = (T_{X,x_i})^+$.

This completes the review of the preparatory material. It is now time to turn to the proof of Theorem 2.1.

Proof. By the above discussion and Theorem 2.2, $\tilde{H}$ has a cellular decomposition. It is well known in this situation that all the homology of $\tilde{H}$ is freely generated by the homology classes of the closures of the cells, and that the cycle maps are isomorphisms [9, Ch. 19.1]. This proves (1) and (2).

For (3), it remains to find the dimensions of the cells in the cell decomposition, or equivalently, to find the number of positive weights of the $\mathbf{G}_m$ action on the tangent space to $\tilde{H}$ at a fixed point. By considering $Z = \cup Z_i$, the calculation is reduced to the analogous calculation for schemes supported at a single p_i, and having arbitrary lengths $d \leq 4$ (the cells corresponding to Z are the products of the cells in $\mathrm{Hilb}^{d_i}\mathbf{P}^3$ associated to the fixed points Z_i in the case that Z is not supported at a single point). The results of the calculation will only be given for some of the fixed points; the remaining cases can be immediately deduced by permutation of the coordinates.

Let the 1 parameter subgroup be defined by $\mathrm{diag}(t^{\gamma_1}, \ldots, t^{\gamma_4})$. Then the weights of the resulting $\mathbf{G}_m$ action at the fixed points considered here are given by the following table.

d	Z	weights
1	(x,y,z)	$\gamma_1-\gamma_2,\ \gamma_1-\gamma_3,\ \gamma_1-\gamma_4$
2	(x^2,y,z)	$\gamma_1-\gamma_2,\ 2\gamma_1-2\gamma_2,\ \gamma_2-\gamma_3,\ \gamma_2-\gamma_4,\ \gamma_1-\gamma_3,\ \gamma_1-\gamma_4$
3	(x^3,y,z)	$\gamma_1-\gamma_2,\ 2\gamma_1-2\gamma_2,\ 3\gamma_1-3\gamma_2,\ \gamma_1-\gamma_3,\ \gamma_2-\gamma_3,\ \gamma_1-\gamma_4,$ $\gamma_2-\gamma_4,\ -\gamma_1+2\gamma_2-\gamma_3,\ -\gamma_1+2\gamma_2-\gamma_4$
3	(x,y^2,yz,z^2)	$\gamma_1-\gamma_2,\ \gamma_3-\gamma_2,\ \gamma_4-\gamma_2,\ \gamma_1-\gamma_3,\ \gamma_1-\gamma_3,$ $\gamma_1-\gamma_4,\ \gamma_1-\gamma_4,\ \gamma_1-2\gamma_3+\gamma_4,\ \gamma_1+\gamma_3-2\gamma_4$
4	(x^4,y,z)	$\gamma_1-\gamma_2,\ 2\gamma_1-2\gamma_2,\ 3\gamma_1-3\gamma_2,\ 4\gamma_1-4\gamma_2,\ \gamma_1-\gamma_3,\ \gamma_2-\gamma_3,$ $-\gamma_1+2\gamma_2-\gamma_3,\ -2\gamma_1+3\gamma_2-\gamma_3,\ \gamma_1-\gamma_4,\ \gamma_2-\gamma_4,$ $-\gamma_1+2\gamma_2-\gamma_4,\ -2\gamma_1+3\gamma_2-\gamma_4$
4	(x,y^2,z^2)	$\gamma_1-\gamma_2,\ \gamma_3-\gamma_2,\ \gamma_4-\gamma_2,\ 2\gamma_1-2\gamma_3,\ \gamma_1-\gamma_3,$ $\gamma_1-2\gamma_3+\gamma_4,\ \gamma_4-\gamma_3,\ 2\gamma_1-2\gamma_4,\ \gamma_1-\gamma_4,\ \gamma_1+\gamma_3-2\gamma_4,$ $\gamma_3-\gamma_4,\ -\gamma_1-\gamma_2+\gamma_3+\gamma_4$
4	(x,y^2,yz,z^3)	$\gamma_1-\gamma_2,\ \gamma_3-\gamma_2,\ \gamma_4-\gamma_2,\ -\gamma_1-\gamma_2+2\gamma_4,\ \gamma_1-\gamma_3,$ $2\gamma_4-2\gamma_3,\ \gamma_1-\gamma_4,\ \gamma_1-\gamma_3,\ \gamma_4-\gamma_3,\ 2\gamma_1+\gamma_3-3\gamma_4$ $2\gamma_1-2\gamma_4,\ \gamma_1-\gamma_4$
4	$(x,y,z)^2$ $(x^2)^*\wedge(xy)^*$	$\gamma_1-\gamma_2,\ \gamma_1-\gamma_3,\ \gamma_1-\gamma_4,\ \gamma_2-\gamma_4,\ 2\gamma_2-2\gamma_3,$ $2\gamma_2-\gamma_3-\gamma_4,\ 2\gamma_2-2\gamma_4,\ \gamma_3-\gamma_4,\ \gamma_2-\gamma_3,\ \gamma_2-\gamma_4,$ $\gamma_2+\gamma_3-2\gamma_4,\ \gamma_1-2\gamma_2+a_4$

4	$(x,y,z)^2$ $(xy)^* \wedge (xz)^*$	$\gamma_1-\gamma_2,\ \gamma_1-\gamma_3,\ \gamma_1-\gamma_4,\ \gamma_3-\gamma_2,\ \gamma_2-\gamma_3,$ $\gamma_2-\gamma_4,\ \gamma_2+\gamma_3-2\gamma_4,\ \gamma_4-\gamma_2,\ \gamma_2-2\gamma_3+\gamma_4,\ \gamma_2-\gamma_3,$ $\gamma_2-\gamma_4,\ \gamma_1-\gamma_2$
4	$(x,y,z)^2$ $(x^2)^* \wedge (y^2)^*$	$\gamma_1-\gamma_2,\ \gamma_1-\gamma_3,\ \gamma_1-\gamma_4,\ \gamma_2-\gamma_3,\ \gamma_2-\gamma_4,$ $2\gamma_2-\gamma_3-\gamma_4,\ 2\gamma_2-2\gamma_4,\ \gamma_3-\gamma_2,\ -\gamma_2+2\gamma_3-\gamma_4,\ \gamma_3-\gamma_4,$ $2\gamma_3-2\gamma_4,\ \gamma_1-\gamma_2-\gamma_3+\gamma_4$
4	$(x,y,z)^2$ $(x^2)^* \wedge (yz)^*$	$\gamma_1-\gamma_2,\ \gamma_1-\gamma_3,\ \gamma_1-\gamma_4,\ \gamma_2-\gamma_3,\ \gamma_2-\gamma_4,$ $2\gamma_2-2\gamma_3,\ 2\gamma_2-2\gamma_4,\ \gamma_4-\gamma_2,\ \gamma_3-\gamma_2,\ \gamma_4-\gamma_3$ $\gamma_3-\gamma_4,\ \gamma_1-\gamma_2$

(2.3)

The calculation leading to the Table (2.3) is straightforward. All but the last four rows may be calculated from the identification of the tangent space to H at $Z(\mathcal{I})$ with $\mathrm{Hom}(\mathcal{I}, \mathcal{O}/\mathcal{I})$ [10]. For example, consider $\mathcal{I} = (x, y^2, yz, z^2)$. Then $\mathrm{Hom}(\mathcal{I}, \mathcal{O}/\mathcal{I})$ has dimension 9. There is indeed the basis $e_1, \ldots, e_9$ defined by

$$e_1(x) = w,\ e_2(x) = y,\ e_3(x) = z,\ e_4(y^2) = yw,\ e_5(y^2) = zw,\ e_6(yz) = yw,$$
$$e_7(yz) = zw,\ e_8(z^2) = yw,\ e_9(z^2) = zw,$$

where e_i is zero on any of the given generators g of $\mathcal{I}$ for which $e_i(g)$ has not been explicitly defined. $\mathbf{G}_m$ acts on $\mathrm{Hom}(\mathcal{I}, \mathcal{O}/\mathcal{I})$ through its action on $\mathcal{I}$ and $\mathcal{O}/\mathcal{I}$ at the point $(1,0,0,0)$; so the weights can be calculated. It turns out that the e_i above are in fact eigenvectors, and a calculation of the weights gives the relevant data in Table (2.3).

For the last four rows of Table (2.3), the tangent space to $\tilde{H}$ at a point of E must be computed. This 12 dimensional space contains the 11 dimensional tangent space to E as a subspace, with a 1 dimensional quotient. The tangent space to E contains the tangent space to a fiber of E over F (isomorphic to $G(2,6)$) and the quotient is the tangent space to F. So the $\mathbf{G}_m$ weight of the following 3 types of tangent eigenvectors must be computed.

1. A tangent vector to F at the image point in $F \subset H$.
2. A tangent vector to the fiber $G(2,6)$.
3. A normal vector to $E \subset \tilde{H}$.

To compute these, (1) has already been done in table (2.3), $d = 1$. For (2), the tangent space to the Grassmannian of subspaces of a vector space P at a subspace S is known to be $\mathrm{Hom}(S, P/S)$, from which these weights may be computed. For (3), recall that a point e of E consists of a fat point $Z(m_p^2)$ together with a normal vector $\vec{n}$ to F. At such an e, it turns out that a normal vector to E in $\tilde{H}$ must project to a scalar multiple of $\vec{n}$ (the projection being the natural projection from $\tilde{H}$ to H. If in addition e is $\mathbf{G}_m$-fixed, then $\vec{n}$ can be chosen to be an eigenvector for $\mathbf{G}_m$, and its weight is the same as the weight of the induced $\mathbf{G}_m$ action on the normal space to $E \subset \tilde{H}$ at e. In fact, such special $\vec{n}$ have already been written down in the course of describing the fixed points of D on $\tilde{H}$. For example, consider the last row of table (2.3). $\vec{n}$ can be taken to be the normal vector corresponding to $(x^2)^* \wedge (yz)^*$ under the isomorphism of Theorem 1.6 (4). Its weight is immediately calculated as $\gamma_1 - \gamma_2$ (note that $\mathbf{G}_m$ acts non-trivially on the section $(x \wedge y \wedge z) \otimes w$ of $\wedge^3\mathcal{K} \otimes \mathcal{O}_F(1)$ used in (1.9) and Lemma 1.8).

It is now a straightforward albeit tedious matter to fix a general 1 parameter subgroup of D and use Table (2.3) to calculate the dimension of each of the cells deduced from Theorem 2.2. The number of cells of complex dimension i is as asserted in (3). As a check, the calculation has been carried out on a computer for several different choices of 1 parameter subgroups of D, and the same answer was found in each computer run. This completes the proof of Theorem 2.1. □

The ranks of the Chow groups of H can be deduced from Theorem 2.1.

Theorem 2.4. *The Chow groups $A_i(H)$ for $0 \leq i \leq 12$ have respective ranks*

$$1, 2, 6, 14, 26, 35, 42, 36, 28, 15, 7, 2, 1.$$

Proof. Put $U = H - F \simeq \tilde{H} - E$. Let $i : F \hookrightarrow H$ and $\tilde{i} : E \hookrightarrow \tilde{H}$ be the natural inclusions. Consider the commutative diagram

$$\begin{array}{ccccccc} A_*(E) & \xrightarrow{\tilde{i}_*} & A_*(\tilde{H}) & \longrightarrow & A_*(U) & \longrightarrow & 0 \\ \downarrow & & \downarrow & & \| & & \\ A_*(F) & \xrightarrow{i_*} & A_*(H) & \longrightarrow & A_*(U) & \longrightarrow & 0 \end{array} \tag{2.5}$$

Claim: i_* and $\tilde{i}_*$ are inclusions.

Proof of claim: For i_*, note that $A_i(F) \simeq \mathbf{Z}$ for $0 \leq i \leq 3$, while $A_i(F) = 0$ otherwise. Now i_* is clearly an injection on $A_0(F)$. If $0 \neq \eta \in A_i(F)$ with $i > 0$, pick a very ample divisor D on H (by projectivity of H [10]). Then $D^i \cdot i_*\eta \neq 0$, hence $i_*\eta \neq 0$.

For the injectivity of $\tilde{i}_*$, first note that $\tilde{i}^*$ is surjective. This is not difficult to see. $A^*(E)$ is generated as a $\mathbf{Z}$-algebra by the classes c_1, c_2, and h, where c_i are the Chern classes of the universal rank 2 subbundle on $E \simeq G_2(S^2\mathcal{K}^*)$, and h is the pullback of the hyperplane class of F. Since $\tilde{i}^*$ is a ring homomorphism, it suffices to show that $\tilde{i}^*$ is surjective in codimension less than or equal to 2. But this may be seen by the explicit knowledge of the cell decomposition of $\tilde{H}$ and the explicit structure of the Chow groups of E.

Now suppose that $\eta \in A_*(E)$ is a nonzero class such that $\tilde{i}_*\eta = 0$. Without loss of generality, assume that η has pure dimension k. Choose a class $\rho \in A^k(E)$ of complementary dimension such that $\rho \cdot \eta \neq 0$, and a class $\tilde{\rho} \in A^k(\tilde{H})$ such that $\tilde{i}^*\tilde{\rho} = \rho$. Then

$$\tilde{i}_*(\rho \cdot \eta) = \tilde{i}_*(\tilde{i}^*\tilde{\rho} \cdot \eta) = \tilde{\rho} \cdot \tilde{i}_*\eta = 0.$$

But $\tilde{i}_*$ is injective on $A_0(E)$, a contradiction. This proves the claim.

The ranks of the groups $A_*(E)$ are easy to calculate, and the first vertical map in (2.5) is surjective. Theorem 2.4 follows immediately from diagram (2.5), the claim, and the calculation of the ranks of $A_*(E)$. □

Appendix A. Local Equations for H near F

Here is a list of a basis for the 15 quadratic obstructions to the deformation of a fat point, in the coordinates specified in (1.3).

$$a_3a_4 - a_1a_6 + a_5a_6 + a_6a_9 - a_2a_{12} - a_3a_{15}$$
$$a_2a_7 - a_1a_8 + a_5a_8 + a_8a_9 - a_2a_{14} - a_3a_{17}$$
$$a_2a_4 - a_1a_5 + a_5^2 - a_3a_7 + a_1a_9 - a_9^2 - a_2a_{11} - a_3a_{14} + a_2a_{15} + a_3a_{18}$$
$$a_8a_{10} - a_5a_{13} + a_9a_{13} - a_6a_{16}$$
$$a_5a_7 - 2a_4a_8 - a_7a_9 + a_8a_{11} + a_2a_{13} - a_5a_{14} + a_9a_{14} + a_3a_{16} - a_6a_{17}$$
$$a_4a_5 + 2a_6a_7 - a_4a_9 - a_2a_{10} + a_8a_{12} - a_3a_{13} - a_5a_{15} + a_9a_{15} - a_6a_{18}$$
$$a_6a_7 + a_8a_{12} - a_3a_{13} - a_6a_{14}$$
$$a_4a_6 - a_3a_{10} - a_6a_{11} + a_5a_{12} - a_9a_{12} + a_6a_{15}$$
$$a_4a_8 - a_2a_{13} - a_8a_{15} + a_6a_{17}$$
$$a_7a_8 + a_8a_{14} - a_2a_{16} - a_5a_{17} + a_9a_{17} - a_8a_{18}$$
$$a_7a_{10} - a_4a_{13} + a_{11}a_{13} - a_{10}a_{14} - a_{13}a_{15} + a_{12}a_{16}$$
$$a_4a_7 - a_1a_{13} + a_5a_{13} + a_9a_{13} - a_4a_{14} - a_7a_{15} + a_{14}a_{15} - a_{12}a_{17}$$
$$a_4^2 - a_1a_{10} + a_5a_{10} + a_9a_{10} - a_4a_{11} - a_7a_{12} - a_{12}a_{14} + a_{11}a_{15} - a_{15}^2 + a_{12}a_{18}$$
$$a_7a_{13} + a_{13}a_{14} - a_4a_{16} + a_{15}a_{16} - a_{10}a_{17} - a_{13}a_{18}$$
$$a_7^2 - a_{14}^2 - a_1a_{16} + a_5a_{16} + a_9a_{16} - a_4a_{17} + a_{11}a_{17} - a_{15}a_{17} - a_7a_{18} + a_{14}a_{18}$$

References

[1] D. Avritzer and I. Vainsencher. $\mathrm{Hilb}^4\mathbf{P}^2$, LNM 1436, Springer-Verlag.

[2.] K. Behnke and J. Christophersen. Hypersurface sections and obstructions (rational surface singularities). Compositio Math. 77 (1991), 233–268.

[3] A. Bialinicki-Birula, Some theorems on actions of algebraic groups. Ann. Math. 98 (1973), 480–497.

[4] A. Bialinicki-Birula, Some properties of the decompositions of algebraic varieties determined by actions of a torus. Bull. Acad. Pol. Sci. Sér. Sci. Math. Astron. Phys. 24 (1976), 667–674.

[5] G. Ellingsrud and S. A. Strømme. On the homology of the Hilbert scheme of points in the plane. Invent. Math. 87 (1987), 343–352.

[6] G. Ellingsrud and S. A. Strømme, The Number of Twisted Cubic Curves on the General Quintic Threefold, Univ. of Bergen Preprint 63-7-2-1992.

[7] G. Ellingsrud and S. A. Strømme, Towards the Chow ring of the Hilbert scheme of $\mathbf{P}^2$. To appear in J. Reine Angew. Math.

[8] J. Fogarty, Algebraic families on an algebraic surface. Amer. J. Math. 90 (1968) 511–521.

[9] W. Fulton, Intersection Theory. Springer-Verlag, Berlin 1984.

[10] A. Grothendieck. Fondements de la Géometrie Algébriques. Extraits du Séminaire Bourbaki, 1962.

[11] A. Iarrobino, Reducibility of the families of 0 dimensional schemes on a variety. Invent. Math. 15 (1972) 72–77.

[12] R. Miro-Roig, G. Trautmann, The moduli scheme $M(0,2,4)$ over $\mathbf{P}_3$, preprint, Universidad de Zaragoza and Universität Kaiserslautern, 1993.

[13] F. Roselló-Llompart. The Chow ring of $\mathrm{Hilb}^3\mathbf{P}^3$. LNM 1436, Springer-Verlag, 225–255.

[14] M. Schlessinger and J. Stasheff, The Lie algebra structure of tangent cohomology and deformation theory. J. Pure Appl. Alg. 38 (1985) 313–332.

[15] J. Stevens, Private communication to the author.

[16] D. Dias-Villet, Private communication to the author.

Some applications of the canonical module of a 0-dimensional scheme

Martin Kreuzer

Abstract. In this paper the author studies some applications of the canonical module of a 0-dimensional scheme to:
a) Hyperplane sections of curves.
b) Cohomologically split schemes.
c) Pure schemes.

1991 Mathematical Subject Classification: 13C05, 14C99, 14N05.

Keywords: canonical module, 0-dimensional scheme, Cayley-Bacharach property, Hilbert function.

1. Introduction

In this note we consider 0-dimensional schemes X together with a fixed embedding into a d-dimensional projective space $\mathbf{P}^d$ over an algebraically closed field k. In [K1] we studied the canonical module ω_R of the projective coordinate ring $R = k[X_0, \dots, X_d]/I_X$ of $X \subseteq \mathbf{P}^d$. First of all, we recall some of the basic definitions and results of [K1]. In particular, we are interested in the applications of those results w.r.t. the Hilbert function of X.

In the sequel, we fix our coordinate system such that $X \subseteq D_+(X_0)$. Thus $x_0 := X_0 + I_X$ is a linear nonzerodivisor of R and ω_R can be described by

$$\omega_R \cong \underline{\mathrm{Hom}}_{k[x_0]}(R, k[x_0])(-1).$$

Definition. We say that X is a *Cayley-Bacharach scheme*, if every hypersurface of degree $\sigma_X := \max\{n \in \mathbf{Z} : H_X(n) < \deg X\}$ containing a subscheme of degree $\deg X - 1$ of X contains X.

Theorem 1.1 (cf. [K1], 2.4). *The scheme X is a locally Gorenstein Cayley-Bacharach scheme if and only if there exists an element $\varphi \in (\omega_R)_{-\sigma_X}$ such that* $\mathrm{Ann}_R(\varphi) = 0$.

Corollary 1.2. *If X is a locally Gorenstein Cayley-Bacharach scheme, then*

$$H_X(n) + H_X(\sigma_X - n) \leq \deg X \quad \textit{for all } n \in \mathbf{Z}.$$

Proof. Since $R(\sigma_X) \cong R\varphi \subseteq \omega_R$, we get $H_X(\sigma_X - n) = \dim_k R(\sigma_X)_{-n} \leq \dim_k(\omega_R)_{-n} = \deg X - H_X(n)$. For the last equality we used [K1], 1.3.c. □

In section 2 we shall apply Theorem 1.1 to give an algebraic proof of the fact that generic hyperplane sections of certain curves are Cayley-Bacharach schemes.

The uniformity of X required in the definition of a Cayley-Bacharach scheme can be extended as follows.

Definition. Let $n \geq 1$. Then X is called $n-$*uniform*, if every subscheme $Y \subseteq X$ of degree $\deg X - n \leq \deg Y \leq \deg X$ has Hilbert function $H_Y = \min\{H_X, \deg Y\}$.

It is easy to see (cf. [K1]) that with this definition, Cayley-Bacharach schemes are exactly the 1-uniform schemes, while ($\deg X - 1$)-uniformity agrees with the usual definition of "in uniform position" .

Theorem 1.3 (cf. [K1], 3.1). *Let* $\Delta_X := H_X(\sigma_X + 1) - H_X(\sigma_X)$. *The scheme* X *is* Δ_X*-uniform if and only if the multiplication map*

$$R_{\sigma_X} \otimes (\omega_R)_{-\sigma_X} \to (\omega_R)_0$$

is biinjective (i.e. $r \in R_{\sigma_X}, \varphi \in (\omega_R)_{-\sigma_X}, r\varphi = 0$ *implies* $r = 0$ *or* $\varphi = 0$ *).*

Corollary 1.4. *If* X *is* Δ_X*-uniform, then*

$$H_X(n) + H_X(\sigma_X - n) \leq \deg X - \Delta_X + 1 \quad \textit{for all } n \in \{0, \ldots, \sigma_X\}.$$

Proof. Since x_0 is a nonzerodivisor of both R and ω_R, also $R_n \otimes (\omega_R)_{-\sigma_X} \to (\omega_R)_{-\sigma_X+n}$ is biinjective. By applying the Biinjective Map Lemma (cf. [K1], section 3), we obtain $\dim_k\,(\omega_R)_{-\sigma_X+n} = \deg X - H_X(\sigma_X - n) \geq \dim_k\,R_n + \dim_k\,(\omega_R)_{-\sigma_X} - 1 = H_X(n) + \Delta_X - 1$. □

In case $\Delta_X \geq 2$, Corollary 1.4 obviously yields stronger inequalities for the Hilbert function of X than Corollary 1.2. In section 3 we examine intermediate uniformities (between 1-uniform and Δ_X-uniform) w.r.t. the question how much uniformity is really needed in order to prove the inequalities in Corollary 1.4 .

Finally, in section 4, we discuss the question under which hypotesis it is possible to strengthen the inequalities of Corollary 1.4. An interesting candidate for a set of even stronger inequalities is given by the following notion.

Definition. We say that H_X is *flawless*, if $\Delta H_X(\sigma_X + 1 - n) \geq \Delta H_X(n) \geq \Delta H_X(n-1)$ for all $0 \leq 2n \leq \sigma_X + 1$.

In our discussion of possible hypoteses yielding flawless Hilbert functions we shall arrive at the following question, which at the same time concludes this introduction:

Question. If X is pure (i.e. if ω_R is generated by its elements of degree $-\sigma_X$) and sufficiently uniform (e.g. in uniform position), does this imply that H_X is flawless?

2. Applications to hyperplane sections of curves

In this section we let $C \subseteq \mathbf{P}^{d+1}$ be an integral curve, and $X := C \cap H \subseteq \mathbf{P}^d$ a generic hyperplane section of C. By S we denote the projective coordinate ring of C, and by $l \in S_1$ the equation of H in S. Then the projective coordinate ring R of X satisfies $R = (S/lS)/H^0_{\mathfrak{m}}(S/lS)$. Furthermore, we let $M := H^1_{\mathfrak{m}}(S) = H^1_*(\mathbf{P}^{d+1}, \mathcal{J}_C)$ denote the Hartshorne-Rao module of C, and consider it as a graded S-module.

Our first result describes the relationship between the canonical modules $\omega_R = H^1_{\mathfrak{m}}(R)^*$ of R and $\omega_S = H^2_{\mathfrak{m}}(S)^*$ of S. Here * denotes graded dualization w.r.t. k, i.e. $\underline{\mathrm{Hom}}_k(-, k)$.

Theorem 2.1 (The Fundamental Exact Sequence).
There is an exact sequence of graded R-modules

$$0 \to (\omega_S/l\omega_S)(1) \to \omega_R \to (M/lM)^* \to 0.$$

Proof. Let $T := S/lS$. Since $R = T/H^0_{\mathfrak{m}}(T)$, we have $H^1_{\mathfrak{m}}(R) \cong H^1_{\mathfrak{m}}(T)$, and therefore $\omega_R \cong \omega_T$ (cf. [GW]). An application of the long exact sequence of local cohomology modules to

$$0 \to S(-1) \xrightarrow{l} S \to T \to 0$$

yields a homogeneous exact sequence

$$0 \to H^0_{\mathfrak{m}}(T) \to H^1_{\mathfrak{m}}(S)(-1) \xrightarrow{l} H^1_{\mathfrak{m}}(S) \to H^1_{\mathfrak{m}}(T) \to H^2_{\mathfrak{m}}(S)(-1) \xrightarrow{l} H^2_{\mathfrak{m}}(S) \to 0.$$

Now we can use $M = H^1_{\mathfrak{m}}(S)$ and apply the functor * to obtain

$$0 \to \omega_S \xrightarrow{l} \omega_S(1) \to \omega_R \to M^* \xrightarrow{l} M^*(1) \to H^0_{\mathfrak{m}}(T)^* \to 0.$$

Since obviously $\ker(M^* \xrightarrow{l} M^*(1)) \cong (M/lM)^*$, splitting this sequence concludes the proof. □

Using Theorems 1.1 and 2.1 we can now give an algebraic proof of the fact that X is a Cayley-Bacharach scheme (regardless of char k) in the case $\sigma_X - 1 = \sigma_C := \max\{n \in \mathbf{Z} : h^1\mathcal{O}_C(n) \neq 0\}$. For this we need a few preparations.

Lemma 2.2. *Let N be a finitely generated graded S-module, and let $l \in S_1$ be a general linear form. Then* $\mathrm{Ann}_N(l) := \{x \in N : lx = 0\} \subseteq H^0_{\mathfrak{m}}(N)$.

Proof. From Grothendieck' s standard graded exact sequence we construct a commutative diagram with exact rows

$$\begin{array}{ccccccc} 0 & \to & H^0_{\mathfrak{m}}(N)(-1) & \to & N(-1) & \to & \Gamma_*(\tilde{N})(-1) \\ & & \downarrow \mu_l & & \downarrow \mu_l' & & \downarrow \mu_l'' \\ 0 & \to & H^0_{\mathfrak{m}}(N) & \to & N & \to & \Gamma_*(\tilde{N}) \end{array}$$

Since $H^0_{\mathfrak{m}}(\Gamma_*(\tilde{N})) = 0$, we have $\mathfrak{m} \notin \operatorname{Ass}(\Gamma_*(\tilde{N}))$, and μ_l'' is injective. Hence $\operatorname{Ann}_N(l) = \ker \mu_l' = \ker \mu_l \subseteq H^0_{\mathfrak{m}}(N)$. □

Lemma 2.3. *In the situation described above we have* $\operatorname{Ass}(\omega_S) = \{(0)\}$.

Proof. Let $\mathfrak{p} \subseteq S$ be a homogeneous prime of height at least one. From the Graded Duality Theorem (cf. [GW]) we get $\operatorname{Ann}_{\omega_S}(\mathfrak{p}) \cong \underline{\operatorname{Hom}}_S(S/\mathfrak{p}, \omega_S) \cong H^2_{\mathfrak{m}}(S/\mathfrak{p})^* = (0)$. Thus $\mathfrak{p} \notin \operatorname{Ass}(\omega_S)$. □

Theorem 2.4. *Let $X \subseteq \mathbf{P}^d$ be the generic hyperplane section of an integral curve $C \subseteq \mathbf{P}^{d+1}$. If $\sigma_X = \sigma_C + 1$, then X is a Cayley-Bacharach scheme.*

Proof. First of all, we choose a nonzero element $\Phi \in (\omega_S)_{-\sigma_C}$. This is possible, since $\sigma_C = \max\{n \in \mathbf{Z} : h^1\mathcal{O}_C(n) \neq 0\} = -\min\{n \in \mathbf{Z} : (\omega_S)_n \neq 0\}$. Using Lemma 2.3 we get $\operatorname{Ass}(S\Phi) \subseteq \operatorname{Ass}(\omega_S) = (0)$, and therefore

$$\dim(S/\operatorname{Ann}_S(\Phi)) = \max\{\dim S/\mathfrak{p} : \mathfrak{p} \in \operatorname{Ass}(S\Phi)\} = 2.$$

This implies $\operatorname{Ann}_S(\Phi) = 0$. Now we construct the following commutative diagram of homogeneous maps with exact rows and columns.

$$\begin{array}{ccccccccc}
0 & \rightarrow & S(\sigma_C) & \xrightarrow{\Phi} & \omega_S & \rightarrow & \omega_S/S\Phi & \rightarrow & 0 \\
 & & \downarrow \mu_l & & \downarrow \mu_l' & & \downarrow \mu_l'' & & \\
0 & \rightarrow & S(\sigma_C+1) & \xrightarrow{\Phi} & \omega_S(1) & \rightarrow & (\omega_S/S\Phi)(1) & \rightarrow & 0 \\
 & & \downarrow & & \downarrow & & \downarrow & & \\
 & & (S/lS)(\sigma_X) & \xrightarrow{\mu_\varphi} & (\omega_S/l\omega_S)(1) & \rightarrow & \omega_S/(S\Phi + l\omega_S)(1) & \rightarrow & 0 \\
 & & \downarrow & & \downarrow & & \downarrow & & \\
 & & 0 & & 0 & & 0 & &
\end{array}$$

Let φ denote the image of Φ in $(\omega_S/l\omega_S)(1)$. From Lemma 2.2 we conclude that $\ker \mu_l'' = \operatorname{Ann}_{\omega_S/S\Phi}(l) \subseteq H^0_{\mathfrak{m}}(\omega_S/S\Phi)$. Next we apply the Snake Lemma to the above diagram. Since μ_l' is injective, we see that $\ker \mu_\varphi \subseteq H^0_{\mathfrak{m}}(S/lS)$. Therefore, if we regard φ via the Fundamental Exact Sequence as an element of ω_R, we get $\operatorname{Ann}_R(\varphi) = 0$. Since $\varphi \in (\omega_S/l\omega_S)(1)_{-\sigma_C-1} \subseteq (\omega_R)_{-\sigma_X}$, an application of Theorem 1.1 yields the claim. □

In case C is arithmetically Cohen-Macaulay, the Fundamental Exact Sequence collapses to an isomorphism $\omega_S \cong (\omega_S/l\omega_S)(1)$, and the proof of Theorem 2.4 reduces to the one given in [GKR], 5.1 . Nontrivial examples of the thorem can be found using arithmetically Buchsbaum curves.

Example. Let $C \subseteq \mathbf{P}^{d+1}$ be an integral, arithmetically Buchsbaum curve, and let $X = C \cap H \subseteq \mathbf{P}^d$ be a generic hyperplane section of C. It is well-known that the Hartshorne-Rao module M of C is generated in degrees $\leq \sigma_C + 2$. Since $\mathfrak{m}M = 0$, this implies $M_n = 0$ for all $n > \sigma_C + 2$. Suppose that also $M_{\sigma_C+2} = 0$. Then we have $(M/lM)^*_{-\sigma_C-2} = 0$ in the Fundamental Exact Sequence, and therefore $-\sigma_X = \min\{n \in \mathbf{Z} : (\omega_R)_n \neq 0\} = \min\{n \in \mathbf{Z} : (\omega_S/l\omega_S)(1)_n \neq 0\} = -\sigma_C -$

1. Hence Theorem 2.4 can be applied and yields that X is a Cayley-Bacharach scheme.

3. Applications to cohomologically split schemes

For this section, we shall assume that $X \subseteq \mathbf{P}^d$ is a reduced 0-dimensional subscheme. In the introduction we saw that Δ_X-uniform schemes X satisfy the inequalities of Corollary 1.4 , while for 1-uniform schemes we obtained only weaker inequalities. The following example demonstrates that the stronger set of inequalities actually fails for some 1-uniform schemes.

Example. Let $L_1, L_2 \subseteq \mathbf{P}^3$ be two skew lines, and let $\{P_1, \ldots, P_4\} \subseteq L_1$ as well as $\{P_5, \ldots, P_8\} \subseteq L_2$ be two subsets of four distinct points each. Then $X := \{P_1, \ldots, P_8\} \subseteq \mathbf{P}^3$ is a Cayley-Bacharach scheme, but $H_X : 1\ 4\ 6\ 8\ 8 \cdots$ does not satisfy the inequalities of Corollary 1.4 .

Intuitively, the failure of those inequalities comes from the fact that X "splits" into two disjoint subsets having high individual uniformities, but little to do with each other. Algebraically, we can formalize this as follows.

Write $X = \{P_1, \ldots, P_s\}$, where $s =$ deg X. For each $i \in \{1, \ldots, s\}$ the ideal of $X \backslash \{P_i\}$ in X defines a unique 1-dimensional subvectorspace V_i of R_{σ_X+1}.

Definition . We say that X *splits cohomologically* , if we can decompose $X = Y \cup Z$ such that $Y \neq \emptyset$, $Z \neq \emptyset$, $Y \cap Z = \emptyset$, and the only elements $r \in \sum_{P_i \in Y} V_i, r' \in \sum_{P_i \in Z} V_i$ with $r + r' \in x_0 R_{\sigma_X}$ are $r = r' = 0$.

In order to make this an intrinsic definition, we have to show that it does not depend on the choice of the linear nonzerodivisor $x_0 \in R_1$. At the same time, we want to explain the choice of the name.

Remark. The local cohomology module $H^1_{\mathfrak{m}}(R)$ is a graded R-module whose grading satisfies $\sigma_X = \max\{n \in \mathbf{Z} : H^1_{\mathfrak{m}}(R)_n \neq 0\}$. Each subscheme $Y_i := X \backslash \{P_i\}$ corresponds to a 1-dimensional subspace $c(Y_i) \subseteq H^1_{\mathfrak{m}}(R)_{\sigma_X}$ in the following way: if $f_i \in V_i \backslash \{0\}$, then the generalized fraction $\begin{bmatrix} f_i \\ x_0 \end{bmatrix}$ defines $c(Y_i)$. Using the rules for computing with generalized fractions, the above definition translates into the following statement:

X splits cohomologically if and only if $X = Y \cup Z$ with $Y \neq \emptyset, Z \neq \emptyset, Y \cap Z = \emptyset$, and $\sum_{P_i \in Y} c(Y_i) \cap \sum_{P_i \in Z} c(Y_i) = (0)$. Since $c(Y_i)$ clearly does not depend on the choice of x_0, neither does the notion of cohomological splitting.

The reader may verify that the example given above constitutes a cohomologically split scheme. Our next theorem characterizes cohomologically non-split schemes in terms of their canonical module. This will allow us to show that they enjoy a uniformity which is indeed intermediate between 1-uniformity and Δ_X-uniformity.

Theorem 3.1. *Let $X = \{P_1, \ldots, P_s\} \subseteq \mathbf{P}^d$ be a reduced 0-dimensional subscheme. The following conditions are equivalent.*

a) *X does not split cohomologically.*

b) *The multiplication map $R_{\sigma_X} \otimes (\omega_R)_{-\sigma_X} \to (\omega_R)_0$ is nondegenerate and surjective.*

The proof of this theorem relies on a careful study of the canonical ideal of X and is contained in [K2]. Here we are interested in the following application.

Corollary 3.2. *For reduced 0-dimensional subschemes $X \subseteq \mathbf{P}^d$, the following implications hold.*

$$\begin{gathered} X \textit{ is } \Delta_X\textit{-uniform} \\ \Downarrow \\ X \textit{ does not split cohomologically} \\ \Downarrow \\ X \textit{ is a Cayley-Bacharach scheme (i.e. 1-uniform)} \end{gathered}$$

Proof. The second implication immediately follows from [K1], Theorem 2.6, which says that X is a Cayley-Bacharach scheme if and only if the multiplication map $\mu : R_{\sigma_X} \otimes (\omega_R)_{-\sigma_X} \to (\omega_R)_0$ is nondegenerate. For the first implication, we only have to make the additional observation that the multiplication map μ is surjective, if it is biinjective (which it is by Theorem 1.3), since the Biinjective Map Lemma (cf. [K1], section 3) yields $\dim_k \operatorname{im} \mu \geq \dim_k (\omega_R)_{-\sigma_X} - 1 = s - 1 = \dim_k (\omega_R)_0$. □

It is interesting to note that cohomological non-splitting rules out all known examples of Cayley-Bacharach schemes violating the inequalities of Corollary 1.4. Therefore we are led to putting forward the following

Question. If $X \subseteq \mathbf{P}^d$ is a reduced 0-dimensional subscheme which does not split cohomologically, does it follow that $H_X(n) + H_X(\sigma_X - n) \leq \deg X - \Delta_X + 1$ for all $n \in \{1, \ldots, \sigma_X\}$?

Although we have not yet been able to decide this question, we shall end this section with a partial result. Recall that a generic basis $\{\varphi_1, \ldots, \varphi_{\Delta_X}\}$ of $(\omega_R)_{-\sigma_X}$ satisfies $\operatorname{Ann}_R(\varphi_i) = (0)$ for $i = 1, \ldots, \Delta_X$, if X is a Cayley-Bacharach scheme (cf. [GKR]). Let us say that X *has property* $(*)$, if $(\omega_R)_0$ has a decomposition of the form

$$(\omega_R)_0 = R_{\sigma_X}\varphi_1 \oplus k x_0^{\sigma_X} \varphi_2 \oplus \ldots \oplus k x_0^{\sigma_X} \varphi_{\Delta_X}.$$

Theorem 3.3. *Let $X \subseteq \mathbf{P}^d$ be a reduced 0-dimensional subscheme.*

a) *If X has property $(*)$, then $H_X(n) + H_X(\sigma_X - n) \leq \deg X - \Delta_X + 1$ for all $n \in \{0, \ldots, \sigma_X\}$.*

b) *If X has property $(*)$, then X does not split cohomologically.*

Proof. a). Since x_0 is not a zerodivisor of ω_R, it follows from property $(*)$ that $(\omega_R)_{-\sigma_X+n}$ contains $R_n\varphi_1 \oplus kx_0^n\varphi_2 \oplus \ldots \oplus kx_0^n\varphi_{\Delta_X}$ for all $n \in \{0,\ldots,\sigma_X\}$. Hence $\deg X - H_X(\sigma_X - n) = \dim_k(\omega_R)_{-\sigma_X+n} \geq H_X(n) + \Delta_X - 1$.

b). Property $(*)$ yields that the multiplication map $R_{\sigma_X} \otimes (\omega_R)_{-\sigma_X} \to (\omega_R)_0$ is nondegenerate and surjective. Then we can apply Theorem 3.1. □

Since in [K2] we also show that Δ_X-uniform schemes have property $(*)$, we have found another condition of uniformity, which is intermediate between Δ_X-uniformity and cohomological non-splitting, and which at the same time implies the inequalities of Corollary 1.4. Therefore the above question really asks:

Question. Do cohomologically non split schemes have property $(*)$?

4. Applications to pure schemes

After having examined the inequalities of Corollary 1.4, we venture further and ask under which circumstances it is possible to prove even stronger inequalities for the Hilbert function of a 0-dimensional scheme $X \subseteq \mathbf{P}^d$. The inequalities describing flawless Hilbert functions appear to be natural candidates because of a result of T. Hibi which we are now going to explain.

From the presentation $R \cong k[X_0,\ldots,X_d]/I_X$ we get $\overline{R} \cong k[X_1,\ldots,X_d]/J_X$, where $\overline{R} := R/(x_0)$ and $J_X := I_X \bmod X_0$. Choose a degree-compatible ordering τ of the monomials of $k[X_1,\ldots,X_d]$, and form the ideal of leading terms $\mathrm{Lt}_\tau(J_X) := (\mathrm{Lt}_\tau f)_{f\in J_X}$. Let M be the set of all monomials of $k[X_1,\ldots,X_d]$ which are not in $\mathrm{Lt}_\tau(J_X)$. It is well-known that the elements of M form an order ideal of monomials (i.e. if $m \in M$ and $m'|m$, then $m' \in M$), and that the Hilbert function of X satisfies $\Delta H_X(n) = \#\{m \in M : \deg m = n\}$ for all $n \in \mathbf{Z}$. We say that X is *combinatorically pure*, if we can choose x_0 and τ in such a way that all maximal elements of M have the same degree (namely $\sigma_X + 1$).

Theorem 4.1 (cf. [H]). *If X is combinatorically pure, then H_X is flawless.*

In the present situation this result is not entirely satisfactory, because the choices involved in the above definition make it hard to check in practice whether a given X is combinatorically pure. As a more algebraic notion, we propose the following.

Definition. We say that X is *pure*, if ω_R is generated by its elements of minimal degree, i.e. by $(\omega_R)_{-\sigma_X}$.

First of all, let us show that this notion generalizes the above notion of combinatorical purety. For that and for other purposes it is useful to characterize pure schemes in a number of different ways.

Theorem 4.2. *The following conditions are equivalent.*

a) *X is combinatorically pure.*

b) *The canonical module* $\omega_{\overline{R}}$ *of* $\overline{R}$ *is generated by its elements of minimal degree (which is* $-\sigma_X - 1$*).*

c) *For all* $m, n \geq 0$ *with* $m + n \leq \sigma_X + 1$*, the multiplication map* $\overline{R}_m \otimes \overline{R}_n \to \overline{R}_{m+n}$ *is nondegenerate.*

d) *The socle* $\mathfrak{S}(\overline{R}) := \{r \in \overline{R} : \mathfrak{m}r = 0\}$ *is precisely* $\mathfrak{S}(\overline{R}) = \overline{R}_{\sigma_X+1}$.

Proof. The equivalence of a) and b) follows from the facts that x_0 is a nonzerodivisor of ω_R and that $\omega_{\overline{R}} \cong (\omega_R/x_0\omega_R)(1)$ by applying the graded version of Nakayama's Lemma. Clearly, b) is equivalent with the multiplication maps $\overline{R}_n \otimes (\omega_{\overline{R}})_{-\sigma_X-1} \to (\omega_{\overline{R}})_{-\sigma_X-1+n}$ being surjective for $n = 0, \ldots, \sigma_X + 1$. Using $\omega_{\overline{R}} \cong \overline{R}^*$, a standard dualization argument shows that it is also equivalent with the multiplication maps $\overline{R}_n \otimes \overline{R}_{\sigma_X+1-n} \to \overline{R}_{\sigma_X+1}$ being nondegenerate for $n = 0, \ldots, \sigma_X + 1$. This in turn is easily seen to be equivalent with c) and d). □

The proof of the following corollary is based on an observation of L. Robbiano.

Corollary 4.3. *If* X *is combinatorically pure, then* X *is pure.*

Proof. If $r \in \mathfrak{S}(\overline{R})$, then $\mathrm{Lt}_\tau(r) \in \mathfrak{S}(\mathrm{gr}_\tau(\overline{R}))$, because $\mathrm{Lt}_\tau(x_i) \cdot \mathrm{Lt}_\tau(r)$ is either zero or it equals $\mathrm{Lt}_\tau(x_i r) = \mathrm{Lt}_\tau(0) = 0$. Now $\mathfrak{S}(\mathrm{gr}_\tau(\overline{R})) = \mathrm{gr}_\tau(\overline{R})_{\sigma_X+1}$, because of our assumption about the maximal elements of M. Hence $r \in \overline{R}_{\sigma_X+1}$. The other inclusion is clear, so that Theorem 4.2.d yields the claim. □

In other words, Theorem 4.2.d means that $\overline{R}$ is a level ring. Using the other conditions of that theorem, we can derive some implications for the Hilbert functions of pure schemes which have been announced first in [S].

Corollary 4.4. *Let* $X \subseteq \mathbf{P}^d$ *be pure.*

a) *For all* $m, n \geq 0$ *with* $m + n \leq \sigma_X + 1$ *we have* $\Delta H_X(m) \leq \Delta H_X(n) \cdot \Delta H_X(m+n)$.

b) *The function* $f(n) := \Delta H_X(\sigma_X + 1 - n)$ *is the sum of* Δ_X *Hilbert functions of quotient rings of* $\overline{R}$.

c) *If* $Y \subseteq X$ *is a subscheme with* $\deg Y = H_X(n)$ *for some* $n \in \{0, \ldots, \sigma_X + 1\}$*, and if* $H_Y = \min\{H_X, \deg Y\}$*, then* Y *is pure again. In particular,* H_Y *satisfies the above restrictions, too.*

Proof. a) follows from Theorem 4.2.c, since that condition implies that the canonical map $\overline{R}_m \to \mathrm{Hom}_k(\overline{R}_n, \overline{R}_{m+n})$ is injective.

b) follows from Theorem 4.2.b by induction on Δ_X. If $\varphi_1, \ldots, \varphi_{\Delta_X} \in (\omega_{\overline{R}})_{-\sigma_X-1}$ is a minimal set of generators of $\omega_{\overline{R}}$, then $\omega_{\overline{R}}/\overline{R}\varphi_1$ is generated by the images of $\varphi_2, \ldots, \varphi_{\Delta_X}$, while $\overline{R}\varphi_1 \cong (\overline{R}/\mathrm{Ann}_{\overline{R}}(\varphi_1))(\sigma_X + 1)$ has (up to a shift) the Hilbert function of a quotient ring of $\overline{R}$. Now note that $f(n)$ is nothing but the Hilbert function of $\omega_{\overline{R}}$ shifted by $\sigma_X + 1$.

c) follows from Theorem 4.2.c,d, since $R/(I_{Y/X} + (x_0)) \cong \bigoplus_{i=0}^{n} \overline{R}_i$ has its socle concentrated in degree n. □

In view of Corollary 4.3 it would be natural to conjecture that any pure scheme has a flawless Hilbert function. Unfortunately, this seems not to be true, since there are examples of 0-dimensional standard graded Grenstein rings $\overline{R}$ (i.e. level rings of type one) whose Hilbert functions are not unimodal. Here H_X (or $H_{\overline{R}}$) is called *unimodal*, if there exists an $n \in \{0, \ldots, \sigma_X + 1\}$ such that $\Delta H_X(0) \leq \ldots \leq \Delta H_X(n) \geq \ldots \geq \Delta H_X(\sigma_X + 1)$. Thus we would like to pose the following more modest

Question. If X is pure and sufficiently uniform (e.g. in uniform position), does this imply that H_X is flawless?

Let us show how one can use the uniformity assumption in a discussion of this question.

Remark. Suppose we can show that any Δ_X-uniform, pure scheme satisfies $\Delta H_X(n) \leq \Delta H_X(\sigma_X + 1 - n)$ for $0 \leq n \leq \sigma_X + 1$. Then any pure scheme in uniform position has a flawless Hilbert function. Namely, we can show $\Delta H_X(n-1) \leq \Delta H_X(n)$ in the specified range by considering a subscheme $Y \subseteq X$ of degree $\deg Y = H_X(2n)$, and by applying Corollary 4.4.c and the assumption to it.

Moreover, the above uniformity assumption seems natural, because any pure scheme has at least some uniformity.

Theorem 4.5. *If X is pure, then X is a Cayley-Bacharach scheme.*

Proof. Let $Y \subseteq X$ be a subscheme of degree $\deg Y = \deg X - 1$, and let $I_{Y/X} \subseteq R$ be its defining ideal. Choose $f \in I_{Y/X} \backslash \{0\}$ of minimal degree. Then $(I_{Y/X})_{\deg f+1} = kx_0 f$ implies that the image of f in $\overline{R}$ is an element of the socle of $\overline{R}$. Hence $\deg f = \sigma_X + 1$, i.e. X is a Cayley-Bacharach scheme (cf. [GKR]). □

Finally, we would like to point out an interesting consequence, if the above question has a positive answer.

Theorem 4.6. *Let* char $k = 0$. *Suppose any pure scheme in uniform position has a flawless Hilbert function. Then the Hilbert function of any n-dimensional standard graded Gorenstein domain S has a unimodal difference function $\Delta^n H_S$.*

Proof. Since the case $\dim S = 1$, $S = k[X]$ is trivial, we can assume $n \geq 2$. Let $d + 1 = \operatorname{embdim}(S)$, and let $V \subseteq \mathbf{P}^d$ be the projective variety with projective coordinate ring S. Since $\dim V = n - 1$, we can take $n - 1$ generic hyperplane sections and arrive at a set of points $X \subseteq \mathbf{P}^{d-n-1}$ which is in uniform position by the Uniform Position Lemma. Notice that $H_X = \Delta^{n-1} H_S$ and X is arithmetically Gorenstein, hence pure. By assumption, H_X is flawless, and therefore $\Delta^n H_S = \Delta H_X$ is symmetrical and unimodal. □

References

[GKR] A.V. Geramita, M. Kreuzer and L. Robbiano, Cayley-Bacharach schemes and their canonical modules, Trans. Amer. Math. Soc. (to appear)

[GW] G. Goto and K. Watanabe, On graded rings I, J. Math. Soc. Japan **30** (1978), 179–213

[H] T. Hibi, What can be said about pure *O*-Sequences?, J. Combin. Theory **50** (1989), 319–322

[K1] M. Kreuzer, On the canonical module of a 0-dimensional scheme, preprint 1992

[K2] M. Kreuzer, On the canonical ideal of a set of points, in preparation (1992)

[S] R.P. Stanley, Cohen-Macaulay complexes, in: Higher Combinatorics, NATO Adv. Study Inst. Ser., Reidel, Boston 1977

Acknowledgements. Part of this research was done while the author was staying at Queen's University, Kingston. He wishes to thank A.V. Geramita for his hospitality and encouragement. Furthermore, the author is indebted to L. Robbiano and L.G. Roberts for stimulating discussions, and to E. Kunz for the opportunity to lecture about a preliminary version of this paper in his seminar. Last but not least, cordial thanks to the organizers of the Ravello conference for providing such a stimulating and warm atmosphere.

Zero-dimensional Schemes on Abelian Surfaces

Antony Maciocia

Abstract. The moduli spaces of semistable torsion-free sheaves with $c_1 = 0$ and $c_2 = -2$ and -3 over a principally polarised complex torus are described explicitly in terms of zero-dimensional subschemes of the torus. The boundary structures are computed in detail. The first moduli space is a compactified family of Jacobians and the second is a Hilbert scheme.

1991 Mathematics Subject Classification: 14J60, 14C17, 14C05, 14H40, 53C07.

0. Introduction

In this paper we shall show how detailed information about zero-dimensional subschemes of a principally polarised complex Abelian variety $(\mathbf{T}, L)$ can be used to give us information about the moduli space of stable bundles. The information we are looking for is existence and connectedness of these moduli spaces. We shall show how the moduli spaces are related to Hilbert spaces of zero-dimensional subschemes of our torus (and its dual) using a combination of Serre's method of constructing vector bundles and Mukai's Fourier transform for tori. Properties of the vector bundles representing points of the moduli spaces can be related to geometrical properties of certain zero-dimensional schemes. The most important properties is whether various zero-dimensional schemes and their subschemes lie on certain divisors. These type of problems have been called 'interpolation problems' by Geramita (see his article in this volume).

We will consider the interpolation problem from an intrinsic viewpoint and call on results from [8] which dealt with many of these questions. However, the problem can also be viewed extrinsically by embedding the torus in some projective space. The most natural one would be $\mathbf{C}P^8$ given by the very ample linear system $|L^3|$. One could also gain some information from the singular Kummer surface in $\mathbf{C}P^3$.

Stability in this context means either μ-stability of Mumford-Takemoto or G-stability of Gieseker. To define these notions we require a polarized variety (X, ℓ) of

The author is grateful to the organisers of the Ravello conference on Zero-dimensional subschemes for their support. He would also like to thank the Seggie-Brown trust for support while this work was carried out.

dimension n (or a complex manifold with a chosen Kähler form ω). Let us denote the Chern characters of sheaves on X by $(n+1)$-tuples: $(r, c_1, \frac{1}{2}c_1^2 - c_2, \ldots)$.

Definition 0.1. We say that a torsion-free sheaf E is μ-stable (respectively, μ-semistable) with respect to ℓ if for all subsheaves F such that E/F is torsion-free we have $\mu(F) < \mu(E)$ (resp. $\mu(E) \leq \mu(F)$), where $\mu(E)$ is the *slope* of E defined by $\mu(E) = c_1(E) \cdot \ell^{n-1}/r(E)$.

Definition 0.2. We say that E is G-stable (resp. G-semistable) with respect to ℓ if for all subsheaves F with E/F torsion-free, we have $P_m(F) < P_m(E)$ (resp. $P_m(F) \leq P_m(E)$) for all m sufficiently large, where $P_m(E) = \chi(E \otimes L^m)/r(E)$ for some representative line bundle L of ℓ.

There is a chain of implications:

$$\mu\text{-stable} \Rightarrow \text{G-stable} \Rightarrow \text{G-semistable} \Rightarrow \mu\text{-semistable}.$$

Further properties of stabilities can be found in [7, Chapter 5] or [15, II.1].

The following is well known:

Theorem 0.3 (Gieseker [4], Maruyama [10,11]). *Let E be a G-stable vector bundle. Then the component of E in the moduli space of G-stable vector bundles of a given Chern character is a quasi-projective variety and its closure $\mathcal{M}(E)$ in the space of G-semistable torsion-free sheaves is projective.*

We shall denote the closure of the μ-stable moduli space of vector bundles in the moduli space of S-equivalence classes of G-semistable torsion-free sheaves with Chern character $(2, 0, -k)$ by $\mathcal{M}_k$. We shall restrict our attention to $\mathcal{M}_2$ and $\mathcal{M}_3$. Our principal results are the following.

Theorem 0.4. *The moduli space $\mathcal{M}_2$ is naturally fibred over a $\mathbf{C}P^3$-bundle over the dual torus $\hat{\mathbf{T}}$ with fibres given by certain compactifications of Jacobians of genus 5 curves.*

Theorem 0.5. *The moduli space $\mathcal{M}_3$ is isomorphic to* $\mathrm{Hilb}^6\hat{\mathbf{T}} \times \mathbf{T}$.

Similar calculations for the case $\mathcal{M}_{(2,L_1,-1)}$ were carried out in [1].

It is also well known that the open subspace of μ-stable bundles coincides with gauge equivalence classes of Hermitian-Einstein connections on the underlying smooth bundle [17], [2], [3]. In complex dimension 2 these connections are called *instantons*.

We shall be interested in the case when (X, ℓ) is a principally polarized Abelian surface $(\mathbf{T}, L)$, where we represent ℓ by symmetric line bundle L, $\chi(L) = 1$ and $0 \in D_L$, the divisor corresponding to a non-trivial section of L. Taubes ([16]) and Gieseker ([5]) have solved the general existence question for instantons or μ-stable bundles which, in our case, provide examples of μ-stable bundles for $c_1(E) = 0$

and $\chi(E) \leq -4$. It is easy to show (see below) that there are no μ-stable bundles with Chern character $(r, 0, -1)$.

It is also known that the moduli space of simple torsion-free sheaves Spl on an Abelian surface or K3 surface is smooth for all E (see [13]) and are symplectic manifolds in a natural way. This contains the moduli space of G-stable torsion-free sheaves. This implies, for example, that $\mathcal{M}_3 \cap \mathrm{Spl}$ is smooth, consistent with Theorem 0.5 above.

We shall make the simplifying assumption that $\mathbf{T}$ is irreducible (i.e., cannot be written as the product of elliptic curves). The reducible case can be treated similarly but the statement of our results and their proofs are more complicated. Our assumption implies that the Neron-Severi group of $\mathbf{T}$ is generated by L.

Our primary tool for studying bundles and sheaves on such surfaces is the *Fourier-Mukai transform* whose definition and basic properties we shall now outline.

1. The Fourier-Mukai Transform

This construction is due to Mukai (see [12], [14]). Let $\hat{\mathbf{T}}$ denote the dual torus of $\mathbf{T}$. This is just $\mathrm{Pic}^0\mathbf{T}$ and we shall denote the correspondence by $\hat{x} \mapsto \mathcal{P}_{\hat{x}}$. Let $\mathcal{P}$ denote the Poincaré bundle on $\mathbf{T} \times \hat{\mathbf{T}}$. Given a sheaf E over $\mathbf{T}$ then we define $\mathbf{R}^i\mathcal{F}(E) = R^i\hat{\pi}_*(\pi^*E \otimes \mathcal{P})$, where $\mathbf{T} \xleftarrow{\pi} \mathbf{T} \times \hat{\mathbf{T}} \xrightarrow{\hat{\pi}} \hat{\mathbf{T}}$ are the projection maps.

Definition 1.1. Following Mukai, we say that E satisfies WIT_n if $\mathbf{R}^i\mathcal{F}(E) = 0$ unless $i = n$ and write $\hat{E} = \mathbf{R}^n\mathcal{F}(E)$. WIT stands for "Weak Index Theorem".

If E satisfies WIT_n then the fibres of $\hat{E}$ over $\hat{x} \in \hat{\mathbf{T}}$ are given by $H^n(\mathbf{T}, E \otimes \mathcal{P}_{\hat{x}})$. This lead us to introduce the following stronger condition on E.

Definition 1.2. We say that E satisfies IT_n if for all $i \neq n$ and $\hat{x} \in \hat{\mathbf{T}}$ we have $H^i(E \otimes \mathcal{P}_{\hat{x}}) = 0$.

Examples.

(1) Any line bundle with non-zero Euler character satisfies IT_n for some n. An ample line bundle always satisfies IT_0.

(2) $\mathcal{P}_{\hat{x}}$ satisfies WIT_2 and $\hat{\mathcal{P}}_{\hat{x}} = \mathbf{C}(-\hat{x})$, the skyscraper sheaf at $\hat{x}$.

(3) If E is μ-stable torsion-free sheaf with $c_1(E) = 0$ then E satisfies IT_1.

Proof (of (3)). Since stability is preserved by twisting by line bundles and dualising it suffices, by Serre duality, to show that $H^0(E) = 0$. If $f : \mathcal{O} \to E$ is a non-trivial section of E then f injects as the kernel must be torsion-free. This contradicts μ-stability of E. □

If E satisfies WIT_n then the Chern character of $\hat{E}$ is given by $\mathrm{ch}(E)_i = (-1)^{i+n}\mathrm{ch}(E)_{2-i}$, where we have abused notation by omitting the isomorphism $H^i(\mathbf{T}) \cong H_i(\hat{\mathbf{T}}) \cong H^{2-i}(\hat{\mathbf{T}})$.

Although we think of $\hat{E}$ as the Fourier transform of E, we can generalise to the derived categories of complexes of coherent sheaves on $\mathbf{T}$ and $\hat{\mathbf{T}}$ to obtain a functor $\mathbf{R}\mathcal{F} : D(\mathbf{T}) \to D(\hat{\mathbf{T}})$. It is remarkable that this obeys a Fourier Inversion Theorem:

Theorem 1.3 [12, Thm.2.2].
$\mathbf{R}\mathcal{F}$ is an isomorphism of categories with inverse given by $(-1_{\mathbf{T}})^*\mathbf{R}\pi_*(\hat{\pi}^* - \otimes\mathcal{P})$.

In other words, $(\hat{E})\check{} \cong (-1_{\mathbf{T}})^*E$ for a sheaf E satisfying WIT. More generally, even if E does not satisfy WIT we obtain a spectral sequence whose E_2 term is $\mathbf{R}^p\mathcal{F}\mathbf{R}^q\mathcal{F}(E)$ and which converges to $E_\infty^{p+q} \cong (-1_{\mathbf{T}})^*E$ when $p+q=2$ and 0 otherwise.

2. Stable sheaves with $c_1 = 0$ and $\chi > -4$

If E is a μ-stable bundle with Chern character $(r, 0, -1)$ then by example (3) above it satisfies IT_1 and hence $\hat{E}$ is a vector bundle with Chern character $(1, 0, -r)$. This is impossible and so no such bundles exist. However, $\mathcal{M}_1$ is not empty but consists of non-locally-free G-stable and G-semistable sheaves.

Our aim will be to construct points in the interior of the moduli space of μ-stable bundles of Chern characters $(2, 0, -2)$ and $(2, 0, -3)$. We shall use the well established Serre/Schwarzenberger method of constructing bundles (see [6, pp720–731]). The boundary of the first moduli space has been studied in detail in [9].

Notation. We use the notation L_x to denote $L \otimes \mathcal{P}_x$ and we drop the tensor product sign when this does not lead to confusion. We use $\hat{}$ to denote objects associated with the dual torus. However, since $\hat{\mathbf{T}} \cong \mathbf{T}$ via the principal polarization, the use of formulae such as $x + \hat{x}$ is permisable and mean that x originated in the torus and $\hat{x}$ originated in the dual torus.

Proposition 2.1. *Suppose that E is μ-semistable torsion-free sheaf of Chern character $(2, 0, -k)$, $k = 2$ or 3. Then, for some $\hat{x}$, $EL_{\hat{x}}$ admits a section. Moreover, if $k = 2$ then $\hat{x}$ can be chosen from a translate of some divisor in $|\hat{L}^2|$ and if $k = 3$ then $\hat{x}$ can be chosen from $V \in \mathrm{Hilb}^6\hat{\mathbf{T}}$.*

Proof. Observe that $\mu(EL_{\hat{x}}) > 0$ and hence $H^2(EL_{\hat{x}}) = 0$ by μ-semistability. Suppose that $H^0(EL_{\hat{x}}) = 0$ for all $\hat{x}$ so that EL satisfies IT_1. If $\chi(E) = -2$ then $\chi(EL_{\hat{x}}) = 0$ and so EL has zero Fourier transform which is impossible. If $\chi(E) = -3$ then $\widehat{EL}$ is a line bundle with Chern character $(1, 2\ell, -2)$, but

$(2\ell)^2/2 = 4 \neq -2$: a contradiction. If $k = 2$ then $\det(EL) \sim \hat{L}^{-2}$ and so $\mathbf{R}^1\mathcal{F}(EL)$ must have non-trivial fibres over at least a divisor in a translate of $|\hat{L}^2|$. On the other hand, if $k = 3$ then the singularity set of $\mathbf{R}^1\mathcal{F}(EL)$ is at least $V \in \mathrm{Hilb}^6\hat{\mathbf{T}}$.□

Now consider $Q = E/L^*_{\hat{x}}$. If this had any torsion, say T, then $c_1(T) \cdot \ell \geq 0$ and E would surject to Q/T which is torsion-free. Now, $\mu(Q) = 2$ and so $\mu(Q/T) \leq 2$. The kernel of $E \to Q/T$ has slope $-\mu(Q/T)$ and is torsion-free. This implies that it is $L^m\mathcal{P}_{\hat{y}}\mathcal{I}_W$ for some $\hat{y} \in \hat{\mathbf{T}}$, $-1 \leq m \leq 0$ and zero-dimensional scheme W. If we assume that E is G-semistable then either $m = -1$ and we have a sequence:

$$0 \longrightarrow L^*_{\hat{x}} \longrightarrow E \longrightarrow L_{\hat{d}+\hat{x}}\mathcal{I}_X \longrightarrow 0, \tag{2.2}$$

where $X \in \mathrm{Hilb}^{2-\chi(E)}\mathbf{T}$ and $\det E \cong \mathcal{P}_{\hat{d}}$, or we have $\mathcal{I}_W\mathcal{P}_{\hat{y}} \to E$ with $|W| = 1$ if $k = 2$ and $|W| = 2$ or 3 if $k = 3$.

Definition 2.3. If $k = 2$ we define $\mathcal{X}_2 \subset \mathcal{M}_2$ to be the set of S-equivalence classes of E's for which there are no extensions of the form 2.2. Similarly, we define $\mathcal{X}_3^2$ and $\mathcal{X}_3^3 \subset \mathcal{M}_3$ in the $k = 3$ case where the superscript refers to the length of W as given above. We shall refer to these elements of $\mathcal{X}$ as *exceptional sheaves.* Notice that they are never μ-stable nor vector bundles nor satisfy WIT_1. We shall still consider extensions of the form

$$0 \longrightarrow L^*_{\hat{x}} \longrightarrow E \longrightarrow Q \longrightarrow 0. \tag{2.4}$$

We are now led to the following question:

Which extensions of the form 2.2 *and* 2.4 *give rise to μ-stable/G-stable vector bundles?*

Before answering this question we shall draw some useful corollaries from the above proposition. First, a definition:

Definition 2.5. We say that a zero-dimensional subscheme X of $\mathbf{T}$ is *collinear* if, for some $x \in \mathbf{T}$, $X \subset D_x$, the translate of $D_L \in |L|$ given by x. We shall denote the scheme of collinear subschemes of length l by $\mathrm{Hilb}^l_c\mathbf{T}$ and its complement in $\mathrm{Hilb}^l\mathbf{T}$ by $\mathrm{Hilb}^l_n\mathbf{T}$.

Proposition 2.6. *If E is μ-semistable torsion-free with Chern character* $(2, 0, -k)$ *for $k = 2$ or 3 then E satisfies* (i) *WIT_1 if and only if E is μ-stable,* (ii) *IT_1 if and only if X (exists and) is not collinear and E is μ-stable.*

Proof. Without loss of generality assume that $\hat{x} = 0 = \hat{d}$. From the remarks in definition 2.3 we may assume that E is not exceptional. Apply $\mathbf{R}\mathcal{F}$ to 2.2. Then we see that $\mathbf{R}^0\mathcal{F}(E) \cong \mathbf{R}^0\mathcal{F}(L\mathcal{I}_X)$ and

$$0 \to \mathbf{R}^1\mathcal{F}(E) \to \mathbf{R}^1\mathcal{F}(L\mathcal{I}_X) \to \mathbf{R}^2\mathcal{F}(L^*) \to \mathbf{R}^2\mathcal{F}(E) \to \mathbf{R}^2\mathcal{F}(L\mathcal{I}_X) \to 0.$$

But if we apply $\mathbf{R}\mathcal{F}$ to the twisted structure sequence of $\mathcal{I}_X$:

$$0 \longrightarrow L\mathcal{I}_X \longrightarrow L \longrightarrow \mathcal{O}_X \longrightarrow 0 \tag{2.7}$$

we obtain the long exact sequence

$$0 \to \mathbf{R}^0\mathcal{F}(L\mathcal{I}_X) \to \mathbf{R}^0\mathcal{F}L \to H_X \to \mathbf{R}^1\mathcal{F}(L\mathcal{I}_X) \to 0 \tag{2.8}$$

and $\mathbf{R}^2\mathcal{F}(L\mathcal{I}_X) = 0$. The Mukai spectral sequence implies that $\mathbf{R}^2\hat{\mathcal{F}}\mathbf{R}^1\hat{\mathcal{F}}(L\mathcal{I}_X) = 0$ but $H_X (= \hat{\mathcal{O}}_X)$ satisfies WIT_2 and so the middle map is not zero. Since $\hat{L}$ and H_X are both locally-free, the kernel of $\mathbf{R}^0\mathcal{F}L \to H_X$ must be zero. Hence, $\mathbf{R}^0\mathcal{F}(E) = 0$.

On the other hand, applying $H^*(\mathcal{P}_{\hat{x}} \otimes -)$ to 2.2 we see that $H^0(E\mathcal{P}_{\hat{x}}) \cong H^0(L_{\hat{x}}\mathcal{I}_X)$ and there is a long exact sequence

$$0 \to H^1(E\mathcal{P}_{\hat{x}}) \to H^1(L_{\hat{x}}\mathcal{I}_X) \to H^2(L^*\mathcal{P}_{\hat{x}}) \to H^2(E\mathcal{P}_{\hat{x}}) \to 0.$$

If E is μ-stable then $H^2(E\mathcal{P}_{\hat{x}}) = 0$ for all $\hat{x}$ and so $\mathbf{R}^2\mathcal{F}(E) = 0$ implying that E satisfies WIT_1. E satisfies IT_1 if also $H^0(E\mathcal{P}_{\hat{x}}) = 0$ but this happens if and only if X is not collinear (this proves part (ii)).

Conversely, suppose that E is not μ-stable. Then we have a short exact sequence $\mathcal{P}_{\hat{y}}\mathcal{I}_{X'} \to E \to \mathcal{P}_{-\hat{y}}\mathcal{I}_{X''}$. But the structure sequence of $\mathcal{I}_{X''}$ implies that $H^2(\mathcal{I}_{X''}) \neq 0$ and so $H^2(E\mathcal{P}_{\hat{y}}) \neq 0$ and hence E does not satisfy WIT_1. □

Observe that if X is collinear then there is some flat line bundle $\mathcal{P}_{\hat{y}}$ mapping to E. This contradicts G-semistability and so such extensions cannot be in the closure of the moduli space. It suffices to consider X's which are non-collinear.

3. Incidence of zero-dimensional subschemes on divisors

In this section we shall recall some information about the incidence of zero-dimensional subschemes on divisors on $\mathbf{T}$. This was computed in [8] and we shall summarise the results below. Consider first $L\mathcal{I}_X$ and notice that $\chi(L\mathcal{I}_X) = 1 - |X|$.

Proposition 3.1. [8, §4–§8]. *For all $X \neq \emptyset$, $L\mathcal{I}_X$ satisfies WIT_1.*
(i) *If $|X| = 1$ then $\mathbf{R}^1\mathcal{F}(L\mathcal{I}_X)$ is a line bundle of zero degree supported on a translate of D_L.*
(ii) *If $|X| = 2$ then $\mathbf{R}^1\mathcal{F}(L\mathcal{I}_X) \cong \hat{L}_x\mathcal{I}_{X'}$, where $X' \in \mathrm{Hilb}^2\hat{\mathbf{T}}$.*
(iii) *If $|X| = 3$ then $\mathbf{R}^1\mathcal{F}(L\mathcal{I}_X)$ is a rank 2 torsion-free sheaf with singularity set equal to a single point or empty.*

Proposition 3.2 [8, §8]. *Suppose $|X| > 3$ then $L^2\mathcal{I}_X$ satisfies WIT_1 if and only if X is not collinear. Suppose that X is not collinear. When $|X| = 4$ then $\mathbf{R}^1\mathcal{F}(L^2\mathcal{I}_X)$ is line bundle of degree 3 supported on a divisor in $|L^2\mathcal{P}_\sigma|$, where*

$\sigma = \sum_{x \in X} x$. *When* $|X| = 5$ *then* $\mathbf{R}^1\mathcal{F}(L^2\mathcal{I}_X) \cong \hat{L}^2\mathcal{P}_y\mathcal{I}_{X'}$ *for some* $y \in \mathbf{T}$ *and* $X' \in \mathrm{Hilb}^5\hat{\mathbf{T}}$.

(Recall that the linear system $|L^2|$ is base-point free and contains only $D_x + D_{-x}$ as reducible divisors.)

Definition 3.3. Given a zero dimensional scheme X we define

$$\Phi_n(X)_{\hat{x}} = H^0(L^{\otimes n}\mathcal{P}_{\hat{x}}\mathcal{I}_X).$$

Over an open subset of $\hat{\mathbf{T}}$, $\Phi_n(X)_{\hat{x}}$ will patch together to form a vector bundle. In any case, we can view $\Phi_n(X)$ as a torsion-free sheaf over $\hat{\mathbf{T}}$. The singularity set of $\Phi_n(X)$ we shall denote by $S_n(X)$ (standing for 'sauf'). This can be given a natural scheme structure (see [8, §3]).

Theorem 3.4.
(i) *If* X *has length* 3 *and is not collinear then* $S_2(X) \cong X$ *as schemes. The isomorphism is given by* $(x, y, z) \mapsto (x + y, y + z, z + x)$. $\Phi_2(X)$ *has generic rank* 0 *and jumps to* 1 *on* $S_2(X)$.
(ii) *If* X *is collinear,* $X \subset D_x$ *say, and length* 3 *then* $S_2(X) = D_{x+\tau}$, *where* $\tau = \sum_{x \in X} x$. $\Phi_2(X)$ *has generic rank* 0 *and jumps to rank* 1 *on* $S_2(X)$.
(iii) *If* $|X| = 4$ *and* X *is not collinear then* $S_2(X) \in |L^2\mathcal{P}_\sigma|$, *where* $\sigma = \sum_{x \in X} x$. *If* $S_2(X)$ *is irreducible then* X *contains no collinear length* 3 *subschemes,* $\Phi_2(X)$ *is generically empty and jumps to a single point over* $S_2(X)$.
(iv) *If* $S_2(X)$ *is reducible then*

either

> (a) $\Phi_2(X)$ *is generically empty and jumps to a single point over* $S_2(X)$, *in which case* X *contains precisely one length* 3 *collinear subscheme,*

or

> (b) $\Phi_2(X)$ *is generically empty and jumps to a single point over* $S_2(X)$, *except that it has rank one over over one of the intersection points of the two components of* $S_2(X)$. *In this case* X *contains precisely two length* 3 *collinear subschemes.*

The importance of these propositions is in determining when extensions of the form 2.2 give rise to vector bundles E. There is a condition on X and the linear system $|L^2\mathcal{P}_{2\hat{x}}| = T_{\hat{x}}|L^2|$ called the *Cayley-Bacharach condition* which determines whether E is locally-free or not. A good exposition of this can be found in [6, chapter 5]. In our case the condition reads:

The Cayley-Bacharach Condition. X satisfies this if and only if for all $X' \subset X$ with $|X'| = |X| - 1$ we have

$$\Phi_2(X)_{2\hat{x}} = \Phi_2(X')_{2\hat{x}}.$$

Observe that $\mathrm{Ext}^1(L_{\hat{x}}\mathcal{I}_X, L^*\mathcal{P}_{-\hat{x}}) \cong H^1(L^2\mathcal{P}_{2\hat{x}}\mathcal{I}_X)^*$ by Serre duality. When $|X| = 4$ the dimension of this equals the rank of $\Phi_2(X)_{2\hat{x}}$ which is generically zero. Hence, we are only concerned with $2\hat{x} \in S_2(X)$. Then X will satisfy the Cayley-Bacharach condition if $2\hat{x} \notin \bigcup S_2(X')$, where the union is over length 3 subschemes X' of X. If $S_2(X)$ is irreducible then this union consists of the six points (counted with multiplicities) $\{p+q, q+r, r+s, p+r, q+s, p+s\}$, where $X = \{p,q,r,s\}$. If $S_2(X)$ is reducible equal to $D_u + D_v$, say, then we must remove the component(s) corresponding to $S_2(X')$, where $X' \subset D_x$. If X contains only one collinear subscheme of length 3 then we must choose $2\hat{x}$ from D_v minus some points. If X contains two collinear subschemes of length 3 then the Cayley-Bacharch condition is only met at a single point (one of the two intersection points of D_u and D_v). In particular, we have the following

Proposition 3.5. *For all $X \in \mathrm{Hilb}^4 T$ there is some $\hat{x}$ and a non-trivial extension of the form 2.2 such that E is locally-free.*

4. Details of the moduli spaces

$\chi(E) = -2$

We shall first consider the case when E has Euler character -2. It is also convenient to relax the condition that $\det E = \mathcal{O}$. Then 2.2 gives rise to a sequence

$$0 \longrightarrow L^*_{\hat{x}} \longrightarrow E \longrightarrow L_{\hat{x}+\hat{d}}\mathcal{I}_Z \longrightarrow 0, \tag{4.1}$$

where $Z \in \mathrm{Hilb}^4\mathbf{T}$. Observe that any such extension has E μ-semistable. Furthermore, E is G-semistable provided Z is not collinear. Furthermore, E fails to be G-stable if there is a map from $\mathcal{P}_{\hat{y}}\mathcal{I}_x \to E$. This must inject into $L_{\hat{x}+\hat{d}}\mathcal{I}_Z$ and hence Z contains a collinear $Y \in \mathrm{Hilb}^3\mathbf{T}$. This, in turn, implies that $S_2(Z)$ is reducible. Observe also that the cohomology jumping divisor $S(E)$ of EL is $T_{-\hat{x}-\hat{d}}S_2(Z)$.

We also know from [8, Thm 7.5] that $L^2\mathcal{I}_Z$ is a line sheaf of degree 3 on $S_2(Z)$ when Z is not collinear. Hence, if we apply $\mathbf{R}\mathcal{F}(L \otimes -)$ to 4.1 we see that $\widehat{EL}$ is a line sheaf of degree 2 on $S(E)$. If $S(E)$ is reducible then we need to be more precise about the degree of such line sheaves (which need no longer be locally-free). This is done by restricting the line sheaf to irreducible factors and writing the degree as a tuple of the restriction degrees. In our case, the restriction degree of $L^2\mathcal{I}_Z$ is $(2,1)$ over $D_u + D_v$ unless there are 2 collinear length 3 subschemes of Z in which case the restriction type is $(1,1)$. It will be convenient to introduce the term *extension type* to denote the degrees of the kernels of the restriction maps plus 2. For example, the restriction type of $L^2\mathcal{I}_Z$ when Z contains one collinear Y is $(2,1)$ and the extension type is $(1,2)$. If Z contains 2 collinear Y's then the restriction type is $(1,1)$ but the extension type is $(2,2)$. Notice that the sum of the components of the extension type is minimised when the line sheaf is locally-free. Notice that $a + c = b + d$ depends only on the Euler character of the sheaf.

Finally, let us introduce the notation $\mathrm{Jac}^{(a,b)}_{(c,d)}(D_u + D_v)$ to denote the space of line sheaves up to isomorphism of restriction type (a, b) and extension type (c, d). We omit the extension type if the line sheaves are line bundles. It will also be convenient to refer to $\left[\begin{smallmatrix}(a,b)\\(c,d)\end{smallmatrix}\right]$ as the *type* of a line sheaf.

We already know from [8] that the possible types of $T = (L^2\mathcal{P}_{\hat{x}+\hat{d}}\mathcal{I}_Z)^{\wedge}$ are $\left[\begin{smallmatrix}(2,1)\\(1,2)\end{smallmatrix}\right]$ and $\left[\begin{smallmatrix}(1,1)\\(2,2)\end{smallmatrix}\right]$. In the latter case we do not allow torsion-free line sheaves over $D_u + D_v$ which are restrictions of $L_\alpha\mathcal{I}_\beta$ (because their Fourier transforms are not torsion-free—see [8, §9]). From this it is possible to determine the types of kernels of maps $T \to \mathcal{O}_x$. These are

$$A = \begin{bmatrix}(1,1)\\(1,1)\end{bmatrix}, \quad B = \begin{bmatrix}(0,2)\\(2,0)\end{bmatrix}, \quad C = \begin{bmatrix}(0,1)\\(2,1)\end{bmatrix}, \quad D = \begin{bmatrix}(0,0)\\(2,2)\end{bmatrix}$$

and B' and C' formed by interchanging D_u and D_v. Notice that if we have a column $0, 2$ in a type then there is an extension $M \to \widehat{EL} \to N$ where M and N are of the form $(L_{\hat{y}}\mathcal{I}_p)^{\wedge}$, for some $p \in \mathbf{T}$. This implies that E is not G-stable. This happens in types B, C and D. If we allow S-equivalence to act then it forces types B, C, D to be equivalent. It also identifies B with B', etc. Type B corresponds to locally-free $\widehat{EL}$.

Lemma 4.2. *If E is exceptional then $\widehat{EL}$ is the restriction of some $L_\alpha\mathcal{I}_Q$, $Q \in \mathrm{Hilb}^2\hat{\mathbf{T}}$, and so $\widehat{EL}$ has type D.*

Proof. Using the notation of 2.4 we see that $\widehat{QL}$ is torsion-free on $D_u + D_v$. Then [8, Lemma 9.3] implies that $\widehat{QL}$ is the restriction of $L_\alpha\mathcal{I}_\beta$. Applying $\mathbf{R}\mathcal{F}$ to $(2.4\otimes L)$ we obtain the short exact sequence $\widehat{EL} \to \widehat{QL} \to \mathcal{O}_{\hat{x}}$. If the composite $L_\alpha\mathcal{I}_\beta \to \widehat{QL} \to \mathcal{O}_{\hat{x}}$ is zero then $L_\alpha\mathcal{I}_\beta$ surjects to $\widehat{EL}$. The kernel is torsion-free with Chern character $(1, -L, 2)$ which is impossible. Hence, we have $L_\alpha\mathcal{I}_Q \to \widehat{EL}$ as required. The last part of the lemma follows from the obvious fact that $Q = D_u \cap D_v$. □

If we assemble the Jacobians together to form a bundle of Jacobians over the space of divisors in translates of the linear system $|L^2|$, which we can abbreviate to $\mathbf{P}\widehat{L^2}$, we obtain the full moduli space:

Theorem 4.3. *The moduli space of G-semistable torsion-free sheaves of Chern character $(2, 0, -2)$ over a principally polarised Abelian surface $(\mathbf{T}, L)$ is given by a fibration $\mathcal{M}_2 \to \mathbf{P}\widehat{L^2}$ whose fibres are $\mathrm{Jac}(D')$ if $D' \in \mathbf{P}\widehat{L^2}$ is smooth and $\mathrm{Jac}^{(1,1)}_{(1,1)}(D_u + D_v) \cup B$ if $D' = D_u + D_v$, where elements of B can be represented by elements of $\mathrm{Jac}^{(0,0)}_{(2,2)}$. Points in B correspond precisely to G-semistable, non-G-stable, torsion-free sheaves and taken together they are isomorphic to $\mathrm{Sym}^2(\mathbf{T}\times\hat{\mathbf{T}})$. The non-$\mu$-stable part coincides with the non-locally-free part and is given by line sheaves over a divisor $D' \in \mathbf{P}\widehat{L^2}$ which are of the form $L_y\mathcal{I}_Q/L^*_z$ for $Q \in \mathrm{Hilb}^2 D'$*

and $z, y \in \mathbf{T}$ and so includes $\mathcal{X}_2$. This part of the moduli space is isomorphic to a $\mathbf{C}P^1$ bundle over $\mathrm{Sym}^2\hat{\mathbf{T}} \times \mathrm{Hilb}^2\mathbf{T}$ *which is collapsed at the G-semistable points. In particular, $\mathcal{M}_2$ is both non-empty and connected.*

The Donaldson moduli space of instantons can be obtained from this by applying S-equivalence corresponding to μ-stability rather that G-stability. This has the effect of blowing-down the $\mathbf{C}P^1$-fibration given in the theorem as well as the diagonal in $\mathrm{Hilb}^2\mathbf{T}$ and provides an explicit complex analytic structure on the compactified moduli space.

$\chi(E) = -3$

We shall now turn our attention to $\mathcal{M}_3$.

Proposition 4.4. *Suppose that* $\mathrm{ch}(E) = (2, 0, -3)$ *and E is G-semistable and torsion-free. Then $\mathbf{R}^1\mathcal{F}(EL)$ is of the form $L^2\mathcal{P}_y\mathcal{I}_V$ for some $V \in \mathrm{Hilb}^6\hat{\mathbf{T}}$ if and only if the cohomology jumping set of EL is zero-dimensional.*

Proof. We express E as an extension of the form 2.2. The assumptions on E imply that X is not collinear since any section of $L_{\hat{y}}\mathcal{I}_X$ lifts to $E\mathcal{P}_{\hat{y}-\hat{x}}$ which would contradict G-semistability. We have a sequence $\mathcal{P}_{-\hat{x}} \to EL \to L^2\mathcal{P}_{\hat{x}}\mathcal{I}_X$ with $|X| = 5$. Apply $\mathbf{R}\mathcal{F}$ and propositions 2.6 and 3.2 to obtain

$$0 \to \mathbf{R}^1\mathcal{F}(EL) \to \mathbf{R}^1\mathcal{F}(L^2\mathcal{P}_{\hat{x}}\mathcal{I}_X) \to \mathcal{O}_{\hat{x}} \to 0.$$

But we know that $\mathbf{R}^1\mathcal{F}(L^2\mathcal{I}_X)$ is isomorphic to $\hat{L}^2\mathcal{P}_y\mathcal{I}_{X'}$ for some $X' \in \mathrm{Hilb}^5\hat{\mathbf{T}}$ if and only if the condition of the jumping set of EL (which equals the cohomology jumping set of $L^2\mathcal{P}_{\hat{x}}\mathcal{I}_X$) given in the proposition holds. □

We must, therefore split our moduli space into those sheaves whose jumping set is in $\mathrm{Hilb}^6\mathbf{T}$ (denoted $\mathbf{M}_3$) and those whose jumping set is a translate of D_L (denoted $\tilde{\mathbf{M}}_3$).

Apply $\mathbf{R}\mathcal{F}$ to $L^2\mathcal{P}_y\mathcal{I}_V \to L^2\mathcal{P}_{\hat{y}} \to \mathcal{O}_V$ to obtain $T_y^*\widehat{L^2} \to H_V \to (-1_{\mathbf{T}})^*EL$. Then $\det(EL) = \mathcal{O}$ implies that $\det(H_V) = \mathcal{P}_{\Sigma_{v\in V}v} = L^{-2}\det T_y^*\widehat{L^2} = \mathcal{P}_{-2y}$. We use here the fact that $(F\mathcal{P}_y)\hat{} = T_y^*\hat{F}$ for any sheaf F satisfying WIT. Hence, the condition $\hat{d} = 0$ becomes

$$\sum_{v\in V} v = -2y. \tag{4.5}$$

Then proposition 4.4 tells us that $\mathbf{M}_3$ is isomorphic to the subset $\mathbf{H}$ of $\mathrm{Hilb}^6\hat{\mathbf{T}} \times \mathbf{T}$ given by (V, y) satisfying the condition that $L^2\mathcal{P}_y\mathcal{I}_V$ has torsion-free Fourier transform. The latter condition is equivalent to the geometric condition that V does not contain any collinear subschemes of length 5 and if we want $\det(E) \cong \mathcal{O}$ then we must also impose condition 4.5.

If $E \in \mathcal{M}_3$ is G-semistable but not μ-stable then there is a short exact sequence

$$0 \longrightarrow \mathcal{P}_{\hat{a}}\mathcal{I}_Y \longrightarrow E \longrightarrow \mathcal{P}_{-\hat{a}}\mathcal{I}_{Y'} \longrightarrow 0, \tag{4.6}$$

where $|Y| + |Y'| = 3$ and $|Y| > 1$. This gives two possibilities. Either $|Y| = 2$ or $|Y| = 3$. Notice that E is forced to be G-stable. This suggests a further splitting of $\mathbf{M}$ into:

$\mathcal{M}_3^0$	μ-stable vector bundles	$\lvert\mathrm{sing}(E)\rvert = 0$
$\mathcal{M}_3^1$	μ-stable non-vector bundles	$\lvert\mathrm{sing}(E)\rvert = 1$
$\mathcal{M}_3^2$	G-stable sheaves with $\lvert Y\rvert = 2$	$\lvert\mathrm{sing}(E)\rvert = 3$
$\mathcal{M}_3^3$	G-stable sheaves with $\lvert Y\rvert = 3$	$\lvert\mathrm{sing}(E)\rvert = 3$

Similarly for $\tilde{\mathbf{M}}$ we define $\tilde{\mathcal{M}}_3^i$ by the same definition.

From [14, Cor.4.5] we know that $\mathcal{M}_1$ consists only of non-locally-free non-μ-stable sheaves and hence it is impossible for E^{**} to lie in $\mathcal{M}_1$ and hence the singularity set of E must have lengths 0, 1 or 3. This explains the last column in the definition above. Since $S_2(V)$ is a translation of the singularity set of E for $E \in \mathbf{M}$ we can deduce the immediate geometrical proposition:

Proposition 4.7. *There is no length 6 subscheme of $\mathbf{T}$ with $|S_2(V)| = 2$. In other words, if V lies on two divisors in $\mathbf{P}\widehat{L^2}$ then it must lie on a third.*

Lemma 4.8. *If a torsion-free sheaf F has Chern character $(2, 2L, -1)$ and satisfies WIT_1 then it must be G-stable.*

Proof. If $L^m\mathcal{P}_{\hat{z}}\mathcal{I}_W \to F \to L^{2-m}\mathcal{P}_{\hat{z}}\mathcal{I}_{W'}$ with $m > 0$ then we must have $2 - m > 0$ since F satisfies WIT_1. Hence, $m = 1$ and so F is μ-semistable. Since L satisfies IT_0 we cannot have $|W| = 0$ and if $|W| = 1$ then $\mathbf{R}^1\mathcal{F}(L\mathcal{I}_W)$ is a torsion sheaf which cannot map to $L^2\mathcal{I}_V$. Hence, $|W| > 1$ and so F is G-stable. □

In particular, $\mathbf{R}^1\mathcal{F}(L^2\mathcal{I}_V)$ is G-stable if V contains no collinear length 5 subschemes.

Lemma 4.9. *$\mathcal{X}_3^2 \subset \mathcal{M}_3^2$ and $\mathcal{X}_3^3 \not\subset \mathbf{M}_3$.*

Proof. Let $E \in \mathcal{X}_3^2$. Refering to 2.4 we have a sequence $S \to LQ \to L_{\hat{y}}\mathcal{I}_q$, where S is a line bundle over a translate of D_L of degree 0. Now, S satisfies WIT_1 and the transform of the above sequence is of the form $L_\alpha\mathcal{I}_\beta \to \widehat{QL} \to S'$ and hence $\widehat{QL}$ is torsion-free. This implies that EL satisfies WIT_1 and $\widehat{EL} \subset \widehat{QL}$ is also torsion-free and hence isomorphic to some $L^2\mathcal{P}_y\mathcal{I}_V$. Hence, using lemma 4.9, $E \in \mathcal{M}_3^2$.

On the other hand, if $\widehat{EL} = L^2\mathcal{P}_y\mathcal{I}_V$ then $\hat{Q} = L^2\mathcal{P}_y\mathcal{I}_{X'}$ for $X' \in \mathrm{Hilb}^5\hat{\mathbf{T}}$. Hence, S must have degree 0 and so $|W| = 2$. □

In fact, the above proof shows that $\tilde{\mathcal{M}}_3^2 = \emptyset$.

Theorem 4.10.
(i) $E \in \mathcal{M}_3^0$ *if and only if V is generic with $S_2(V) = \emptyset$.*

(ii) $E \in \mathcal{M}_3^1$ *if and only if* $S_1(V)$ *has length* 1.
(iii) $E \in \mathcal{M}_3^2$ *if and only if* V *contains at least one collinear length* 4 *subscheme (but no collinear length* 5 *subscheme).*
(iii)' $E \in \mathcal{X}_3^2$ *if and only if* V *contains* 3 *collinear length* 4 *subschemes (but no collinear length* 5 *subscheme).*
(iv) $E \in \mathcal{M}_3^3$ *if and only if* V *satisfies* $\Phi_3(V)_y = \Phi_3(X')_y$ *for some* y *and* $H^1(L^3\mathcal{P}_y\mathcal{I}_V) \neq 0$ *and such that* V *contains no collinear length* 5 *subschemes. The codimension of* $\mathcal{M}_3^n$ *in* $\mathcal{M}_3$ *is* n *and the codimension of* $\mathcal{X}_3^2$ *is* 6.

Proof. (i) and (ii) are trivial.
(iii) EL fits into extensions of the form $L\mathcal{I}_Y \to EL \to L\mathcal{I}_p$ if and only if there is an extension $L\mathcal{I}_{Y'} \to L^2\mathcal{P}_y\mathcal{I}_V \to (L\mathcal{I}_p)\hat{}$. This happens if and only if $V \setminus Y'$, which has length 4, is collinear.
(iii)' $E \in \mathcal{X}_3^2$ if and only if each X' contains a collinear length 4 subscheme. If there were four collinear subschemes then there would have to be a collinear length 5 subscheme in V.
(iv) Applying $\mathbf{R}\mathcal{F}$ to $L_\alpha\mathcal{I}_Y \to EL \to L_\beta$ gives the short exact sequence $\hat{L}^*\mathcal{P}_{-\beta} \to G \to L^2\mathcal{P}_y\mathcal{I}_V$, where G is torsion-free, has Chern character $(2, L, -1)$ and satisfies WIT_1. Hence, G is μ-stable. If G were not locally-free then there is some $L_\gamma\mathcal{I}_p \to G$ and the induced map $L_\gamma\mathcal{I}_p \to L^2\mathcal{P}_y\mathcal{I}_V$ implies that $L^2\mathcal{I}_V$ does not have a torsion-free Fourier transform; a contradiction. Hence, G is locally-free.

Conversely, suppose that V does satisfy Cayley-Bacharach with respect to $L^3\mathcal{P}_y$ for some y so that $H^1(L^3\mathcal{P}_y\mathcal{I}_V) \neq 0$. Then the resulting extension gives rise to G and hence to $L_\alpha\mathcal{I}_Y \to EL \to L_\beta$ for suitable α, β.

To compute the dimensions of $\mathcal{M}_3^n$ observe that $\dim \mathcal{M}_3^0 = 14$. Now $\dim \mathcal{M}_3^2 = 2 + 4 + 4 + 2 = 12$ given by 2 dimensional choice of D_u, a 4 dimensional choice of $Z \subset D_u$, a 4 dimensional choice of 2 other points and then a 2 dimensional choice of y. If V contains 3 collinear length 4 subschemes then the freedom of choice is given by choosing 4 points on any translate of D_L and the other two points are determined by the intersection of a pair of translates of D_L each passing through 2 of these four points. Hence, $\dim \mathcal{X}_3^2 = 2 + 4 + 2 = 8$.

For $\mathcal{M}_3^1$ we observe that Cayley-Bacharach fails for X when $2\hat{x} + \hat{d}$ lies on $S_2(Z) \setminus S_2(X)$ for some $Z \subset X$. This is a codimension 1 condition. For $\mathcal{M}_3^3$ we require $2\hat{x} + \hat{d}$ to lie on the intersection of $S_2(Z_i)$ for three $Z_i \subset X$. This imposes two conditions on $\hat{x}$ and a further condition on X giving codimension 3. □

We can likewise classify sheaves in $\tilde{\mathbf{M}}_3$. Points $E \in \tilde{\mathbf{M}}_3 \setminus \mathcal{X}_3^3$ are characterised by the fact that if $EL/\mathcal{P}_{-\hat{x}} = L^2\mathcal{P}_{\hat{x}+\hat{d}}\mathcal{I}_X$ then X contains a collinear length 4 subscheme which we shall denote by Z. For the moment we shall assume that $\hat{x} = 0 = \hat{d}$. Then $L^2\mathcal{I}_X$ satisfies WIT_1 but its Fourier transform A, say, is not torsion-free. Applying $\mathbf{R}\mathcal{F}$ to $L\mathcal{I}_p \to L^2\mathcal{I}_X \to T$ we obtain $S \to A \to \hat{L}_y\mathcal{I}_q$, where $\mathrm{ch}(S) = (0, L, -1)$. If the composite $S \to A \to \mathcal{O}_0$ vanishes then $S \to \widehat{EL}$ injects which implies $L\mathcal{I}_p \to EL$ injects contradicting the G-semistability of E. Hence,

$\widehat{EL}$ fits into an extension of the form

$$0 \longrightarrow R \longrightarrow \widehat{EL} \longrightarrow L_y \mathcal{I}_q \longrightarrow 0, \tag{4.11}$$

where R is a degree -1 line bundle over some translate $D_{\hat{y}}$ of $D_{\hat{L}}$. Extensions 4.11 cannot split.

On the other hand, if $E \in \mathcal{X}_3^3$ then we have a sequence $\mathcal{P}_{-\hat{x}} \to EL \to Q$ for some torsion sheaf Q. We also have (4.6) with $|Y| = 3$ which shows that EL satisfies WIT_1. Then $L_{\hat{a}} \mathcal{I}_Y \to Q$ and Q satisfies WIT_1. Since $\widehat{EL}$ is not torsion-free we see that $\hat{Q}$ is not torsion-free. Applying $\mathbf{R}\mathcal{F}$ to the short exact sequences gives $\hat{L}_x^{-1} \to K' \to \widehat{EL}$ and $K' \to K \to \mathcal{O}_{\hat{x}}$, where K is locally-free and μ-stable of Chern character $(2, l, 0)$. Let T denote the torsion subsheaf of $\hat{Q}$. Then $\hat{Q}/T \cong \hat{L}_y \mathcal{I}_p$ for some p as K maps to $\hat{Q}$. This implies that $\widehat{EL}$ fits into an extension of the form 4.11 again.

Now suppose that we are given a non-trivial extension $R \to A \to L_y \mathcal{I}_q$. Then A satisfies WIT_1 and its Fourier transform fits into an exact sequence $K \to \hat{A} \to S$, where $\mathrm{ch}(K) = (2, L, 0)$, K is locally-free (because R is IT_1) and S is a degree 0 line bundle over a translate of D_L. If T is the torsion subsheaf of $\hat{A}$ then it must be a line bundle over $\mathrm{supp}(S)$ of degree less than 0. But S/T is supported in dimension 0 and $K \to \hat{A}/T \to S/T$ is short exact. This implies that $\hat{A}/T \cong K \oplus S/T$ which contradicts the fact that $\hat{A}/T$ is torsion-free. Hence, $\hat{A}$ is torsion-free.

We know from lemma 4.8 that $\hat{A}$ must be G-stable and so lies in $\tilde{\mathbf{M}}_3$. We also see easily that extensions 4.11 are uniquely determined by E. So we have proved:

Theorem 4.12. *There is an isomorphism between non-trivial extensions of sheaves of the form $L_y \mathcal{I}_q$ and degree -1 line bundles over translates of $D_{\hat{L}}$, and $\tilde{\mathbf{M}}_3$.*

Using the fact that R can be written as $L_x \mathcal{I}_Y / \mathcal{O}$ with $Y \subset \mathrm{supp}(R)$ and the structure sequence $L_y \mathcal{I}_q \to L_y \to \mathcal{O}_q$ we can compute $\mathrm{Ext}^1(L_y \mathcal{I}_q, R)$ which is isomorphic to $H^1(R \otimes L_y) \cong \mathbf{C}^4$ if $y \notin \mathrm{supp}(R)$ and isomorphic to

$$\mathrm{coker}(\mathrm{Ext}^1(\mathcal{O}_q, R) \to H^1(R \otimes L_y)) \oplus \mathrm{Hom}\,(R, \mathcal{O}_q)^* \cong \mathbf{C}^4$$

if $y \in \mathrm{supp}(R)$. This provides us with an isomorphism between $\tilde{\mathbf{M}}_3$ and a $\mathbf{C}P^3$ fibration over $\hat{\mathbf{T}} \times \hat{\mathbf{T}} \times \hat{\mathbf{T}} \times \mathbf{T}$. This latter factor consists of $(\hat{a}_1, \hat{a}_2, q, y)$ where $(\hat{a}_1, \hat{a}_2) \mapsto R$ given by $R \cong T^*_{\hat{a}_2}(\ker(\mathcal{O}(D_{\hat{L}}) \to \mathcal{O}_{\hat{a}_1}))$. This implies that $\dim \tilde{\mathbf{M}}_3 = 11$.

Theorem 4.13. $\mathcal{X}_3^3 = \tilde{\mathcal{M}}_3^3$ *and* $\tilde{\mathbf{M}}_3 = \tilde{\mathcal{M}}_3^1 \cup \tilde{\mathcal{M}}_3^3$.

Proof. Suppose $E \in \tilde{\mathbf{M}}_3 \setminus \mathcal{X}_3^3$ and we assume, as before, that $\hat{x} = 0 = \hat{d}$. Then there is a collinear length 3 subscheme of X and hence a sequence $L_\alpha \mathcal{I}_N \to L^2 \mathcal{I}_X \to Q$ where $N \in \mathrm{Hilb}^2\mathbf{T}$ and Q is a line bundle over a translate of D_L of degree 1. The composite $EL \to L^2 \mathcal{I}_X \to Q$ has kernel K' which has Chern character $(2, l, -1)$ and this factors through $K = \ker(\widehat{EL} \to L^2 \mathcal{I}_X / L_\alpha \mathcal{I}_p)$ which is locally-free. This implies that K' is torsion-free and hence $\tilde{\mathcal{M}}_1^0 = \emptyset$.

We also have part of a long exact sequence

$$\mathrm{Ext}^1(Q,\mathcal{O}) \longrightarrow \mathrm{Ext}^1(L^2\mathcal{I}_X,\mathcal{O}) \longrightarrow \mathrm{Ext}^1(L_\alpha\mathcal{I}_N,\mathcal{O}).$$

Then $EL \in \tilde{\mathcal{M}}_3^2$ only if the representative of K' in the last Ext group vanishes. But this is impossible as $K' \subset K$. Hence, $\tilde{\mathcal{M}}_3^2 = \emptyset$. We can repeat this with a length 3 subscheme Y instead of N to show that $E \notin \tilde{\mathcal{M}}_3^3$. This proves the theorem. □

We shall now relate $\tilde{\mathbf{M}}_3$ to the part of $\mathrm{Hilb}^6\hat{\mathbf{T}} \times \mathbf{T}$ which is not isomorphic to $\mathbf{M}_3$. Consider an extension of the form 4.11 above and consider a suitable structure sequence for R: $L^{-1}\mathcal{P}_x \to \mathcal{P}_w\mathcal{I}_p \to R$. Without loss of generality assume that $x = 0 = w$. The $\mathrm{Ext}^1(R, L^{-1}) \cong \mathbf{C}$ and this choice of extension gives a canonical map (up to a scalar) $\phi : \mathrm{Ext}^1(L_y\mathcal{I}_q, R) \to \mathrm{Ext}^2(L_y\mathcal{I}_q, L^{-1})$ via the cup product. This map forms part of a long exact sequence. The previous terms are $0 \to \mathrm{Ext}^1(L_y\mathcal{I}_q, \mathcal{I}_p)$ because $H^1(L^2\mathcal{P}_y\mathcal{I}_q) = 0$. Suppose that 4.11 was the image of $\mathcal{I}_p \to G \to L_y\mathcal{I}_q$. Then we obtain $L^{-1} \to G \to \widehat{EL}$. This implies that $E \in \mathcal{X}_3^3$.

Conversely, suppose that $E \in \mathcal{X}_3^3$. Then we have $L^{-1}\mathcal{P}_x \to G \to \widehat{EL}$ and the induced map $G \to L_y\mathcal{I}_q$ has kernel $\mathcal{P}_w\mathcal{I}_p$ for some w and p. This implies that 4.11 is in the image of some element in $\mathrm{Ext}^1(L_y\mathcal{I}_q, \mathcal{P}_w\mathcal{I}_p)$. Hence, the kernels of ϕ as we vary x, w, p and q give rise precisely to $\mathcal{X}_3^3$.

Serre duality implies that $\ker(\Phi)$ is isomorphic to the dual of the cokernel of $\psi : \mathrm{Hom}\,(L^{-1}, L_y\mathcal{I}_q) \to \mathrm{Ext}^1(R, L_y\mathcal{I}_q)$.

Claim. Points of $(\mathrm{coker}(\psi))^*$ correspond to collinear elements $V \in \mathrm{Hilb}^6\hat{\mathbf{T}}$.
Proof (of claim). Points of the image of ψ correspond to injections $\mathcal{I}_p \to B$ where B fits in $L_y\mathcal{I}_q \to B \to R$. If B has torsion then the torsion must take the form $T \to B \to L_\gamma$ and there must be an induced injection $\mathcal{I}_p \to L_\gamma$. Let the quotient be S. Applying $\mathbf{R}\mathcal{F}$ we obtain $\hat{L}_\gamma \to \mathbf{R}^0\mathcal{F}(S) \to \mathcal{P}_p$ and $\mathbf{R}^1\mathcal{F}(S) = \mathcal{O}_0$. But $\mathrm{Ext}^1(\mathcal{P}_p, \hat{L}) = 0$ and so $\mathbf{R}^0\mathcal{F}(S) \cong \hat{L}\oplus\mathcal{P}_p$. The Mukai spectral sequence for S gives $\mathcal{O} \to L_\gamma \oplus \mathcal{O}_p \to S$ and hence S has torsion over its support. But $S = L_y\mathcal{I}_q/L^{-1}$ which implies that S is torsion-free; a contradiction. Hence, B must be torsion-free of the form $L^2\mathcal{P}_y\mathcal{I}_V$ with V not collinear but containing a collinear length 5 subscheme. The converse argument also holds showing that points of the image of ψ correspond to non-collinear V's. This implies that the dual of the cokernel corresponds precisely to the collinear V's as desired. □

This implies that there is a canonical isomorphism between the collinear part of $\mathrm{Hilb}^6\hat{\mathbf{T}} \times \mathbf{T}$ and $\mathcal{X}_3^3$. This shows that $\dim \mathcal{X}_3^3 = 8$ and provides a canonical section of the $\mathbf{C}P^3$-fibration $\tilde{\mathbf{M}}_3 \to \hat{\mathbf{T}}^3 \times \mathbf{T}$. We have also shown in the proof of the claim that the dual fibration $\tilde{\mathbf{M}}_3^*$ is naturally isomorphic to the subvariety of $\mathrm{Hilb}^6\hat{\mathbf{T}} \times \mathbf{T}$ consisting of V's with collinear length 5 subschemes. On the other hand, we know that $\mathcal{M}_3$ and $\mathrm{Hilb}^6\hat{\mathbf{T}} \times \mathbf{T}$ are birationally isomorphic via $\mathbf{M}_3$ and the there is a canonical isomorphism between two sections of the $\mathbf{C}P^3$-fibrations. This shows that the $\mathbf{C}P^3$-fibration is holomorphically self-dual and proves:

Theorem 4.14. *There is a canonical isomorphism* $\mathcal{M}_3 \cong \mathrm{Hilb}^6\hat{\mathbf{T}} \times \mathbf{T}$. *In particular,* $\mathcal{M}_3$ *is connected.*

In general, one might expect that $\mathcal{M}_{2p+1} \cong \mathrm{Hilb}^{4p+2}\hat{\mathbf{T}} \times \mathbf{T}$.

References

[1] Braam, P.J., Maciocia, A., Todorov, A., Instanton Moduli as a Novel Map from Tori to K3-Surfaces, Invent. Math. 108 (1992) 419–451.

[2] Donaldson, S.K., Anti-self-dual Yang-Mills Connections over Complex Algebraic Surfaces and Stable Vector Bundles, Proc. London Math. Soc. 50 (1985) 1–26.

[3] ———, Infinite Determinants, Stable Bundles and Curvature, Duke Math. J. 54 (1987) 231-247.

[4] Gieseker, D., On the moduli of vector bundles on an algebraic surface, Ann. Math. 106 (1977) 45–60.

[5] ———, A Construction of Stable Bundles on an Algebraic Surface, J. Differential Geom. 27 (1988) 137–154.

[6] Griffiths, P., Harris, J., Principles of Algebraic Geometry, Wiley (1978).

[7] Kobayashi, S., Differential geometry of complex vector bundles, Princeton University Press (1987).

[8] Maciocia, A., Incidence of points on divisors in a principally polarised abelian surface, Edinburgh University Preprint (1992).

[9] ———, Classifying sheaves in the Maruyama boundary for surfaces with trivial canonical bundle, Edinburgh University Preprint (1992).

[10] Maruyama, M., Moduli of stable sheaves, I, J. Math. Kyoto Univ. 17 (1977) 91–126.

[11] ———, Moduli of stable sheaves, II, J. Math. Kyoto Univ. 18(1978) 557–614.

[12] Mukai, S., Duality between $D(X)$ and $D(\hat{X})$ with its application to Picard sheaves, Nagoya Math. J. 81 (1981) 153–175.

[13] ———, Symplectic structure of the moduli space of sheaves on an abelian or K3 surface, Invent. Math 77 (1984) 101-116.

[14] ———, Fourier Functor and its Application to the Moduli of Bundles on an Abelian Variety, in: Algebraic Geometry, Sendai, 1985, 515–550. Adv. Stud. Pure Math. 10, North-Holland, 1985.

[15] Okonek, Ch., Schneider, M., Spindler, H., Vector Bundles on Complex Projective Spaces, Birkhäuser (1980).

[16] Taubes, C.H., Self-dual connections on 4-manifolds with indefinite intersection matrix, J. Differential Geom. 19 (1984) 517–560.

[17] Uhlenbeck, K.K., Yau, S-T., On the Existence of Hermitian-Yang-Mills Connections in Stable Vector Bundles, Comm. Pure Appl. Math. Supp. 39 (1986) 257–293.

Hypersurface Sections of Curves

Juan C. Migliore

Abstract. If X is a subscheme of projective space $\mathbb{P}^n$ and F is a general hypersurface of degree d, F cuts out on X a subscheme Z which is also a subscheme of F and hence of $\mathbb{P}^n$. We are primarily concerned with the case where X is a curve, and we study the question of when it is true that if Z is a complete intersection (resp. arithmetically Gorenstein) then the same must be true of X. We show how to extend the approach of Huneke and Ulrich to allow any degree d, and in particular we show that for $d \geq 2$ then without exception if Z is a complete intersection it follows that X must also be a complete intersection. (Strano and Re had proved this for $d = 1$, but there are some natural exceptions in this case. We recover their result.) Following Huneke and Ulrich (who studied $d = 1$ for reduced, connected curves not lying on a quadric hypersurface) we also attack the Gorenstein problem, trying to eliminate these hypotheses, but only partial results are obtained and a complete classification is still open.

1991 Mathematics Subject Classification: 14M05, 13C40, 14M10, 13H10

0. Introduction

If X is a subscheme of projective space $\mathbb{P}^n$ and F is a general hypersurface of degree d, F cuts out on X a subscheme Z which is also a subscheme of F and hence of $\mathbb{P}^n$. A very natural and important question is to study the properties that either Z or X force on the other. One such property which is of interest is the question of being arithmetically Cohen-Macaulay. This question has been studied especially when $d = 1$.

It is a classical result that if X is arithmetically Cohen-Macaulay then so is Z. In the other direction, it is not hard to show that if $\dim Z \geq 1$ then the Cohen-Macaulayness of Z implies that X is also arithmetically Cohen-Macaulay, as long as X is at least locally Cohen-Macaulay and equidimensional. (For the case $d = 1$ see **[M]** Remark 1.10, **[R]** Teorema 2. The case $d \geq 2$ is proved similarly. More generally, **[HU]** Proposition 2.1 proves this result without assuming that X is locally Cohen-Macaulay or even homogeneous.) So the interesting situation is when Z is a zero-scheme in $\mathbb{P}^n$. Are there any situations in which the (automatic) Cohen-Macaulayness of Z forces the curve X to be arithmetically Cohen-Macaulay?

To date, most work in this area has focused on hyperplane sections (i.e. $d = 1$). In **[GM]** the following question is asked: *If C is reduced and irreducible and lies* *on no quadric surface, and if a general hyperplane section $C \cap H$ is a complete intersection, must C be a complete intersection?*

This question was answered in the affirmative by Strano in **[S1]**. His clever proof has spawned a great deal of further research in this area. One important contribution is the work of Re (**[R]**), who proved the analogous result for complete intersections $Z \subset \mathbb{P}^{n-1}$, even without the assumption of integral. Specifically, he proves that if W is a locally Cohen-Macaulay equidimensional subscheme of $\mathbb{P}^n$ whose general hyperplane section is a complete intersection, and if $n > 3$ or else if $n = 3$, deg $W > 4$ and W does not lie on a quadric surface, then W is a complete intersection. (Strano also proves the second part of this result in **[S2]**.)

Finally, the recent paper of Huneke and Ulrich **[HU]** extends this result: let W be a reduced, connected subscheme of $\mathbb{P}^n$, $n \geq 4$, which does not lie on a quadric hypersurface. If the general hyperplane section Z of W is arithmetically Gorenstein then so is W.

Some of the ideas in these papers are similar, as are some of the results in **[K]**. However, the results of Huneke and Ulrich are based on a very nice result, their so-called Socle Lemma, which seems to be really the central idea behind these results.

In this paper, we follow the approach of Huneke and Ulrich to extend these results. Our generalizations are along two lines: we remove the assumption of being reduced and connected, and we pass to general hypersurface sections (i.e. we allow any degree $d \geq 1$).

§1 provides the technical results we will need later. It is essentially a restatement of the preparatory results in **[HU]**, but we have to find the analogous statements since theirs are only for the case $d = 1$. Nevertheless, apart from this rephrasing of their statements in greater generality, the material (especially the ideas for the proofs) is essentially taken from their paper. The goal is to prove Lemma 1.4, which is the analogue of their "Socle Lemma."

§2 begins the study of curves and their hypersurface sections. The first result (Proposition 2.2 and Remark 2.3) is again an analogue of one in **[HU]**. It gives a condition on Z which forces X to be arithmetically Cohen-Macaulay. The main result of this section is Theorem 2.4, which generalizes the results of Strano and Re. That is, it is shown that if Z is a complete intersection but X is not then in fact d must be 1. We then recover Re's result (except we note that X may also be a double line in $\mathbb{P}^n$). As a corollary, suppose X is a non-degenerate subscheme of any dimension in $\mathbb{P}^n$ and Y is a "general" complete intersection of complementary dimension. Assume that either deg $Y > 1$ or that codim $X \geq 3$ and deg $X > 2$. If X and Y meet in a zero-scheme Z which is a complete intersection in $\mathbb{P}^n$ then X must be a complete intersection.

§3 passes to the situation of **[HU]**, namely the case where Z is a Gorenstein set of points. The problem is to determine when X must also be Gorenstein. Supposing that X is not Gorenstein, we analyze this from the point of view of the Hilbert function of Z and (roughly) the least degree of a hypersurface containing Z which does not lift to X. We give a number of examples of non-Gorenstein X with Gorenstein hypersurface section, and show that these are very rare, even in the non-reduced case. Our main result (Corollary 3.5) extends that of [**HU**]: if X is reduced and does not lie on a quadric hypersurface, and Z is Gorenstein, then X is Gorenstein provided that either $n \geq 5$ or $d \geq 2$. We also give a corollary similar to that given for complete intersections: assume that X has dimension r and does not lie on a quadric hypersurface, and that Y is a "general" complete intersection of codimension r also not lying on a quadric hypersurface. Then if they meet in a zero-scheme Z which is Gorenstein it follows that X must be Gorenstein.

1. Preparatory Results

Let k be an algebraic closed field of characteristic zero. Let $R = k[X_0, ..., X_n]$. Let M be a non-zero finitely generated graded R-module. For a sheaf $\mathcal{F}$ on $\mathbb{P}^n$ we sometimes write $H^i(\mathcal{F})$ for $H^i(\mathbb{P}^n, \mathcal{F})$ (and similarly for the dimension $h^i(\mathcal{F}) = \dim_k H^i(\mathcal{F})$).

The goal of this section is to re-state some of the preparatory results in [**HU**] in greater generality. We stress that the ideas for the statements and proofs of the results in this section are based entirely on [**HU**].

Lemma 1.1 ([**HU**] Lemma 3.1). *Suppose that $V = V_r, V_{r-1}, ..., V_0 = W$ are finite-dimensional vector spaces over an infinite field k. Assume that we have two sequences of homomorphisms,*

$$V_r \xrightarrow{\alpha_r} V_{r-1} \xrightarrow{\alpha_{r-1}} \dots \xrightarrow{\alpha_1} V_0,$$

and

$$V_r \xrightarrow{\beta_r} V_{r-1} \xrightarrow{\beta_{r-1}} \dots \xrightarrow{\beta_1} V_0,$$

which satisfy the condition that

$$\beta_1 \circ \dots \circ \beta_{i-1}(\operatorname{Im} \alpha_i) \subseteq \operatorname{Im} \alpha_1$$

for all $1 \leq i \leq r$. Write $\Phi = \alpha_1 \circ \dots \alpha_r, \Psi = \beta_1 \circ \dots \circ \beta_r$, and

$$\Delta = \Delta_\lambda = (\alpha_1 + \lambda\beta_1) \circ \dots \circ (\alpha_r + \lambda\beta_r)$$

for $\lambda \in k$. Then either
i) $\Psi(\operatorname{Ker} \Phi) \subseteq \operatorname{Im} \alpha_1$,
or
ii) $\operatorname{rank} \Delta_\lambda > \operatorname{rank} \Phi$ *for general* λ.

(A similar result is proved in [**K**] Theorem 1.1.)

Corollary 1.2. *Let $F \in R_d$ be a general homogeneous polynomial of degree d and let $G \in R_d$ be arbitrary. F and G each induce a homomorphism from M_i to M_{i+d} for any i. Then for any $r > 0$ we have*

$$G^r \cdot [\operatorname{Ker}(\times F^r)] \subseteq \operatorname{Im}(\times F).$$

Proof. Let $x \in \mathrm{Ker}(\times F^r)$. Say x occurs in a component M_j of M. In Lemma 1.1 set $V_r = M_j$, $V_{r-1} = M_{j+d}$, etc. and set α_r, α_{r-1}, etc. each to be multiplication by F and β_r, β_{r-1}, etc. each to be multiplication by G. It is a basic property of modules that the condition of Lemma 1.1 is satisfied, and so the result follows from the generality of F. □

The next result is the translation of **[HU]** Proposition 3.5 to our situation, and the proof is exactly a re-wording of their proof in this new context. We include the proof here only for completeness, for the convenience of the reader; the main difference is that instead of looking at all degrees less than t (in the case $d = 1$) we jump d steps each time.

Proposition 1.3. *Let $m \subset R$ be the maximal homogeneous ideal $(X_0, ..., X_n)$. Fix $t \in \mathbf{Z}$. Let $F \in R_d$ be a general homogeneous polynomial of degree d. Denote by ϕ the multiplication induced by F on M and by ϕ^i the multiplication induced by F^i for any $i > 0$. Let*

$$0 \rightarrow K \rightarrow M(-d) \xrightarrow{F} M \rightarrow C \rightarrow 0$$

be exact. Then

(a) $\dim\, [\mathrm{Ker}\, \phi^i]_{t-id} \geq \sum_{j=0}^{i-1} \dim\, K_{t-jd} - \sum_{j=1}^{i-1} \dim\, [0 :_C m^{jd}]_{t-jd}$

(b) $\sum_{j \geq 0} \dim\, K_{t-jd} \leq \sum_{j \geq 1} \dim\, [0 :_C m^{jd}]_{t-jd}$

where "dim" refers to the vector space dimension over k.

Proof. We prove (a) by induction on i. Note that $K(d) = \mathrm{Ker}\, \phi^1 \subseteq \mathrm{Ker}\, \phi^i$ for any $i > 1$. For $i = 1$, then, the statement is trivial. So assume that $i > 1$ and that (a) holds for $i - 1$. Hence we have an exact sequence

$$0 \rightarrow K_{t-(i-1)d} \rightarrow [\mathrm{Ker}\, \phi^i]_{t-id} \xrightarrow{\phi} [\mathrm{Ker}\, \phi^{i-1}]_{t-(i-1)d} \rightarrow L_{t-(i-1)d} \rightarrow 0$$

where $L = [\mathrm{Ker}\, \phi^{i-1}] \;/\; [\mathrm{Im}\, \phi \cap \mathrm{Ker}\, \phi^{i-1}]$.

We first claim that $L_{t-(i-1)d}$ is contained in $[0 :_C m^{(i-1)d}]_{t-(i-1)d}$. Indeed, let $G \in R_d$ be any homogeneous polynomial of degree d. By Corollary 1.2 we have that the image of $[0 :_M F^\ell]_{t-\ell d}$ in C is contained in $[0 :_C G^\ell]_{t-\ell d}$ for all $\ell > 0$. But then recall that m^ℓ is generated by y^ℓ as y ranges over elements of R_1. Then by considering the d-uple embedding of $\mathbb{P}^n$ in $\mathbb{P}^N$ ($N = \binom{n+d}{n} - 1$) one can use this to show that $m^{\ell d}$ is generated by polynomials of the form G^ℓ as G ranges over R_d. Thus combining these two facts, we get that the image of $[0 :_M F^\ell]_{t-\ell d}$ in C is contained in $[0 :_C m^{\ell d}]_{t-\ell d}$. Then set $\ell = i - 1$ and we have finished the claim.

Now, we have from the above exact sequence

$$\dim\,[\operatorname{Ker}\phi^{i-1}]_{t-(i-1)d} + \dim\,K_{t-(i-1)d} = \dim\,[\operatorname{Ker}\phi^{i}]_{t-id} + \dim\,L_{t-(i-1)d}$$
$$\leq \dim\,[\operatorname{Ker}\phi^{i}]_{t-id} + \dim\,[0 :_C m^{(i-1)d}]_{t-(i-1)d}$$

The result (a) then follows from the inductive hypothesis, making the appropriate substitution in the above inequality, and simplifying. The statement (b) follows from (a) by letting $i \gg 0$. □

The paper **[HU]** isolates their so-called "Socle Lemma" as the key point in arriving at their results. In our situation it is not the socle (the submodule of C annihilated by the maximal ideal m) but rather $[0 :_C m^d]$ which is important. Following **[HU]**, for any graded module N we denote by $a_-(N)$ the least degree i for which N_i is non-zero. The Socle Lemma then translates to

Lemma 1.4. *Let M be a non-zero finitely generated graded R-module. Let $F \in R_d$ be a general homogeneous polynomial of degree d. Let*

$$0 \to K \to M(-d) \xrightarrow{F} M \to C \to 0$$

be exact. If $K \neq 0$ then

$$a_-(K) \geq a_-([0 :_C m^d]) + d.$$

Proof. The idea of this proof is to derive a "degree d divisibility result," as was done in **[K]** for degree 1. Let $t = a_-(K)$, so $\operatorname{Ker}(\times F)$ begins in degree $t-d$. Suppose that $[0 :_C m^d]_i = 0$ for all $i \leq t-d$. We will derive a contradiction.

By 1.3(b), $[0 :_C m^{jd}]_{t-jd} \neq 0$ for some $j \geq 1$. Let $0 \neq z \in [0 :_C m^{jd}]_{t-jd}$. That is, $m^d(m^{(j-1)d}z) = 0$. Now, z occurs in degree $t-jd$ so $m^{(j-1)d}z \subseteq C_{t-d}$. But $[0 :_C m^d]_i = 0$ for all $i \leq t-d$, so $m^{(j-1)d}z = 0$. Similarly one shows that $m^{(j-2)d}z = 0$ and so on, concluding that $z = 0$. Contradiction. □

2. Curves and Complete Intersections

Let X be a curve in $\mathbb{P}^n$ with homogeneous ideal I_X. We assume only that X is locally Cohen-Macaulay and equidimensional. This guarantees that its deficiency module $M(X) = \bigoplus_{i\in\mathbf{Z}} H^1(\mathbb{P}^n, \mathcal{I}_X(i))$ (where $\mathcal{I}_X$ is the ideal sheaf of X in $\mathbb{P}^n$) is a graded R-module of finite length, and hence that the results of §1 apply to $M = M(X)$.

Let F be a general form of degree d. We may take F to be smooth, hence defining a smooth hypersurface. Let Z be the zero-scheme in $\mathbb{P}^n$ cut out on X by F. Let I_Z be its homogeneous ideal in R and let $\mathcal{I}_Z$ be its ideal sheaf in $\mathbb{P}^n$. Let $I_{Z|F}$ be the homogeneous ideal of Z in F (i.e. an ideal in the ring $R/(F)$, but we will view it as an R-module) and let $\mathcal{I}_{Z|F}$ be its sheafification.

We have the following two exact sequences of sheaves:

$$0 \to \mathcal{O}_{\mathbf{P}^n}(-d) \xrightarrow{F} \mathcal{I}_Z \to \mathcal{I}_{Z|F} \to 0 \tag{1}$$

and

$$0 \to \mathcal{I}_X(-d) \xrightarrow{F} \mathcal{I}_X \to \mathcal{I}_{Z|F} \to 0 \tag{2}$$

In particular, from (1) we conclude that

$$\bigoplus_{i \in \mathbf{Z}} H^1(\mathbb{P}^n, \mathcal{I}_{Z|F}(i)) \cong \bigoplus_{i \in \mathbf{Z}} H^1(\mathbb{P}^n, \mathcal{I}_Z(i))$$

as graded R-modules. We will denote this module by $H^1_*(\mathcal{I}_Z)$ and use this isomorphism without further comment.

From (2) we have the exact diagram

$$\begin{array}{ccccccccccccc}
0 & \to & I_X(-d) & \xrightarrow{F} & I_X & \to & I_{Z|F} & \longrightarrow & M(-d) & \xrightarrow{F} & M & \longrightarrow & H^1_*(\mathcal{I}_Z) \\
 & & & & & & \searrow & \nearrow & & & \searrow & \nearrow & \\
 & & & & & & & K & & & & C & \\
 & & & & & & \nearrow & \searrow & & & \nearrow & \searrow & \\
 & & & & & & 0 & & 0 & & 0 & & 0
\end{array} \tag{3}$$

which we will use to apply the results of §1.

Remark 2.1.
(a) Let $t = a_-(\text{soc } H^1_*(\mathcal{I}_Z))$. Then

$$a_-([0 :_{H^1_*(\mathcal{I}_Z)} \mathfrak{m}^d]) \geq t - d + 1.$$

This is because if $0 \neq x \in [0 :_{H^1_*(\mathcal{I}_Z)} \mathfrak{m}^d]_i$ for $i \leq t - d$ then there exists some polynomial G of degree $\leq d - 1$ such that $Gx \neq 0$ but $Gx \in [\text{soc } H^1_*(\mathcal{I}_Z)]_i$ for some $i < t$. Contradiction.

(b) Note that F is a minimal generator for I_Z. Otherwise, suppose that $F = \sum X_i G_i$ where $G_i \in I_Z$ all have degree $d - 1$. By Bezout's theorem, $G_i \in I_X$ for all i. But then $F \in I_X$ as well. Contradiction.

(c) Suppose Z has minimal free resolution

$$0 \to F_n \to \ldots \to F_2 \to F_1 \oplus R(-d) \to I_Z \to 0$$

where $R(-d)$ is the summand corresponding to the minimal generator F. Then applying the mapping cone to the exact diagram

$$\begin{array}{ccccccccc}
 & & & & \vdots & & & & \\
 & & & & F_3 & & & & \\
 & & & & \downarrow & & & & \\
 & & 0 & & F_2 & & & & \\
 & & \downarrow & & \downarrow & & & & \\
 & & R(-d) & \rightarrow & R(-d) \oplus F_1 & & & & \\
 & & \downarrow & & \downarrow & & & & \\
0 & \rightarrow & R(-d) & \xrightarrow{F} & I_Z & \rightarrow & I_{Z|F} & \rightarrow & 0 \\
 & & \downarrow & & \downarrow & & & & \\
 & & 0 & & 0 & & & &
\end{array}$$

gives that the R-module $I_{Z|F}$ has a minimal free resolution

$$0 \rightarrow F_n \rightarrow \ldots \rightarrow F_2 \rightarrow F_1 \rightarrow I_{Z|F} \rightarrow 0$$

(d) If $F_n = \bigoplus_{i=1}^{b_n} R(-m_i)$ then it is shown in **[HU]** that

$$\text{soc } H^1_*(\mathcal{I}_{Z|F}) = \bigoplus_{i=1}^{b_n} k(-m_i + n + 1) \qquad \square$$

Proposition 2.2. *Assume that X is not arithmetically Cohen-Macaulay, so that $K \neq 0$. Let $b = a_-(K)$, the least degree in which $I_{Z|F}$ contains an element which does not lift to X. Then $b + n \geq \min\{m_i\}$ (where m_i are defined in* 2.1 (d)*).*

Proof.

$$\begin{aligned}
b = a_-(K) &\geq a_-([0 :_C m^d]) + d \quad \text{(by 1.4)} \\
&\geq a_-([0 :_{H^1_*(\mathcal{I}_Z)} m^d]) + d \quad (\text{since } C \subset H^1_*(\mathcal{I}_Z)) \\
&\geq a_-(\text{soc } H^1_*(\mathcal{I}_Z)) - d + 1 + d \quad \text{(by 2.1(a))} \\
&= 1 + \min\{m_i - n - 1\} \quad (\text{by 2.1}(d))
\end{aligned}$$

from which the result follows. $\square$

Remark 2.3.

(a) The R-module $I_{Z|F}$ must have a minimal generator in degree b (see 2.2), and the minimal generators of $I_{Z|F}$ correspond to those of I_Z other than F (2.1 (c)). Thus we conclude from 2.2 that if min $\{m_i\} - n$ is strictly greater than the degree of any generator of I_Z other than F then the curve X must be arithmetically Cohen-Macaulay. This is the basic idea behind the results of Strano, Re and Huneke-Ulrich in lifting the complete intersection or Gorenstein properties from the hyperplane section to the curve, but now we see that it holds for hypersurface sections as well.

(b) Let X be a curve in $\mathbb{P}^3$ and let $d = 1$. Then Z is a zero-scheme in $\mathbb{P}^2$, and so as such it has a Hilbert-Burch matrix and corresponding degree matrix. Now, Remark 2.3(a) refers to the resolution of Z as a subscheme of $\mathbb{P}^3$. Comparing this resolution with the corresponding one in $\mathbb{P}^2$, the conclusion of (a) can be translated to say that if the entries of the degree matrix of Z in $\mathbb{P}^2$ are all strictly greater than 2, then X must be arithmetically Cohen-Macaulay. This gives a nice tie-in with Strano's theorem (essentially 2.4 (a) below): a complete intersection in $\mathbb{P}^2$ has a degree matrix of size 1×2, and those complete intersection Z which do *not* force X to also be a complete intersection are those Z with a 2 in the degree matrix, i.e. Z lying on a conic. □

Theorem 2.4. *Let X be a non-degenerate locally Cohen-Macaulay equidimensional one-dimensional subscheme of $\mathbb{P}^n$. Let F be a general homogeneous polynomial of degree $d \geq 1$ cutting out a zero-scheme Z on X. Assume that Z is a complete intersection in $\mathbb{P}^n$. Then X is a complete intersection, unless one of the following holds:*
(a) *$n = 3, d = 1, X$ lies on a quadric surface and has even degree;*
(b) *$n = 3, d = 1, \deg X = 4$ and X does not lie on a quadric;*
(c) *$n > 3, d = 1, X$ is a double line (i.e. X has degree 2 and is supported on a line).*

Proof. Say Z is the complete intersection of hypersurfaces (in $\mathbb{P}^n$) of degree $a_1, ..., a_n$ with $a_1 \leq ... \leq a_n$. Since Z is a complete intersection, the last free part of the minimal free resolution for I_Z (and also for $I_{Z|F}$) has rank one and is twisted by $\sum a_i$. Suppose X is not a complete intersection. Then X is not arithmetically Cohen-Macaulay, so by 2.2 we have $b+n \geq \sum a_i$. Notice that one of the a_i is d and one is b. We will denote by $\sum^\bullet a_i$ the sum $a_1 + ... + a_n - b$, so we have $\sum^\bullet a_i \leq n$.

We have two possibilities: either $\sum^\bullet a_i = n-1$ or $\sum^\bullet a_i = n$. In the first case all the a_i other than b are 1. But b is the least degree in which $I_{Z|F}$ has a generator which does not lift to X, so here $b = 1$ as well since X is non-degenerate, and so deg $Z = 1$. It follows that deg $X = 1$. Impossible (X is non-degenerate).

Suppose $\sum^\bullet a_i = n$. Then $n-2$ of the a_i are 1, one is 2 and one is b. Furthermore, one of these is d. If $n \geq 4$ then at least two of the a_i are 1. At most one of these corresponds to F (if $d = 1$) so at least one of the generators of $I_{Z|F}$ has degree 1. Hence by definition $b = 1$ and so deg $Z = 2$. Then the non-degeneracy of X forces $d = 1$, deg $X = 2$, X non-reduced.

The last possibility is $n = 3$. Then Z is the complete intersection of surfaces of degree 1,2 and b. If $b = 1$ then $d = 1$ and Z is a double line, as above. So assume that $b \geq 2$. The definition of b then forces $d = 1$: d is one of 1,2 or b, and non-degeneracy of X means $I_{Z|F}$ cannot have an element of degree 1 since $b \geq 2$. So Z is a zero-scheme in $\mathbb{P}^2$ which is the complete intersection of a conic (not necessarily reduced) and a plane curve of degree b. If $b = 2$ then deg Z = deg X = 4 and X may or may not not lie on a quadric. If $b > 2$ then by definition of b X lies on a quadric and clearly has even degree. □

Remark 2.5. Some of the above analysis was essentially done by Strano and by Re in the case $d = 1$ ([**R**], [**S1**], [**S2**]). However, they do not seem to have considered the case of a double line. In any case, they note that the exceptions (a) and (b) actually occur. Of course (c) does as well. □

Corollary 2.6. *Let X be a non-degenerate locally Cohen-Macaulay equidimensional subscheme of dimension r in $\mathbb{P}^n$. Let Y be a general complete intersection of codimension r (this makes sense once the degrees of the generators of I_Y are chosen). Assume that either of the following holds:*

(a) $\deg Y > 1$, *or*

(b) $\operatorname{codim} X \geq 3$ *and* $\deg X > 2$.

If Y cuts out a zero-scheme on X which is a complete intersection in $\mathbb{P}^n$, then X must be a complete intersection. □

We leave the details to the reader (but see Corollary 3.7, which is similar). Note that without either of the assumptions (a) or (b) this corollary is false: take for example X to be the projection of the Veronese to $\mathbb{P}^4$ and Y to be a general 2-plane. The intersection may be viewed as the intersection of a rational quartic and a plane in $\mathbb{P}^3$, and hence is a complete intersection. But of course X is not a complete intersection.

Lemma 2.7. *Assume that $X \subset \mathbb{P}^n$ is non-degenerate. Let F be a general homogeneous polynomial of degree $d \geq 1$, cutting out on X a subscheme Z of $\mathbb{P}^n$. As before let $b = a_-(K)$, the least degree in which $I_{Z|F}$ contains an element which does not lift to X.*

(a) *If X is reduced and connected then $b > d$. In particular, Z is non-degenerate in $\mathbb{P}^n$ if $d \geq 2$ and Z is non-degenerate in the hyperplane F if $d = 1$.*

(b) *If X is reduced then $b > d-1$. In particular, if $d \geq 2$ then Z is non-degenerate in $\mathbb{P}^n$.*

Proof. The case $d = 1$ of (a) was observed in **[HU]** but the idea is the same for the general case. The hypothesis of (a) guarantees that $h^1(\mathcal{I}_X) = 0$ while the hypothesis of (b) guarantees only that $h^1(\mathcal{I}_X(-1)) = 0$. Consider the exact sequence

$$0 \to (I_X)_i \to (I_X)_{i+d} \to (I_{Z|F})_{i+d} \to H^1(\mathcal{I}_X(i)) \to \dots .$$

Then taking $i = 0$ or -1 respectively proves (a) and (b). □

3. Gorenstein Hypersurface Sections of a Curve

Now we begin an analysis of the case where Z is Gorenstein. The first part of the argument follows that in **[HU]**. Assume that I_Z has a minimal free resolution

$$0 \to R(-m) \to \bigoplus_i R(-m+a_i) \to \dots \to \bigoplus_i R(-a_i) \to I_Z \to 0$$

Then by 2.1(c), $I_{Z|F}$ has the same minimal free resolution except that the direct summand $R(-d)$ corresponding to F is removed. We will denote by $\{a_i\}^\bullet$ the set of degrees of generators of $I_{Z|F}$. Suppose that X is not Gorenstein. Let $a = \min\{a_i\}^\bullet$ and as before let $b = a_-(K)$, the least degree in which $I_{Z|F}$ contains an element which does not lift to X. By 2.2, $b + n \geq m$ and by 2.3, $b \leq \max\{a_i\}^\bullet$. On the other hand, the smallest twist of any free module F_i in the resolution of $I_{Z|F}$ is at least one more than the smallest twist in the module F_{i-1}. Hence $m - \max\{a_i\} \geq \min\{a_i\}^\bullet + n - 2 = a + n - 2$. Then

$$\begin{aligned} \max\{a_i\}^\bullet + n &\geq b + n \\ &\geq m \\ &\geq a + n - 2 + \max\{a_i\}. \end{aligned}$$

Consequently, $0 \geq a - 2$. So we have our first result:

(3.1) *If Z is Gorenstein and its ideal has no generator in degree* 1 *or* 2 *other than (possibly) F then X must be Gorenstein.*

Now we must consider the cases $a = 1, 2$. First suppose $a = 1$. Since X is non-degenerate, this means that $b = 1$ as well. Then from above we get $1 + n \geq m \geq 1 + n - 2 + \max\{a_i\}$. In particular $\max\{a_i\} = 2$ and $m = n + 1$.

By 2.1(d), then, soc $H^1_*(\mathcal{I}_Z) = k$ (in degree 0) and hence $h^1(\mathcal{I}_Z) = 1$. Consider the exact sequence

$$0 \to H^0(\mathbb{P}^n, \mathcal{I}_Z) \to H^0(\mathbb{P}^n, \mathcal{O}_{\mathbb{P}^n}) \to H^0(Z, \mathcal{O}_Z) \to H^1(\mathcal{I}_Z) \to 0 \qquad (1)$$

This gives $d(\deg Z) = 1 + 1 = 2$. If $d \geq 2$ this is impossible. So the only possibility is $d = 1$ and $\deg Z = \deg X = 2$, and again X must be a double line:

(3.2) *If Z is Gorenstein and is degenerate in $\mathbb{P}^n$ then X is Gorenstein unless X is a double line and $d = 1$.*

We now consider the case $a = 2$. Now we have equality in all the steps of the calculation above. In particular, $b = \max\{a_i\}^\bullet = \max\{a_i\} = m - n$. From 2.1(d) again, soc $H^1_*(\mathcal{I}_Z) = k(-b+1)$, so $h^1(\mathcal{I}_Z(b-1)) = 1 = h^1(\mathcal{I}_{Z|F}(b-1))$. Consider the exact sequence

$$\begin{array}{ccccccccc} \cdots & \to & H^1(\mathcal{I}_{Z|F}(b-1)) & \to & H^2(\mathcal{I}_X(b-1-d)) & \to & H^2(\mathcal{I}_X(b-1)) & \to & 0 \\ & \searrow \ \nearrow & & & & & & & \\ & C_{b-1} & & & & & & & \\ & \nearrow \ \searrow & & & & & & & \\ 0 & & 0 & & & & & & \end{array}$$

Suppose that $\dim C_{b-1} = 0$. We know that $h^1(\mathcal{I}_{Z|F}(i)) = 0$ for $i \geq b$. Hence $C_i = 0$ for $i \geq b - 1$. Thus $a_-([0 :_C m^d]) \leq b - d - 1$. But we have seen that $b + n = m$. Hence all the inequalities in the proof of 2.2 are equalities, so $a_-([0 :_C m^d]) = b - d$. Contradiction. Therefore $\dim C_{b-1} = 1$ and so $h^2(\mathcal{I}_X(b-1-d)) = 0$, and hence

$$h^2(\mathcal{I}_X(i)) = 0 \text{ for all } i \geq b - 1 - d. \qquad (2)$$

Suppose first that $b = 2$, so $h^1(\mathcal{I}_Z(1)) = 1$. From the exact sequence (1) (twisted by 1) we get

$$\deg Z = d(\deg X) = n + 2 - h^0(\mathcal{I}_Z(1)).$$

If $d = 1$, this gives $\deg X = n + 1$. Notice that this can always happen, for any n: take a general set of $n + 1$ lines in $\mathbb{P}^n$. Its general hyperplane section is a general set of $n + 1$ points in $\mathbb{P}^{n-1}$, which is Gorenstein (see Example 3.4).

Now assume that $a = b = 2, d \geq 3$. Then in particular $h^2(\mathcal{I}_X(-2)) = 0$ by (2). This is impossible for any curve; this should be a standard fact, but we are not aware of a proof in the literature so we include it.

Claim: For any curve X in $\mathbb{P}^n$, $h^2(\mathcal{I}_X(-2)) \neq 0$.

Proof of Claim. Suppose otherwise. First suppose that X is reduced. From the exact sequence of sheaves

$$0 \to \mathcal{I}_X(t) \to \mathcal{O}_{\mathbb{P}^n}(t) \to \mathcal{O}_X(t) \to 0 \tag{3}$$

and the fact that X is a curve it follows that $h^i(\mathcal{I}_X(t)) = 0$ for all t and all i, $2 < i < n$. Furthermore, $h^n(\mathcal{I}_X(-n) = 0$ from the same sequence, and $h^1(\mathcal{I}_X(-1)) = 0$ since X is reduced. But then Castelnuovo-Mumford implies that the ideal I_X is generated in degree 0, an impossibility.

Now suppose that X is non-reduced. Let Y be the curve on which X is supported. We have an exact sequence of sheaves

$$0 \to \mathcal{I}_X \to \mathcal{I}_Y \to \mathcal{I}_Y/\mathcal{I}_X \to 0$$

and taking cohomology on this sequence (with any twist) gives (since $\dim X = 1$) $h^2(\mathcal{I}_X(t)) \geq h^2(\mathcal{I}_Y(t))$ for any t. Therefore, $h^2(\mathcal{I}_X(-2)) > 0$. This proves the claim.

We are left with the case $a = b = d = 2$. Now $\deg Z = 2 \cdot \deg X = n + 2$. If n is odd then this is impossible. If n is even then Riemann-Roch gives (using (2)) that $h^0(\mathcal{O}_X(-1)) = -(\frac{n+2}{2}) - g + 1 \geq 0$ (where g is the arithmetic genus). So we have shown:

(3.3) *Assume that Z is Gorenstein and non-degenerate in $\mathbb{P}^n$, and that $a = b = 2$. Then X is Gorenstein unless*

(a) $d = 1$ *and* $\deg X = n + 1$,

(b) $d = 2, n$ *is even and* $\deg X = \frac{n+2}{2}$, $p_a(X) \leq 1 - (\frac{n+2}{2})$. *In particular, if X is reduced then it can only be a disjoint union of $\frac{n+2}{2}$ skew lines. (See Example 3.4.)*

Finally we consider the case $a = 2, b \geq 3$. Here there seems to be less to say. Notice that this condition forces X to lie on a quadric hypersurface. Since $h^2(\mathcal{I}_X(b-1-d)) = 0$, as above we have that $d \leq b$. (If X is reduced and connected then $d < b$ by Lemma 2.7.) Also, we still have that $h^1(\mathcal{I}_Z(b-1)) = 1$. Hence from (1) (twisted by $b - 1$), $\deg Z = d(\deg X) = \binom{n+b-1}{n} + 1 - h^0(\mathbb{P}^n, \mathcal{I}_Z(b-1))$. Finally, notice that if $d = b \geq 3$ and X is reduced then X must be a union of skew lines (since $h^2(\mathcal{I}_X(-1)) = 0$).

Example 3.4. The following curves X are not arithmetically Cohen-Macaulay, but for the given values of d the hypersurface sections are Gorenstein.

$\boxed{a = b = d = 1}$ For any n we may take X to be a double line. This is the only possible example (see (3.2)).

$\boxed{a = b = 2, d = 1}$ For any n take X to be a general set of $n+1$ skew lines or a smooth rational curve of degree $n+1$. To check that Z is Gorenstein, note that Z is a set of $n+1$ points in $\mathbb{P}^{n-1}$ with the uniform position property, and consequently with the Cayley-Bacharach property. But considering the coordinate ring of Z and passing to the zero-dimensional case (taking a quotient with a general linear form), one gets that the quotient has a symmetric Hilbert function: 1 $n-1$ 1 0 ... . Therefore Z is Gorenstein ([**DGO**]). (I am grateful to A.V. Geramita for pointing out this interpretation of Gorenstein rings.)

$\boxed{a = b = d = 2, n \text{ even}}$ Take X to be a general set of $\frac{n+2}{2}$ skew lines. (Notice that X lies on a quadric hypersurface– see Corollary 3.5 as well.)

$\boxed{a = 2, b = 3, d = 1}$ [**HU**] Example 3.26 gives an example of a monomial curve in $\mathbb{P}^4$ of degree 8 lying on a quadric hypersurface.

$\boxed{a = 2, b = 3, d = 2}$ We will show that no reduced irreducible X will give an example. Notice that deg Z must be $2n+2$, so deg $X = n+1$. We know from (2) that $h^2(\mathcal{I}_X) = 0$ so by Riemann-Roch $h^0(\mathcal{O}_X) = -g+1$ (where g is the arithmetic genus) and by the hypothesis of reduced, irreducible that $h^1(\mathcal{I}_X) = 0$. Then by (3) we get that $g = 0$. By the main theorem of [**GLP**], $h^1(\mathcal{I}_X(i)) = 0$ for all $i \geq (n+1)+1-n = 2$. Since $h^2(\mathcal{I}_X(2)) = 0$, Riemann-Roch gives $h^0(\mathcal{O}_X(2)) = 2(n+1)-g+1 = 2n+3$. Then (3) gives $h^0(\mathcal{I}_X(2)) = \binom{n+2}{2} - 2n - 3$.

On the other hand, the conditions $a = 2, b = 3$ force the Hilbert function of Z (after taking a quotient with a general linear form) to be 1 n n 1 0 ... . One can deduce from this that $h^0(\mathcal{I}_Z(2)) = \binom{n+2}{2} - 2n - 1$, and hence $h^0(\mathcal{I}_{Z|F}(2)) = \binom{n+2}{2} - 2n - 2$. Then since $b = 3 (> 2)$ we get that $h^0(\mathcal{I}_X(2)) = \binom{n+2}{2} - 2n - 2$, contradicting the previous paragraph.

Notice that if $g = 1$ then Z is Gorenstein, but X is as well. (So we do not have the condition that $h^2(\mathcal{I}_X) = 0$.)

We do not know any examples of non-Gorenstein X with Gorenstein Z satisfying $a = 2, b = 3, d = 2$.

$\boxed{a = 2, b = 3, d = 3}$ We consider reduced curves. As in the last example, the Hilbert function must be 1 n n 1 0 ... and so deg $Z = 2n+2$. Then as in (3.3), the only possibility is a set of $\frac{2n+2}{3}$ skew lines; (the main point is that $b = d$– then consider the arithmetic genus and use the hypothesis of reducedness). Those n congruent to 0 or 1 (mod 3) are eliminated by the condition 3(deg X) = deg Z = $2n+2$. So consider for example $n = 5$ and consider a non-degenerate set of 4 skew lines. We compute $h^0(\mathcal{I}_{Z|F}(2)) = h^0(\mathcal{I}_Z(2)) = 10$ and $h^0(\mathcal{I}_X(2)) = 9 + h^1(\mathcal{I}_X(2))$. So the condition that $b = 3$ means that $h^1(\mathcal{I}_X(2)) = 1$. A general set of 4 lines in $\mathbb{P}^5$ does not satisfy this, but E. Ballico has shown us a simple proof that a "general" set of 4 lines possessing a 4-secant line does. (One simply shows directly that such a set of lines imposes 10 conditions on quadrics.) He also notes that the same idea works in higher projective space, so this gives an example for any n congruent to 2 (mod 3), and is the only reduced example.

$\boxed{a = 2, b \geq 3, d = 1}$ Recall that X can be a curve of even degree on a smooth quadric surface in $\mathbb{P}^3$ (see §2). □

Corollary 3.5. *Suppose that $X \subset \mathbb{P}^n$ does not lie on a quadric hypersurface and that its intersection Z with a general hypersurface of degree d is Gorenstein. Assume furthermore that X is reduced.*

(a) *If $n \geq 5$ then X is Gorenstein.*

(b) *If $d \geq 2$ then X is Gorenstein.*

Proof. If $a \geq 3$ then by (3.1) X is Gorenstein. Since X is reduced, by (3.2) we may assume that $a = 2$ and by hypothesis $b = 2$. Then (3.3) gives the only possibilities, and one checks that both (a) and (b) force X to lie on a quadric if it is reduced. □

Remark 3.6. We do not know if for n sufficiently large the condition that it not lie on a quadric and that Z be Gorenstein are enough to force X to be Gorenstein even in the non-reduced case. □

We remark that under reasonable conditions we can get results along the lines of Corollary 2.6, under the weaker Gorenstein hypothesis. For instance,

Corollary 3.7. *Let X be a reduced locally Cohen-Macaulay equidimensional subscheme of $\mathbb{P}^n$ of dimension r not lying on a quadric hypersurface and let Y be a "general" complete intersection of codimension r not lying on a quadric hypersurface. If the points Z of intersection of X and Y are Gorenstein then X is Gorenstein.*

Proof. We have that Y is a complete intersection of hypersurfaces F_i each of degree $d_i \geq 3$. Take them one at a time: let $X = X_0, X_1 = X_0 \cap F_1, \ldots, X_r = X_{r-1} \cap F_r$. By the generality assumption, each X_i is reduced and we may use (2) of §2 with $d = d_i$. We have in particular that

$$0 \to H^0(\mathcal{I}_X(2 - d_1)) \xrightarrow{F_1} H^0(\mathcal{I}_X(2)) \to H^0(\mathcal{I}_{X_1|F_1}(2)) \to 0$$

and

$$0 \to H^0(\mathcal{O}_{\mathbb{P}^n}(2 - d_1) \xrightarrow{F_1} H^0(\mathcal{I}_{X_1}(2)) \to H^0(\mathcal{I}_{X_1|F_1}(2)) \to 0$$

(the latter is the analogue of (1) of §2, where the second term is the cohomology of the ideal sheaf of X_1 in $\mathbb{P}^n$). Then by hypothesis we have $0 = h^0(\mathcal{I}_X(2)) = h^0(\mathcal{I}_{X_1|F_1}(2)) = H^0(\mathcal{I}_{X_1}(2))$. Continuing inductively, we arrive at a reduced locally Cohen-Macaulay equidimensional curve X_{r-1} not lying on a quadric hypersurface, and $Z = X_r$ is a Gorenstein set of points. But $d_r \geq 3$ so by Corollary 3.5 (b), X_{r-1} is Gorenstein. But for a locally Cohen-Macaulay equidimensional subscheme X_i of dimension ≥ 2, the Cohen-Macaulayness of the general hypersurface section X_{i+1}

implies the Cohen-Macaulayness of X_i. (This is a straightforward result. It was noted for $d = 1$ in **[M]** Remark 1.10, **[R]** Teorema 2, and **[HU]** Proposition 2.1, and the proof in the higher degree case is similar.) Therefore X is Cohen-Macaulay, hence Gorenstein. □

References

[BE] Ballico, E. and Ellia, Ph., On Postulation of Curves in $\mathbb{P}^4$, Math. Z. **188** (1985), 215–223.

[DGO] Davis, E., Geramita, A.V. and Orecchia, F., Gorenstein Algebras and the Cayley- Bacharach Theorem, Proc. Amer. Math. Soc. **93** (1985), 593–597.

[GM] Geramita, A.V. and Migliore, J., On the Ideal of an Arithmetically Buchsbaum Curve, J. Pure Appl. Algebra **54** (1988), 215–247.

[GLP] Gruson, L., Lazarsfeld, R. and Peskine, C., On a theorem of Castelnuovo, and the equations defining space curves, Invent. Math. **72** (1983), 491–506.

[HU] Huneke, C. and Ulrich, B., General Hyperplane Sections of Algebraic Varieties, to appear in J. Algebraic Geom.

[K] Kreuzer, M., An introduction to a theorem of Strano, Curves Seminar at Queen's, Queen's Papers in Pure and Applied Mathematics VIII.

[M] Migliore, J., Curves and their Hyperplane Sections, lecture notes for the Notre Dame Algebraic Geometry Seminar (1989).

[R] Re, R., Sulle sezioni iperpiane di una varietá proiettiva, Le Matematiche XLII, 211-218.

[S1] Strano, R., A characterization of complete intersection curves in $\mathbb{P}^3$, Proc. Amer. Math. Soc. **104** (1988), 711-715.

[S2] —, Sulle sezioni iperpiane delle curve, Rend. Sem. Mat. Fis. Milano LVII (1987).

On the computation of the height of an ideal in a polynomial ring

Roberto Notari

Abstract. In this paper, an algorithm to compute the height of an ideal is presented. It is based on the new concept of c-decomposition of a monomial ideal: this decomposition yields important information on independent sets of variables, and is used also to construct the saturation of a monomial ideal. The algorithm is implemented in the MACSYMA system, and has been tested successfully on various examples.

1991 Mathematics Subject Classification: 13F20, 13C15 .

0. Introduction

One of the basic problems of the algorithmic theory of polynomial ideals is the computation of the dimension $\dim(I)$ of an ideal I in the polynomial ring $\mathcal{R} = k[x_1, \ldots, x_r]$.

$\dim(I)$ can be characterized as the degree of the Hilbert polynomial of I, and this was used by Möller & Mora ([MM]), together with the algorithmic technique of Gröbner bases, introduced by Buchberger ([B]).

$\dim(I)$ can be also characterized as the largest number of elements in $\mathcal{R}$ that are independent modulo I in the sense specified by Gröbner ([Gr]). The Gröbner bases method can be combined with this characterization, too, as showed by Kandry-Rody ([K]), but the number of Gröbner bases calculations involved in this method makes it useless for practical uses in most cases.

Kredel & Weispfenning ([KW]) have showed that $\dim(I)$ can be computed together with maximal independent sets of variables from a single Gröbner base of I.

In this paper, an approach to the computation of $\dim(I)$ throught $\mathrm{ht}(I)$ is made. It is well known that, in $\mathcal{R}$, $\dim(I)+\mathrm{ht}(I) = r$, and that $\mathrm{ht}(I) = \mathrm{ht}(\sqrt{T(I)})$, where $\sqrt{T(I)}$ is the radical of the monomial ideal generated by the leading terms $T(f)$ of the polynomials of a Gröbner base of I(a proof of the last equality is given in Section 3(a) of the paper using the characterization of $\dim(I)$ given in [KW]).

It is possible to compute the height $\mathrm{ht}(J)$ of a monomial ideal J by introducing the c-decomposition of a monomial ideal, which avoids to calculate the primary decomposition of the given monomial ideal. In fact, a c-decomposition gives J as intersection of monomial ideals generated by coprime terms.

Such a decomposition allows also the computation of maximal independent sets of variables modulo J, and the computation of the saturation of J.

The c-decomposition of a monomial ideal can be algorithmically computed, as showed in Section 4, and the corresponding programs have been successfully implemented on a Vax Station 2000, using MACSYMA: the computation of the height of an ideal, using the c-decomposition, is faster than the analogous calculation performed using the algorithm in [KW], as implemented by the author on the same Vax Station.

The paper is organized as follows: in Section 1, some properties of monomial ideals are focused; in Section 2, the c-decomposition is defined and it is showed that every monomial ideal can be c-decomposed, and how to compute the height of the given monomial ideal in terms of the calculated decomposition; in Section 3, the c-decomposition of monomial ideals is used to compute the height of an ideal of $\mathcal{R}$, to compute strongly independent sets of variables modulo the given ideal, and to compute the saturation of a monomial ideal; in Section 4, algorithms to compute a c-decomposiotion and the height of an ideal are presented and discussed, together with examples.

The author would like to thank Prof. F. Orecchia for useful discussions concerning the paper.

1. Monomial ideals

Let $\mathcal{R} = k[x_1, \ldots, x_r]$ be a polynomial ring over the field k.

Definition 1. A *term* T is a product of powers of the variables, i.e. $T = {x_1}^{a_1} \ldots {x_r}^{a_r}$, with $a_i \geq 0$.

The number $\deg(T) = a_1 + \cdots + a_r$ is called *degree* of T, each a_i is also denoted by $\deg(T, x_i)$, while the r-tuple $(a_1, \ldots, a_r)$ is the r-degree of T.

Finally, let supp(T) be the set supp$(T) = \{x_i | \deg(T, x_i) > 0\}$.

The ring $\mathcal{R}$ is a graded ring with respect to the r-fold grading induced by the r-degree ([H], Ch. 4).

Definition 2. An ideal $I \subseteq \mathcal{R}$ is a *monomial ideal* if it is generated by terms.

Monomial ideals generated by coprime terms will be of particular importance in the following.

Definition 3. A monomial ideal generated by coprime terms is a *c-monomial ideal.*

By ([ZS], Vol. II, Ch. VII, §2, Th. 7,8), a monomial ideal is homogeneous in the graded ring $\mathcal{R}$, and therefore, if I, J are monomial ideals, $I+J, I\cdot J, I\cap J, I:J, \sqrt{I}$ are monomial ideals.

Furthermore, let $\mathrm{Supp}(f)$ be the set $\mathrm{Supp}(f) = \{T | c(f,T) \neq 0\}$, where $c(f,T)$ is the coefficient of T in f. Then $f \in I$ if, and only if, $\mathrm{Supp}(f) \subseteq I$.

Let σ be a term-ordering. If f is a polynomial, then $T(f), lc(f)$ are the maximal term in $\mathrm{Supp}(f)$ with respect to σ and the coefficient of $T(f)$, respectively.

In the following proposition, generators of the previously mentioned ideals are calculated from generators of I, J.

Proposition 4. *Let I, J be generated by $\{S_1, \ldots, S_m\}$ and $\{T_1, \ldots, T_n\}$, respectively. Then*

(1) $\{S_1, \ldots, S_m, T_1, \ldots, T_n\}, \{S_1T_1, \ldots, S_1T_m, \ldots, S_nT_m\}$ generate $I+J, I\cdot J$, respectively ;

(2) $I\cap J$ is generated by $\{M_{11}, \ldots, M_{1m}, \ldots, M_{nm}\}$, where M_{ij} is the least common multiple of S_i, T_j;

(3) If J is principal, then $I:J$ is generated by $\{\frac{S_1}{\gcd(S_1,T_1)}, \ldots, \frac{S_m}{\gcd(S_m,T_1)}\}$;

(4) Let S_i' be the term such that

(i) $\mathrm{supp}(S_i') = \mathrm{supp}(S_i)$;

(ii) $\deg(S_i', z) = 1 \quad \forall z \in \mathrm{supp}(S_i)$.

Then, $\sqrt{I}$ is generated by $\{S_1', \ldots, S_m'\}$.

Proof.

(1) The claim is true for every polynomial ideal.

(2) $M_{ij} \in I\cap J$ because $S_i | M_{ij}, T_j | M_{ij}\ \forall i,j$. Conversely, if $S \in I\cap J$, then there exist i,j such that $S_i|S, T_j|S$ and so $M_{ij}|S$.

(3) $S_i | \frac{S_iT_1}{\gcd(S_i,T_1)}$, and therefore $\frac{S_i}{\gcd(S_i,T_1)} \in I:J$.

Conversely, if $S \in I:J$, then there exists i such that $S_i|ST_1$, and so $\frac{S_i}{\gcd(S_i,T_1)} | S \cdot \frac{T_1}{\gcd(S_i,T_1)}$, but $\gcd(\frac{S_i}{\gcd(S_i,T_1)}, \frac{T_1}{\gcd(S_i,T_1)}) = 1$, and so $\frac{S_i}{\gcd(S_i,T_1)} | S$.

(4) $(S_i')^p \in I$, where $p = \max\{\deg(S_i,z) | z \in \mathrm{supp}(S_i)\}$, and so $\mathrm{id}(S_1', \ldots, S_m') \subseteq \sqrt{I}$.

Let f be a polynomial in $\sqrt{I}$. Then $f^q \in I$, and so $T(f)^q \in I$, i.e., there exists i such that $S_i | T(f)^q$, and so $S_i' | T(f)$. Then $f - lc(f)T(f) \in \sqrt{I}$, and, after a finite number of steps, $f \in \mathrm{id}(S_1', \ldots, S_m')$, and hence the claim follows. □

By ([ZS], Vol. II, Ch. VII, §2, Th. 9), every monomial ideal admits a primary decomposition, and the associated prime ideals are monomial ideals.

It is not difficult to prove that the only prime monomial ideals are those generated by $\{x_{i_1}, \ldots, x_{i_s}\}$, and the height $\mathrm{ht}(\mathcal{P})$ of such a prime ideal is the number s of generators.

In fact, the following result holds.

Proposition 5. *If I is a c-monomial ideal, generated by $\{S_1, \ldots, S_m\}$, then* $\mathrm{ht}(I) = m$.

Proof. By Proposition 4(3), $\{S_1, \ldots, S_m\}$ is a $k[x_1, \ldots, x_r]$-regular sequence in I. Let f be a polynomial in I such that $\mathrm{id}(S_1, .., S_m) : \mathrm{id}(f), = \mathrm{id}(S_1, ..., S_m)$

$$\mathrm{id}(S_1, \ldots, S_m) \ni f = f1 \Rightarrow 1 \in \mathrm{id}(S_1, \ldots, S_m) : \mathrm{id}(f) \Rightarrow I = (1)$$

and that is absurd.

But $k[x_1, \ldots, x_r]$ is a C-M ring, and so $m = \mathrm{gr}(I) = \mathrm{ht}(I)$ (for properties of Cohen-Macauley rings, see for examples [GS], Ch. 11). □

2. Decomposition of monomial ideals

Let I be a monomial ideal generated by $\{S_1, \ldots, S_m\}$.

Remark. Given a generating set of I, a minimal generating set can be calculated from the given one by eliminating the terms which can be divided by other terms of the set.

Definition 6. A *c-decomposition* of I is an expression of I as finite intersection

$$I = \bigcap_{j=0}^{t} I_j$$

where $I_0, \ldots, I_t$ are c-monomial ideals.

Proposition 7. *Every monomial ideal can be c-decomposed.*

Proof. Let $\{S_1, \ldots, S_m\}$ be a minimal generating set of the monomial ideal I, subscripted in a way such that

$$1 \neq T_1 = \gcd(S_1, \ldots, S_{i_1}) \text{ with } 1 \leq i_1 \leq m, \gcd(T_1, S_j) = 1 \text{ for } j \geq i_1$$
$$1 \neq T_2 = \gcd(S_{i_1+1}, \ldots, S_{i_2}) \text{ with } i_1 < i_2 \leq m, \gcd(T_2, S_j) = 1 \text{ for } j \geq i_2$$
$$\vdots$$
$$1 \neq T_h = \gcd(S_{i_{h-1}+1}, \ldots, S_{i_h}) \text{ with } i_{h-1} < i_h = m$$

Let us define

$$I_0 = \mathrm{id}(T_1, \ldots, T_h)$$
$$I_j = \mathrm{id}(T_1, \ldots, T_{j-1}, S'_{i_{j-1}+1}, \ldots, S'_{i_j}, S_{i_j+1}, \ldots, S_m) \quad \text{for } j = 1, \ldots, h$$

where $S'_k = S_k | S''_k, S''_k = \prod_{i=1}^{r} x_i^{\alpha_i}$, $\alpha_i = \begin{cases} \deg(T_j, x_i) & \text{if } \deg(T_j, x_i) = \\ & = \deg(S_k, x_i) \\ 0 & \text{otherwise.} \end{cases}$

Claim: $I = I_0 \cap I_1 \cap \ldots \cap I_h$.

If S_j is a generator of I, then there exists $h(j)$ such that $i_{h(j)-1} < j \le i_{h(j)}$, and so $T_{h(j)}|S_j$.

Then, $S_j \in I_0$, and $S_j \in I_l$, with $l > h(j)$ (because $T_{h(j)} \in I_l$), while $S_j \in I_l$ with $l < h(j)$(because S_j is one of the generators of I_l).

But $S_j'|S_j$ and hence $S_j \in I_{h(j)}$.

Conversely, let S be a term of $I_0 \cap I_1 \cap \ldots \cap I_h$. Then $S \in I_0$. Let $\overline{h}$ be the smallest integer such that $T_j|S$.

But $S \in I_{\overline{h}}$ and thus there exists a term $\overline{S}_j$ which divides S.

There are two possibilities:

(1) $j \le i_{\overline{h}}$;

(2) $j > i_{\overline{h}}$.

If (1) holds, then $\overline{S}_j = S_j' = \frac{S_j}{S_j''}$, $\gcd(S_j', S_j'') = 1$ by construction. But $S_j''|T_{\overline{h}}$ and so $S_j''|S$, and then $S_j|S$.

If (2) holds, then $\overline{S}_j = S_j$, i.e., $S \in I$.

By construction, I_0 is a c-monomial ideal.

Furthermore, I_j is generated by terms smaller than the ones of I for $l < j$, and not greater than the ones of I, for $l \ge j$.

Taking a minimal generating set of I_j, and decomposing each $I_j, j > 0$, in the same way, after a finite number of steps the c-decomposition of I is completely computated . □

From now on the considered c-decompositions will be minimal in the usual sense.

Remark. The c-decomposition is not uniquely determined. For example, the ideal $I = \mathrm{id}(x_1x_2, x_1x_3, x_2x_3)$ can be c-decomposed as:

$$I = \mathrm{id}(x_1, x_2x_3) \cap \mathrm{id}(x_2, x_3)$$

$$I = \mathrm{id}(x_2, x_1x_3) \cap \mathrm{id}(x_1, x_3).$$

Proposition 8. *If $I = \cap_{j=0}^t I_j$ is a c-decomposition of I, then*

$$\mathrm{ht}(I) = \min\{\mathrm{ht}(I_j)|j = 0, \ldots, t\}.$$

Proof. If $\mathcal{P}$ is an associated prime ideal of I, then $\mathcal{P} \supseteq \bigcap_{j=0}^t I_j$.

By ([AM], Ch. 1, Prop. 11(ii)), there exists h such that $\mathcal{P} \supseteq I_h$, and so $\mathrm{ht}(\mathcal{P}) \ge \mathrm{ht}(I_h)$.

By definition, $\mathrm{ht}(I) = \min\{\mathrm{ht}(\mathcal{P})|I \subseteq \mathcal{P}\}$, and hence $\mathrm{ht}(I) \ge \min\{\mathrm{ht}(I_j)|j = 0, \ldots, t\}$.

But $I \subseteq I_j$, for all j, and so $\mathrm{ht}(I) \le \mathrm{ht}(I_j), \forall j$, and the claim follows. □

Corollary 9. *If $I = \bigcap_{j=0}^t I_j$ is a c-decomposition of I, then*

$$\mathrm{ht}(I) = \min\{\textit{number of generators of } I_j|j = 0, \ldots, t\}.$$

Proof. Evident, by Proposition 5 and Proposition 8. □

Now, let us study the behaviour of the c-decomposition of monomial ideals generated by terms of two disjoint sets of variables.

Proposition 10. *Let I_1, I_2 be ideals generated by $\{S_1, \dots, S_m\}, \{T_1, \dots, T_n\}$, respectively, with $(\cup_i \operatorname{supp}(S_i)) \cap (\cup_j \operatorname{supp}(T_j)) = \emptyset$, and let I be generated by $\{S_1, \dots, S_m, T_1, \dots, T_n\}$. If $I_1 = \bigcap_{i=0}^{t_1} I_{1i}, I_2 = \bigcap_{j=0}^{t_2} I_{2j}$ are c-decompositions of I_1, I_2, respectively, then $I = \cap_{i,j}(I_{1i} + I_{2j})$ is a c-decomposition of I.*

Proof. At first, $I_{1i} + I_{2j}$ is a c-monomial ideal, because I_{1i}, I_{2j} are c-monomial ideals, generated by terms which have no common variable.

By Definition 3, $I_1 \subseteq I_{1i}, I_2 \subseteq I_{2j}$, and so $I = I_1 + I_2 \subseteq I_{1i} + I_{2j}, \forall i, j$.

Conversely, let S be a term of $\cap_{i,j}(I_{1i} + I_{2j})$ with $S \notin I_1$.

Then there exists h such that $S \notin I_{1h}$, and so $S \in I_{2j}$ $\forall j$, because $S \in I_{1h} + I_{2j}$ $\forall j$. Hence, $S \in I_2 \subseteq I$. □

Corollary 11. *Let I, I_1, I_2 be monomial ideals as in Proposition* 10. *Then* $\operatorname{ht}(I) = \operatorname{ht}(I_1) + \operatorname{ht}(I_2)$.

Proof. By Corollary 9, there exist h_1, h_2 such that $\operatorname{ht}(I_1) = \operatorname{ht}(I_{1h_1}), \operatorname{ht}(I_2) = \operatorname{ht}(I_{2h_2})$ and I_{1h_1}, I_{2h_2} are generated by a minimal number of terms.

Then $I_{1h_1} + I_{2h_2}$ is generated by a number of generators smaller than or equal to the number of generators of $I_{1i} + I_{2j}$ $\forall i, j$.

By Corollary 9, $\operatorname{ht}(I) = \operatorname{ht}(I_{1h_1}) + \operatorname{ht}(I_{2h_2}) = \operatorname{ht}(I_1) + \operatorname{ht}(I_2)$. □

3. Applications of c-decompositions

From now on, I is an ideal of the polyomial ring $\mathcal{R}$, and $T(I)$ is the monomial ideal generated by the leading terms of the polynomials of I, with respect to the fixed term-ordering σ. If $\mathcal{G}$ is a Gröbner base of I with respect to σ, then $T(I)$ is generated by the leading terms of the polynomials of $\mathcal{G}$.

(a) *Computation of the height of an ideal*

Let us recall the definitions of an independent and a strongly independent set of variables modulo I ([KW]).

Let $\mathbf{X}$ be the set $\{x_1, \dots, x_r\}$, and let $\mathbf{Y}$ be a subset of $\mathbf{X}$.

Definition 12. (1) $\mathbf{Y}$ is *independent modulo* I if $I \cap k[\mathbf{Y}] = (0)$.

(2) $\mathbf{Y}$ is *strongly independent modulo* I if $T(I) \cap k[\mathbf{Y}] = (0)$.

Remark. If I is a monomial ideal, then $\mathbf{Y}$ is strongly independent modulo I if, and only if, $\mathbf{Y}$ is independent modulo I.

Proposition 13. *$\mathbf{Y}$ is strongly independent modulo J if, and only if, $\mathbf{Y}$ is strongly independent modulo $\sqrt{J}$.*

Proof. The isolated prime ideals of J and $\sqrt{J}$ are the same. The claim follows from Lemma 1.7(2) in [KW]. □

The dimension $\dim(I)$ of an ideal I can be characterized ([KW], Th. 1.13) as

$$\dim(I) = \max\{|\mathbf{Y}| \mid \mathbf{Y} \text{ is strongly independent modulo } I\}$$

and the fixed term-ordering does not need to be a pure lexicographic one.

By definition of a strongly independent set of variables, I and $T(I)$ have the same dimension, and therefore they have the same height ([M], Ch. 5, §14).

Because $T(I)$, $\sqrt{T(I)}$ have the same height, $\mathrm{ht}(I) = \mathrm{ht}(\sqrt{T(I)})$, and the last number can be computed if a c-decomposition of $\sqrt{T(I)}$ is known, as showed in Corollary 9.

Finally, let us prove Buchberger's criterion for 0-dimensionality ([B]), using c-decompositions.

Proposition (Buchberger's criterion). *Let I be a proper ideal of $\mathcal{R}$. Then $\dim(I) = 0$ if, and only if, $\{x_1, \ldots, x_r\} \subseteq \sqrt{T(I)}$.*

Proof. Let I be a zero dimensional ideal: $\mathrm{ht}(I) = r$.

Then $\mathrm{ht}(\sqrt{T(I)}) = r$, but there is just a radical c-monomial ideal generated by r terms, which is $I_1 = \mathrm{id}(x_1, \ldots, x_r)$, and there is no c-monomial ideal generated by more than r terms. Then, a c-decomposition of $\sqrt{T(I)}$ is given by I_1, and so $\sqrt{T(I)} = I_1$.

Conversely, if $\{x_1, \ldots, x_r\} \subseteq \sqrt{T(I)}$, then $\mathrm{ht}(\sqrt{T(I)}) = r$, and the claim is completely proved. □

(b) *Computation of strongly independent sets of variables*

Proposition 14. *$\mathbf{Y}$ is strongly independent modulo I if, and only if, there exists an isolated prime ideal of $\sqrt{T(I)}$ that is not generated by the variables of $\mathbf{Y}$.*

Proof. By the remark to Definition 12 and Proposition 13 with J equal to $T(I)$, $\mathbf{Y}$ is strongly independent modulo I if, and only if, $\mathbf{Y}$ is strongly independent modulo $\sqrt{T(I)}$.

The claim follows from Lemma 1.7(2) in [KW], because the associated prime ideals of $\sqrt{T(I)}$ are monomial ideals, and hence are generated by variables. □

Now, to calculate the strongly independent sets of variables modulo I, it is enough to show how to compute the isolated prime ideals of $\sqrt{T(I)}$.

Proposition 15. *Let $\sqrt{T(I)} = \bigcap_{j=0}^{t} I_j$ be a c-decomposition of $\sqrt{T(I)}$, and let I_j be generated by $\{S_1^j, \ldots, S_{m_j}^j\}$. The minimal ideals among those generated by picking only a variable up from* $\mathrm{supp}({S^j}_i)$*, for every i, j, are the isolated prime ideals of $\sqrt{T(I)}$.*

Proof. Let $\mathcal{P}$ be an associated prime ideal of $\sqrt{T(I)}$. Then there exists f such that $\mathcal{P} = \sqrt{\sqrt{T(I)} : \mathrm{id}(f)}$ ([AM], Ch. 4, Lemma 4.4). But $\sqrt{\sqrt{T(I)} : \mathrm{id}(f)} = \bigcap_{j=0}^{t} \sqrt{I_j : \mathrm{id}(f)}$, and then there exists j such that $\mathcal{P} = \sqrt{I_j : \mathrm{id}(f)}$.

By ([AM], Ch. 4, Th. 4.5), $\mathcal{P}$ is an associated prime ideal of I_j.

Now, if $\mathcal{P}$ is a prime ideal containing I_j, then ${S^j}_i \in \mathcal{P}$ for every i, and $\mathcal{P} \cap \mathrm{supp}({S^j}_i)$ is not empty, for every i.

Conversely, let $\mathcal{P}$ be generated by $\{x_1^j, \ldots, x_{m_j}^j\}$ with ${x^j}_i \in \mathcal{P} \cap \mathrm{supp}({S^j}_i)$, $\forall i$. Let M be the term $M = \frac{x_1 \cdots x_r}{{x^j}_1 \cdots {x^j}_m}$. Then $\mathcal{P} = I_j : \mathrm{id}(M)$ by Proposition 4(3), and so $\mathcal{P}$ is an associated prime ideal of I_j.

Finally, let $\mathcal{U}$ be the set of prime ideals generated by picking only a variable up from $\mathrm{supp}({S^j}_i)$ for every i, j.

If $\mathcal{P}$ is minimal in $\mathcal{U}$, then, in particular, $\mathcal{P}$ is a prime ideal containing $\sqrt{T(I)}$. Then there exists an isolated prime ideal $\mathcal{P}_1$ of $\sqrt{T(I)}$ contained in $\mathcal{P}$. But $\mathcal{P}_1 \in \mathcal{U}$ and so $\mathcal{P} = \mathcal{P}_1$ by minimality of $\mathcal{P}$. □

(c) *Computation of the saturation of a monomial ideal*

Let us recall the definition of saturation of a homogeneous ideal ([H1], Ch. 2, §5, Ex. 5.10):

$$\mathrm{sat}(I) = \{f \in \mathcal{R} | {x_i}^{m_i} f \in I \ \forall i\}.$$

For other properties of the saturation, see also [T].

Proposition 16. *The saturation of a monomial ideal is a monomial ideal.*

Proof. Let f be a polynomial of $\mathrm{sat}(I)$. Then, for every i, there exists m_i such that ${x_i}^{m_i} f \in I$, and so, I being a monomial ideal, ${x_i}^{m_i} T \in I \ \forall T \in \mathrm{Supp}(f)$, i.e. $T \in \mathrm{sat}(I) \ \forall T \in \mathrm{Supp}(f)$.

Since I is homogeneous in the r-fold grading of $\mathcal{R}$, $\mathrm{sat}(I)$ is homogeneous in the same graded ring, and therefore it is a monomial ideal, c.f. ([ZS] , Vol. II, Ch. VII, §2, Th. 7). □

Now, let us compute the saturation of a c-monomial ideal.

Proposition 17. *Let I be a c-monomial ideal, generated by $\{S_1, \ldots, S_m\}$. Then*

$$\mathrm{sat}(I) = \begin{cases} I, & \text{if } \mathrm{ht}(I) < r \\ (1), & \text{if } \mathrm{ht}(I) = r. \end{cases}$$

Proof. If $\mathrm{ht}(I) = r$, then the S_i's will have the form $x_j{}^{m_j}$, and so $1 \cdot x_j{}^{m_j} \in I\ \forall j$, that is to say, $\mathrm{sat}(I) = (1)$.

If $\mathrm{ht}(I) < r$, then there are two possibilities:

(1) $\cup_i \mathrm{supp}(S_i) \subset \{x_1, \ldots, x_r\}$;

(2) $\cup_i \mathrm{supp}(S_i) = \{x_1, \ldots, x_r\}$.

If (1) holds, then there exists a variable, say x_1, such that $x_1 \notin \mathrm{supp}(S_j)\ \forall j$, while, if (2) holds, then the support of one of the generators of I, say S_1, has cardinality greater than 1, i.e., $\{x_1, x_2\} \subseteq \mathrm{supp}(S_1)$, where the subscripts of the variables are chosen for convenience.

Let T be a term of $\mathrm{sat}(I)$: $x_1{}^{m_1}T \in I$, and so there exists S_h such that $S_h | x_1{}^{m_1}T$.

If (1) holds, or if (2) holds and $h \neq 1$, then $x_1 \notin \mathrm{supp}(S_h)$, and hence $S_h|T$.

If (2) holds and S_1 is the only generator of I which divides $x_1{}^{m_1}T$, then $\deg(S_1, z) \leq \deg(T, z)\ \forall z \in \mathrm{supp}(S_1)\backslash\{x_1\}$. But $x_2{}^{m_2}T \in I$, and S_1 is the only term which divides it. Thus $\deg(S_1, x_1) \leq \deg(T, x_1)$, and then $S_1|T$, and the proof is complete. □

Proposition 18. *If I is a monomial ideal, then* $\mathrm{sat}(I)$ *can be computed.*

Proof. Let $I = \bigcap_{j=0}^{t} I_j$ be a c-decomposition of I.

Claim. $\mathrm{sat}(I) = \bigcap_{j=0}^{t} \mathrm{sat}(I_j)$.

If $T \in \mathrm{sat}(I)$, then $x_i{}^{m_i}T \in I\ \forall i$, and then $x_i{}^{m_i}T \in I_j\ \forall i, j$, i.e., $T \in \mathrm{sat}(I_j)\ \forall j$.

Conversely, if $T \in \mathrm{sat}(I_j)\ \forall j$, then there exists an integer m_i such that $x_i{}^{m_i}T \in I_j\ \forall j$, and therefore $T \in \mathrm{sat}(I)$.

But $\mathrm{sat}(I_j)$ can be computed as showed in Proposition 17, and now the claim follows by Proposition 4(2). □

4. Description of algorithms

(a) *Algorithms to calculate a c-decomposition of a monomial ideal*

The following algorithms calculate a c-decomposition of a monomial ideal: the first one is used to build a stack containing sets of generators of monomial ideals which will be decomposed; the second one acts according to the proof of Proposition 7, separating first the generators and then building new ideals; the last one is used to extract a minimal generating set from the generators of a monomial ideal.

Algorithm CDECOMP(G): C-decomposition of the monomial ideal generated by G.

Input A minimal generating set G of the monomial ideal $\mathrm{id}(G)$.

Output A list L containing the generators of the ideals of a c-decomposition of id(G).

CDECOMP(G)
comment Inizialization of variables
$D \leftarrow \{G\}$, $L \leftarrow \emptyset$
while $|D| > 0$ **do begin**
Select P, first of D, $D \leftarrow D \setminus \{P\}$
$Q =$ DECOMPOSITION(P)
$L \leftarrow L \cup \{\text{first of } Q\}$, $Q \leftarrow Q \setminus \{\text{first of } Q\}$
$D \leftarrow D \cup Q$
endwhile
return(L)

Algorithm DECOMPOSITION(P): Decomposition of the monomial ideal generated by P according to the proof of Proposition 7.

Input A minimal generating set P of the monomial ideal id(P).
Output A set Q containing the generators of the ideals of the decomposition of id(P).

DECOMPOSITION(P)
comment Separation of terms and computation of gcd's
$h \leftarrow 1; D[h] \leftarrow \{\text{first of } P, \text{ first of } P\}; P \leftarrow P \setminus \{\text{ first of } P\}$
for $T \in P$ **do begin**
$j \leftarrow 1$; flag $\leftarrow 0$
repeat
$M \leftarrow \gcd(T, \text{ first of } D[j])$
if $M \neq 1$ **then**
$D[j] \leftarrow D[j] \cup \{T\}$
$D[j] \leftarrow$ change first of $D[j]$ with M
flag $\leftarrow 1$
endif
$j \leftarrow j + 1$
until $j > h$ or flag $= 1$
if ($\gcd(T, \text{ first of } D[j]) = 1 \ \forall j$) **then**
$h \leftarrow h + 1; D[h] \leftarrow \{T, T\}$
endif
endfor
comment Test if id(P) is generated by coprime terms
if $h = |P| + 1$ **then**
return($\{P\}$)
else

```
comment Construction of the ideals of a decomposition of id(P)
    k ← 0
    Q[0] ← {first of D[j]/1 ≤ j ≤ h}
    for j = h step −1 to 1 do begin
        Q ← { first j − 1 terms of Q[0]}
        T ←  first of D[j]; D[j] ← D[j]\{T}
        for S ∈ D[j] do begin
            Q ← Q ∪ {S'}
            comment S' is computed as in the proof of Proposition 7
        endfor
        for i = j + 1 to h do begin
            Q ← Q ∪ D[i]
        endfor
        Q = MINIMALIZE(Q)
        if 1 ∉ Q then
            k ← k + 1
            Q[k] ← Q
        endif
    endfor
    return({Q[0], ..., Q[k]})
endif
```

Algorithm MINIMALIZE(R): Extracts a minimal generating set from a set of generators of a monomial ideal.

Input Generating set of a monomial ideal
Output Minimal generating set of the given monomial ideal

```
MINIMALIZE(R)
    u ← |R|
    for i = u − 1 step −1 to 1 do begin
        flag ← 0; j ← u
        while (j > i and flag  = 0)do begin
            T ← gcd(R[i], R[j])
            if T = R[i] then
                R ← R\{R[j]}
                u ← |R|
            else
                if T = R[j] then
                    flag ← 1
                    R ← R\{R[i]}
                    R[i] ← T
                    u ← |R|
                endif
```

```
            endif
            j ← j − 1
        endwhile
    endfor
    return(R)
```

(b) *Algorithm to calculate the height of an ideal*

Let I be an ideal in the polynomial ring $\mathcal{R}$, and let $\mathcal{G}$ be a Gröbner base of I with respect to the fixed term-ordering σ.

The following algorithm acts according to Paragraph 3(b) :

Algorithm HEIGHT($\mathcal{G}$): Calculates the height of $I = \mathrm{id}(\mathcal{G})$.

> **Input** A Gröbner base of I
> **Output** Height of I

```
HEIGHT(𝒢)
    T(𝒢) ← {T(f)/f ∈ 𝒢}
    √T(𝒢) ← radical ideal of T(𝒢) calculated according to Proposition 4(4)
    Q = CDECOMP(√T(𝒢))
    ht ← | first of Q|
    Q ← Q\{ first of Q}
    for P ∈ Q do begin
        if ht > |P| then
            ht ← |P|
        endif
    endfor
    return(ht)
```

Remark: It is possible to give $T(\mathcal{G})$ or $\sqrt{T(\mathcal{G})}$ in input, and then to go on from that point.

An algorithm like this one can be written to calculate the strongly independent sets of variables modulo I, according to Proposition 15.

(c) *Complexity of the illustrated algorithms*

In the algorithm MINIMALIZE, the only performed operation is the computation of gcd' s of terms, and, in the worst case, i.e., when the terms are coprime, it is performed $\frac{u(u-1)}{2}$ times, with $u = |R|$.

The algorithm DECOMPOSITION can be divided in two parts: in the first part, gcd's are calculated and terms are separated (in the worst case, $\frac{u(u-1)}{2}$ gcd's);

in the second one, the ideals of the decomposition are calculated (at most u divisions of terms).

The algorithm CDECOMP, in most applications, is used to decompose radical ideals, and so the complexity will be computated only for particular radical ideals.

Proposition 19. *If a radical monomial ideal I is generated by all the terms of degree d, then* CDECOMP *calls* DECOMPOSITION *exactly* $\binom{r-1}{d-1}$ *times.*

Proof. By induction on r. The claim is true for $r = 3$ and every degree d (directly verified).

Let $M_{i_1,\ldots,i_d}$ be the term $M_{i_1,\ldots,i_d} = x_{i_1} \cdots x_{i_d}$.
I is generated by $\{M_{1,\ldots,d}, \ldots, M_{r-d+1,\ldots,r}\}$.
The terms are subscripted in such a way that:

$$\begin{aligned}
x_1 &= \gcd(M_{1,\ldots,d}, \ldots, M_{1,r-d+1,\ldots,r}) \\
x_2 &= \gcd(M_{2,\ldots,d+1}, \ldots, M_{2,r-d+2,\ldots,r}) \\
&\vdots \\
x_{r-d} &= \gcd(M_{r-d,\ldots,r-1}, M_{r-d,r-d+2,\ldots,r}) \\
&M_{r-d+1,\ldots,r}
\end{aligned}$$

DECOMPOSITION returns the following ideals :

$$\begin{aligned}
Q_0 &= \mathrm{id}(x_1, \ldots, x_{r-d}, M_{r-d+1,\ldots,r}) \\
Q_1 &= \mathrm{id}(M_{2,\ldots,d}, \ldots, M_{r-d+2,\ldots,r}) \\
&\vdots \\
Q_{r-d} &= \mathrm{id}(x_1, \ldots, x_{r-d-1}, M_{r-d+1,\ldots,r-1}, M_{r-d+2,\ldots,r})
\end{aligned}$$

Q_1 is generated by all the terms of degree $d-1$ in $r-1$ variables, Q_2 is generated by all the terms of degree $d-1$ in $r-2$ variables, and so on.

By induction hypothesis, DECOMPOSITION is called exactly

$$1 + \binom{d-1}{d-2} + \cdots + \binom{r-2}{d-2} = \binom{r-1}{d-1}$$

times. □

Finally, the complexity of HEIGHT depends on the complexity of CDECOMP, because there is no other operation performed in it.

(d) *Examples*

Example 1 (Macauley' s curve) ([MM]).
Input: $T(\mathcal{G}) = \{x_0^8 x_2, x_0 x_3, x_1^8 x_3, x_1^7 x_3^2, x_1^6 x_3^3, \ldots, x_1 x_3^8\}$
$\sqrt{T(\mathcal{G})} = \{x_0 x_2, x_0 x_3, x_1 x_3\}$
$Q = \{\{x_0, x_1 x_3\}, \{x_2, x_3\}\}$

ht $= 2$
Isolated prime ideals :$\{x_0, x_1\}, \{x_0, x_3\}, \{x_2, x_3\}$
Strongly independent sets : $\{x_2, x_3\}, \{x_1, x_2\}, \{x_0, x_1\}$

Example 2 (Macauley' s curve) ([MM]).
Input: $T(\mathcal{G}) = \{x_0^{28}x_2, x_0x_3, x_1^{28}x_3, x_1^{27}x_3^2, \ldots, x_1x_3^{28}\}$
$\sqrt{T(\mathcal{G})} = \{x_0x_2, x_0x_3, x_1x_3\}$
See example 1

Example 3 (Gerdt) ([BGK]).
Input: $T(\mathcal{G}) = \{L_1^3L_2^3, L_1^2L_2^4, L_1L_2^5, L_1^4L_4, L_1^3L_2L_4, L_1L_2^2L_4, L_1L_4^2,$

$L_1^3L_5, L_1^2L_2L_5, L_1L_2^2L_5, L_1L_4L_5, L_2L_4^2L_5, L_1L_5^2, L_4L_5^2, L_1L_6, L_4L_6, L_2L_6^2,$

$L_5L_6^2, L_1^3L_3, L_1L_2L_3, L_4L_5, L_1L_5L_3, L_2L_6L_3, L_1L_3^2, L_7\}$

$\sqrt{T(\mathcal{G})} = \{L_1L_2, L_1L_4, L_1L_5, L_4L_5, L_1L_6, L_4L_6, L_2L_6, L_5L_6, L_1L_3, L_7\}$
$Q = \{\{L_1, L_4, L_6, L_7\}, \{L_1, L_4, L_2, L_5, L_7\}, \{L_1, L_5, L_6, L_7\},$
$\{L_2, L_4, L_5, L_6, L_3, L_7\}\}$
ht $= 4$
Isolated prime ideals : Q
Strongly independent sets :$\{\{L_2, L_3, L_5\}, \{L_3, L_6\}, \{L_1\}, \{L_2, L_3, L_4\}\}$

Example 4 (Computation of the saturation of a monomial ideal).
Input :$I = \mathrm{id}(x^2y^2, x^4z, yz, x^2z^2)$
$Q = \{\{x^2, yz\}, \{y^2, z\}, \{y, x^4, z^2\}\}$
$\mathrm{sat}(I) = \mathrm{id}(x^2, yz) \cap \mathrm{id}(y^2, z) = \mathrm{id}(x^2y^2, x^2z, yz)$

References

[AM] M.F. Atiyah, I.G. Macdonald, Introduction to Commutative Algebra, Addison-Wesley Publishing Company, 1969.

[B] B. Buchberger, Ein algorithmisches Kriterium für die Lösbarkeit eines Algebraischen Gleichungssystems, Aequationes Math. **4** (1970), 374-383.

[BGK] W. Böge, R. Gebauer, H. Kredel, Some examples for solving systems of algebraic equations by calculating Gröbner bases, J. Symbolic Comput. **1** (1986), 83-98.

[Gr] W. Gröbner, Algebraic Geometrie I, II, Bibliographisches Institut, Mannheim, 1968.

[GS] A. Geramita, C. Small, Introduction to Homological Methods in Commutative rings, Queen's Paper in Pure and Applied Mathematics, Kingston, Ontario, 1979.

[H] R. Hartshorne, Connectedness of Hilbert schemes, Inst. Hautes Études Sci. Publ. Math. **29** (1969), 7-48.

[H1] R. Hartshorne, Algebraic Geometry, Springer Verlag, 1977.

[K] A. Kandry-Rody, Dimension of ideals in polynomial rings, Proc. Combinatorial Algorithms in Algebraic Structures (1985).

[KW] H. Kredel, V. Weispfenning, Computing dimension and independent sets, J. Symbolic Comput. **6** (1986), 231-247.

[M] H. Matsumura, Commutative Rings, Benjamin Cummings Publishing Company Inc., 1970.

[MM] H.M. Möller, F. Mora, The computation of the Hilbert function, in: Computer Algebra. EUROCAL '83, Lecture Notes in Comput. Sci. **162**, Springer-Verlag, 1983.

[T] B. Teissier, Variétés polaires II. Multiplicités polaires, sections planes, et conditions de Whitney, in: Proceedings La Rabida (1981), Lecture Notes in Math. **961**, Springer-Verlag, 1982.

[ZS] O. Zariski, P. Samuel, Commutitive Algebra, vol. II, Van Nostrand, 1960.

Generators of ideals of curves: some polynomial algorithms for their computation

Isabella Ramella

Abstract. We want to show how to apply known polynomial algorithms, for constructing generators of points, to the computation of a minimal set of generators of ideals of curves which are the union of lines or have generic ordinary singularities; this yields an algorithm to determine also these last sets, in polynomial time.

1991 Mathematics Subject Classification: Primary 13P10; Secondary 14Q99

1. Introduction

It is well known that, in general, the algorithms of Computer Algebra relative to the computation of generators of algebraic varieties are of exponential type, since they rely on the construction of Gröbner bases. Yet, in the particular case of ideals of points, polynomial algorithms have been recently constructed for the computation of minimal sets of generators. In this paper, we show that it is possible to make use of these last procedures, also for the computation of minimal generators of ideals of lines. For this, we make use of the fact that the ideal of a union of s lines is generated by forms of degree $\leq s$. Then we consider the local ideal I of a curve C at a generic ordinary singularity P. The elements of a minimal generating set of I have leading forms which are a minimal generating set of the points of the tangent cone of C at P. Then, also in this case, we can make use of the polynomial algorithm relative to points to have results about the generators of ideals of curves C. We hope that this paper can be the starting point for the construction, in polynomial time, of generators of other classes of varieties.

2. The ideal of lines in $\mathbf{P}^n$

Let $I = \bigcap\limits_{i=1}^{s} \wp_i \subseteq R = K[X_0, \ldots, X_n]$ be the ideal of the union of s lines $\bigcup\limits_{i=1}^{s} r_i$, $r_i = V(\wp_i)$, where $\wp_i = (L_{i,1}, \ldots, L_{i,n-1})$ and $L_{i,j}$ are independent linear polynomials of R. Let $(F_1, \ldots, F_\nu)$ be a minimal set of homogeneous generators of I , such that $\partial F_1 \leq \ldots \leq \partial F_\nu$ ($\partial F_i =$degree).

Theorem 1. *If the lines r_i are disjoint then $\partial F_\nu \le s$, that is, I can be generated by forms of degree $\le s$.*

Proof. Let π be a hyperplane of $\mathbf{P}^n$ that does not contain any r_i , $i = 1,...,n$. By a suitable change of coordinates, we can assume that π is $x_0 = 0$.

Let $r_i \cap \pi = P_i = V(\wp_i + (X_0))$ and $J = \bigcap_{i=1}^{s} (\wp_i + (X_0))$. We claim $(I + (X_0))_s = J_s$. Clearly $I + (X_0) = (\bigcap_{i=1}^{s} \wp_i) + (X_0) \subset J$. Hence the claim is proved if we show that $J_s \subseteq (I + (X_0))_s$. Let $F \in J_s$ be a form of degree s in J and $S = V(F)$ be the hypersurface corresponding to F. We have that $S \cap r_i$ contains a finite set of points $P_i, P_i^{(1)}, \dots, P_i^{(s-1)}$, which may coincide. Let $\pi_1^{(1)}$ be a hyperplane that contains r_2 and $P_1^{(1)}$ but does not contain r_1; $\pi_1^{(2)}$ a hyperplane containing r_3 and $P_1^{(2)}$ but not r_1; ...; $\pi_1^{(s-1)}$ a hyperplane containing r_n and $P_1^{(s-1)}$ but not r_1. The union $\bigcup_{i=1}^{s-1} \pi_1^{(i)} = V(G_1)$ contains $\bigcup_{i=2}^{s} r_i$ and $\bigcup_{i=1}^{s-1} P_1^{(i)}$ but not r_1. Then $G_1 \in \wp_2 \cap \ldots \cap \wp_s$, $\partial G_1 = s - 1$. By repeating this construction to the lines $r_2, \dots, r_n$, we get forms $G_2, \dots, G_s$ of degree $s - 1$ such that $V(G_i)$ contains the lines $r_1, \dots, r_{i-1}, r_{i+1}, \dots, r_n$ and the points $P_i^{(j)}$ $(j = 1, \dots, s-1)$ but not the line r_i. Then, if $G = F - \lambda_1 G_1 X_0 - \lambda_2 G_2 X_0 - \cdots - \lambda_s G_s X_0$ $(\lambda_i \in K)$, we have

$$G(P_i) = G(P_i^{(j)}) = 0 \qquad \text{for any } i, j.$$

Let us consider a point $P_i^{(s)}$ on the line r_i such that $P_i^{(s)} \notin \{P_1, P_1^{(1)}, \dots, P_1^{(s-1)}\}$ for any $i = 1, \dots, s$. Then $G(P_i^{(s)}) = F(P_i^{(s)}) - \lambda_i G_i(P_i^{(s)})$ and if we set $\lambda_i^0 = \frac{F(P_i^{(s)})}{G_i(P_i^{(s)})}$ $(G_i(P_i^{(s)}) \neq 0$, otherwise G_i, of degree s, would vanish on $s+1$ points of r_i and then $V(G_i)$ would contain r_i). The form

$$G_0 = F - \lambda_1^0 G_1 X_0 - \cdots - \lambda_n^0 G_n X_0$$

of degree s vanishes on all the points $P_i, P_i^{(j)}$, $i = 1, \dots, s$, $j = 1, \dots, s$, which are $s+1$ on any line r_i. Hence $V(G_0)$ contains $\bigcup_{i=1}^{s} r_i$, that is, $G_0 \in \wp_1 \cap \ldots \cap \wp_s$. It follows that $F = \lambda_1^0 G_1 X_0 + \ldots + \lambda_s^0 G_s X_0 + G_0 \in (X_0) + \wp_1 \cap \ldots \cap \wp_s = (X_0) + I$ as claimed. Now J is an ideal of points $P_1, \dots, P_s$ and then, as is well known, it is generated by forms of degree $\le s$. We want to prove, by induction, that I is also generated by forms of degree s. Suppose that for a fixed $m > s$ and for any $G \in I_{m-1}$, we have $G = \sum_{i=1}^{t} H_i F_i$, where $F_i \in I_s$. Then, if $F \in I$ and $\partial F = m > s$, we have $F \in J_m$. Hence, $F = \sum_{i=1}^{r} H_i' F_i'$, where $F_i' \in J_s = (I + (X_0))_s$, that is, $F_i' = F_i'' + X_0 G_i$ where $F_i'' \in I_s$, $\partial G_i = s - 1$. We have $F = \sum_{i=1}^{r} H_i' F_i'' + X_0 (\sum_{i=1}^{r} H_i' G_i)$ and $X_0(\sum_{i=1}^{r} H_i' G_i) = F - \sum_{i=1}^{r} H_i F_i'' \in I = \wp_1 \cap \ldots \cap \wp_s$. But $X_0 \notin \wp_i$, $i = 1, \dots, s$, and so $G = \sum_{i=1}^{r} H_i' G_i \in I_{m-1}$ and therefore, by induction, $G = \sum_{i=1}^{r} H_i F_i$ where $F_i \in I_s$. Hence $F = \sum_{i=1}^{r} H_i' F_i\prime + \sum_{i=1}^{r} H_i F_i$ where $F_i, F_i' \in I_s$, thus proving the claim.

Conjecture. We presume that the assumption of disjointness in Theorem 1 is unnecessary.

Lemma 2. *Let I be the ideal of s disjoint lines in $\mathbf{P}^n$. Choose $s+1$ points on any line r_i, $i = 1, \ldots, s$. If J is the ideal of these $(s+1)s$ points, then $I_m = J_m$ for any $m \leq s$.*

Proof. We recall that $I = \wp_1 \cap \ldots \cap \wp_s$ where $r_i = V(\wp_i)$. Clearly, $I \subset J$. Then it is enough to prove that $J_m \subset \wp_i$ $i = 1, \ldots, s$. If $F \in J_m$ $(m \leq s)$ then the hypersurface $V(F)$ should contain $s+1$ points and then the line r_i, i.e., $F \in \wp_i$.

From Theorem 1 and Lemma 2 it follows immediately that:

Corollary 3. *With the hypotheses of Lemma 2, if $(F_1, \ldots, F_\rho)$ is a minimal set of generators of J and $\nu = \mathrm{Max}\{i \mid \partial F_i \leq s\}$ then $(F_1, \ldots, F_\nu)$ is a minimal set of I.*

Remark. Corollary 3 says that for computing a minimal set of generators of the ideal of a union of s lines, it is enough to compute a minimal set of generators of the ideal of $(s+1)s$ points up to degree s. Various polynomial algorithms are known for computing generators of ideals of points [Ramella], [Robbiano, Mora], [Cioffi] and hence the same algorithms can be used for computing, with polynomial computational cost, the generators of an ideal of lines.

3. The local ideal of curves of ordinary singularities

Let (A, M) be the local ring at a singular point P of an algebraic reduced curve C of embedding dimension emdim $A = m+1$ and multiplicity, $e(A) = s$, over a field K. Suppose A has k-rational normalization $\bar{A}$, i.e., $\bar{A}/M_i = k$ for any maximal ideal $M_1, \ldots, M_h$ of $\bar{A}$. Let $J = M_1 \cap \ldots \cap, M_h$ be the Jacobson radical of $\bar{A}$. For any semilocal ring B, $G(B) = \bigoplus_{n \geq 0} (J^n/J^{n+1})$ denotes the associated graded ring (with respect to its Jacobson radical). The scheme Spec$(G(A))$ is the tangent cone and Proj$(G(A)) \subseteq \mathbf{P}^m$ the projectivized tangent cone to C at P.

Definition 4. The point P is an *ordinary singularity of C*, if Proj$(G(A))$ is reduced.

We now want to characterize geometrically the condition "Proj$(G(A))$ is reduced". Let n be a positive integer and $M^n : M^n = \{b \in \bar{A} \mid bM^n \subset M^n\}$. The subring $B = \bigcup_{n>0} (M^n : M^n)$ of $\bar{A}$ is the ring obtained by blowing up the maximal ideal M of A [4]. It is well known that there is a natural isomorphism of schemes

(1) $\mathrm{Proj}(G(A)) \cong \mathrm{Spec}(B/MB)$ [4, 19.4.2].

From (1) we have easily the following:

Lemma 5. *Proj$(G(A))$ is reduced if and only if $B = \bar{A}$ and $M\bar{A} = J$.*

Proof. If $B = \bar{A}$ and $M\bar{A} = J$, then it follows from (1) that Proj$(G(A))$ is reduced. If $\mathrm{Proj}(G(A)) \cong \mathrm{Spec}(B/MB)$ is reduced, then MB is a radical ideal. Furthermore, MB is a principal ideal [4, Proposition 1.1]. Then B is a locally discrete valuation ring. Hence $B = \bar{B} = \bar{A}$ and $M\bar{A} = J$.

Now we examine the geometric meaning of the condition $M\bar{A} = J$. We recall that $M\bar{A} = M_1^{e_1}, \dots, M_h^{e_h}$, where $e_1 + \dots + e_h = e(A)$. Furthermore, the maximal ideals M_i of $\bar{A}$ are in natural bijective correspondence with the minimal primes $\wp_i$ of the completion $\hat{A}$ of A with respect to M. In fact M_i corresponds to $\wp_i = \Re_i \cap \hat{A}$ where $\Re_i$ is the unique minimal prime ideal of $\hat{\bar{A}} = \bar{\hat{A}}$ contained in $\hat{M}_i$.

The scheme $\mathrm{Spec}(\hat{A}/\wp_i)$ is a branch of the curve at P and we will call it the i-th branch. It is well known that $e_i = e(\hat{A}/\wp_i)$. Thus

(2) $e_i = 1$ if and only if the i-th branch is non singular. $M\bar{A} = J$ if and only if all the branches are non singular. If $M\bar{A} = J$, there are $s = e(A)$ branches at P.

Let now t_i be a basis of the k-vector space (of dimension 1) M_i/M_i^2. Let $x_0, \dots, x_m$ be a set of generators of M and suppose that the natural homomorphism $d_i : M/M^2 \to M_i/M_i^2$ is given by $d_i(\bar{x}_j) = a_{ij}t_j$ $(0 \le j \le m,\ a_{ij} \in k)$. It is easily seen that

(3) the branches at P are non singular if and only if, for any i, there exists an integer j such that $a_{ij} \neq 0$. In this case, from now on, we set:

$$p_1 = (a_{10}, \dots, a_{1m}), \dots, p_s = (a_{s0}, \dots, a_{sm}) \ \in \mathbf{P}^m.$$

Proposition 6. *The following conditions are equivalent:*

i) P is ordinary;
ii) $\mathrm{Proj}(G(A)) = \{P_1, \dots, P_s\} \subset \mathbf{P}^m$;
iii) the branches at P are non singular and $p_i \neq p_j$ if $i \neq j$.

Proof. i) $\Leftrightarrow$ ii). Proj$(G(A))$ is reduced if and only if $B = \bar{A}$ and $MB = M\bar{A} = J = M_1 \cap \dots \cap M_s$ [Lemma 5]. Furthermore, Proj$(G(A))$ has s points if and only if $B \subset \bar{A}$ has s maximal ideals, i.e., $B = \bar{A}$ and $M\bar{A} = J$.

ii) $\Leftrightarrow$ iii). It is easy to check that if $p_i = (a_{i0}, \ldots, a_{im})$ and $\delta_i : G(A) \to G(\bar{A}_{M_i})$, then $\ker \delta_i = (a_{i1}\bar{x}_0 - a_{i0}\bar{x}_1, \ldots, a_{im}\bar{x}_0 - a_{i0}\bar{x}_m)$. Furthermore, if $M\bar{A} = J$, then δ_i is bijective ($1 \leq i \leq s$), hence $\ker \delta_i$ is a minimal prime of $G(A)$, i.e., p_i is a point of $\mathrm{Proj}(G(A))$. Then the equivalence follows from Lemma 5 and (2).

Example. Let $\operatorname{Spec} R$ be a curve given by $R = K[g_0, \ldots, g_m]$ where $g_i \in K[t]$ $1 \leq i \leq m$, t an indeterminate. Suppose that $g = \gcd(g_0, \ldots, g_m)$ is not a constant and $\bar{R} = K[t]$. Let $f_i = g_i/g$. Let $A = (R_N, M)$ be the local ring of R at the point $P \leftrightarrow N = (g_0, \ldots, g_m)$. The branches of $\operatorname{Spec} R$ at P are non singular if and only if $M\bar{A} = J$ [cf.(2)], that is, $g = \prod_{i=1}^{s}(t - a_i)$ where $a_i \neq a_j$, if $i \neq j$. Then the curve has non singular branches if and only if the $m+1$-tuples $(g_0'(a_i), \ldots, g_m'(a_i))$ are non null for any i, $1 \leq i \leq s$ (g_j' is the derivative of g_j). Let $N_i = (t - a_i)$. We have $g_j(t) - g_j'(a_i)(t - a_i) \in N_i^2$. Then $\overline{g_j(t)} = g_j'(a_i)\overline{(t - a_i)}$ in N_i/N_i^2. If M_i is the maximal ideal of $\bar{A}$ corresponding to N_i, the following natural isomorphisms hold: $M/M^2 \cong N/N^2$ and $M_i/M_i^2 \cong N_i/N_i^2$. Further, $\overline{(t - a_i)}$ is a basis of N_i/Ni^2. Hence, if the branches are non singular, then

$$p_i = (g_0'(a_i), \ldots, g_m'(a_i)) = (f_0(a_i), \ldots, f_m(a_i)) \in \mathbf{P}^m$$

are the points of (3), that is, the points of $\mathrm{Proj}(G(A))$ [Proposition 6]. Then P is an ordinary point if and only if $p_i \neq p_j$ for any $i \neq j$ [Proposition 6].

It is clear that: $\mathrm{Proj}(G(A))$ reduced $\Rightarrow$ $G(A)$ reduced.
The converse holds for points in generic position.

Definition 7. Let $P_1, \ldots, P_s$ be points of $\mathbf{P}^m$ over a field k, and let (S, M) be the homogeneous coordinate ring of the variety $\{P_1, \ldots, P_s\}$. Then $P_1, \ldots, P_s$ are in generic position if and only if the Hilbert function $H(S, n)$ of S is maximal, for any $n \in \mathbf{N}$, that is, $H(S, n) = \dim_K(M^n/M^{n+1}) = \mathrm{Min}\{\binom{n+m}{m}, s\}$.

Theorem 8. *If P is an ordinary singularity of a curve C and* $\mathrm{Proj}(G(A)) = \{P_1, \ldots, P_s\}$ *consists of points in generic position, then $G(A)$ is reduced.*

Proof. [6, Theorem 3.3].

If (A, M) is the local ring of a curve $C = \operatorname{Spec}(R/I)$ of embedding dimension $n+1$, then $A = R_N/IR_N$, where $R = K[X_0, \ldots, X_m]$, $N = (X_0, \ldots, X_m)$. If $f \in R$, then there is a unique integer n for which $f \in N^n - N^{n+1}$. Thus the class of f, $f^* \in N^n/N^{n+1}$ is nonzero. This integer n is called the *order* of f and f^* is called the *leading form* of f.

Let I^* be the ideal generated by the leading forms of all the elements in IR_N. Then it is well known that $G(A) = G(R_N)/I^* = K[X_0, \ldots, X_m]/I^*$. We denote by $\nu(I^*)$ and $\nu(IR_M)$ the minimal number of generators of the ideals I^* and IR_M.

Theorem 9. *Let P be an ordinary singularity of $C = \mathrm{Spec}(R/I)$ and suppose that* $\mathrm{Proj}(G(A))$ *consists of s points in generic position. Let $d = \mathrm{Min}\{d \mid \binom{n+d}{d} > s\}$, then IR_N is minimally generated by the elements $(f_1, \ldots, f_\nu)$ of order d and $d+1$ such that $(f_1^*, \ldots, f_\nu^*)$ is a minimal set of generators of the ideal I^* of the points $P_1, \ldots, P_s$ of* $\mathrm{Proj}(G(A))$. *In particular, the order of f_i and the degrees of f_i^* are the same and $\nu(IR_M) = \nu(I^*)$.*

Proof. First we remark that $G(A)$ is reduced by Theorem 8 and then I^* is the ideal of points of $\mathrm{Proj}(G(A))$. In [2, Theorem 15] it is proved that $\nu(IR_M) = \nu(I^*)$ and it is easy to see that the same arguments prove the claim of the theorem .

Remark. Theorem 9 asserts that the minimal number of local generators of a curve C at a generic point P is the same as the minimal number of generators of the ideal of points of the tangent cone of C at P. Since there are polynomial algorithms for computing generators of points, these can be applied to the computation of local generators of curves with generic ordinary singularities.

Remark. As shown in [7], the condition of generic position for points with coordinates in $\mathbf{Q}$ can be checked in the finite field $\mathbf{Z}_p$, thus improving the computational time if compared with the classical algorithms for computing generators of curves which use Gröbner bases. We have implemented these algorithms on an IBM compatible computer using Pascal and on a Vax using the computer algebra system Macsyma.

Example. Let $g = (t^2-1)(t^2-4)$ and let $A = R_N = K[X,Y,Z]_{(X,Y,Z)}/I$ be the local ring at the point $P \leftrightarrow N = (g, tg, t^2g)$ of $R = K[g, tg, t^2g]$. Then $\bar{A} = K[t]$ and, by using the remarks of the previous example, it is easily checked that P is an ordinary singularity and $\mathrm{Proj}(G(A)) = \{P_1, P_2, P_3, P_4\}$, $P_1 = (1,1,1)$, $P_2 = (1,-1,1)$, $P_3 = (1,2,4)$, $P_4 = (1,-2,4)$. Furthermore, since P_1, P_2, P_3, P_4 are not collinear, they are in generic position and their ideal is generated by two forms of degree 2. Then also the ideal I is generated by two elements of order two, i.e., the curve $\mathrm{Spec}\, R$ is a complete intersection at P.

Example. Let $g = t(t^2-1)(t^2-4)$ and let $A = R[g, tg,, t^3g]_N$. Then $\bar{A} = K[t]$, $P \to N$ is an ordinary singularity and $\mathrm{Proj}(G(A)) = \{P_1, P_2, P_3, P_4, P_5\}$, where $P_1 = (1,0,0)$, $P_2 = (1,1,1)$, $P_3 = (1,-1,-1)$, $P_4 = (1,4,64)$, $P_5 = (1,-4,-64)$, are 5 points in generic position. It is easily checked that their ideal is generated by a conic and three cubics, so that the local generators of $C = \mathrm{Spec}\, R$ at P are 4, one of order two and three of order three.

References

[1] F. Cioffi, Algorithms for computing minimal generators of ideals of projective points in polynomial time, These proceedings.

[2] A.V. Geramita, F. Orecchia, Minimally generating ideals defining certain tangent cones, J. Algebra 78 (1982), 36-57.

[3] A. Grothendieck, J. Dieudonné, Élements de Géometrie Algébrique IV, Quatrième Partie, I.H.E.S., Publ. Math. 32, Paris, 1967.

[4] J. Lipman, Stable ideals and Arf rings, Amer. J. Math. 93 (1971), 649-685.

[5] T. Mora, L. Robbiano, Points in affine and projective spaces, Preprint, Dip. di Mat., Univ. di Genova, 1992.

[6] F. Orecchia, Ordinary singularities of algebraic curves, Canad. Math. Bull. 24 (1981), 423-431.

[7] I. Ramella, An algorithmic approach to ideal generation conjecture of points, Rend. Accad. Sci. Fis. Mat. Napoli 56 (1989), 72-81.

[8] I. Ramella, Ideals of points in generic position: a polynomial algorithm for computing a minimal set of generators, Preprint no. 26, Dip. di Mat. e Appl., Univ. di Napoli, 1992.

Remarks on Graded Integral Domains of Dimension One Over a Finite Field

*Leslie G. Roberts**

Abstract. I describe how to produce explicit examples of one-dimensional graded integral domains over the integers mod p, and how to compute the Hilbert function of these domains. The results obtained suggest (at least for $p = 2$, and quotients of $F_p[X_0, X_1, X_2]$) that the Hilbert functions of such domains are exactly the Hilbert functions of sets of rational closed points in 1-uniform position in the projective plane.

1991 Mathematics Subject Classification: 12E20, 13A02

1. Introduction

In this note I discuss how to produce examples of one dimensional graded integral domains over $F_p = \mathbb{Z}/p\mathbb{Z}$ (p a prime), how to compute their Hilbert functions, and the results of some computations that I have performed with $p = 2$. I first did calculations of this type several years ago in support of the remarks in the paragraph before example 4.6 in [RR]. Several times since then I have wanted to reconstruct my examples and had difficulty in doing so. Therefore I would like to get them written down more carefully. In the process of writing things up I have extended the calculations somewhat. In this project I was also motivated by the recent papers of Galligo [G] and Hansen [H], in that the former makes use of Hilbert functions of non-rational closed points, and the latter of Hilbert functions of sets of points over a finite field. Unfortunately, however, I have more observations than explanations, and I offer the challenge of explaining some of the observed phenomena.

My original motivation for these calculations was the problem of characterizing the Hilbert functions of one-dimensional graded integral domains over a field k (especially the case where k is finite).

* This work was supported by an NSERC grant.

2. The construction of examples and method of calculation

Let $F = F_q$ be a finite field with $q = p^d$ elements, p a prime. In this note I discuss the possible Hilbert functions of (reduced) closed points P of degree d in $\mathbb{P}^n_{F_p} = \text{Proj } F_p[X_0, \ldots, X_n]$. (Equivalently, such Hilbert functions are the Hilbert functions of integral standard graded F_p algebras of dimension 1.) The point $P = \text{Spec}(F)$ has to lie in one of the open sets $D_+(X_i)$ for some i, say $i = 0$, and hence there is a surjection $\phi : S := F_p[x_1, \ldots, x_n] \to F$ sending $x_j = X_j/X_0$ to $\alpha_j \in F$. Since ϕ is onto we have $F = F_p[\alpha_1, \ldots, \alpha_n]$. Let $J = \ker(\phi)$ be the prime ideal of P. Let I be the ideal generated by all homogeneous forms f in $R := F_p[X_0, \ldots, X_n]$ such that $f(1, \alpha_1, \ldots, \alpha_n) = 0$. Clearly I is prime, $X_0 \notin I$, and $\phi(I) = J$. Thus I is the homogenization of J with regard to X_0, and I is the homogeneous ideal of P. With some abuse of notation I will write $P = [1 : \alpha_1 : \cdots : \alpha_n]$.

The Hilbert function H of P can now be calculated by linear algebra over F_p. Let h be an irreducible polynomial of degree d over F_p. Then $F \cong F_p[X]/h(X)$, and we can represent elements of F by polynomials of degree $\leq d - 1 \pmod{h}$. The latter can be represented by vectors of length d. Now we have that $H(i) = \dim_{F_p}(R/I)_i$ and ϕ maps the latter isomorphically onto the F_p subspace of F spanned by the monomials of degree $\leq i$ in the α_j. Thus $H(i)$ is the rank (over F_p) of the matrix whose rows are the coefficients of the polynomials of degree $\leq d-1$ corresponding to the various monomials of degree $\leq i$ in the α_j. Note that as we increase i we just add more rows to our matrix.

Calculations were done using APL programs written when I was working on [RR]. The properties of H described below, as well as the cases that can be done by hand, provide (I hope!) a reasonable check on the accuracy of the programs.

3. Properties of the Hilbert function

I will concentrate mostly on the case $n = 2$. For any sequence $S = \{s_i\}_{i \geq 0}$ define the difference sequence $\Delta S = \{s'_i\}_{i \geq 0}$ in the usual way by $s'_0 = s_0$ ($= 1$ in our Hilbert function cases) and $s'_i = s_i - s_{i-1}$ for $i \geq 1$. Then since R/I is a quotient of $R = F_p[X_0, X_1, X_2]$ one shows easily that if a is the lowest degree of a non-zero element of I, then ΔH starts out $\Delta H(i) = i + 1$ $0 \leq i \leq a - 1$ and that if $i \geq a$ then $\Delta H(i) \leq a$ $(a \geq 1)$. Furthermore it follows from Macaulay's growth condition [S] that for $i \geq a - 1$ ΔH decreases monotonically (but not necessarily strictly) to 0. However, in the domain case, once ΔH starts to decrease it decreases strictly, as we see in the next paragraph.

Suppose we extend the ground field from F_p to F. Then the point P splits into a reduced subscheme $X = \{P_1, \ldots, P_d\}$ which consists of d distinct F-rational points in $\mathbb{P}^n_F$. The Galois group G of F/F_p (which is cyclic of order p) permutes the points of X transitively, and hence acts transitively on subsets of $d-1$ points. Thus all subsets of $d-1$ elements have the same Hilbert function (necessarily equal to $\min(d-1, H_X)$, since by [GMR, Theorem 2.5(c)] some subset of $d-1$ elements has this Hilbert function). Therefore X is a Cayley-Bacharach scheme (that is, X is 1-uniform in the language of [K, section 3]. One might wonder if X can have a greater degree of uniformity). If $n = 2$ the Hilbert functions of Cayley-Bacharach 0-dimensional subschemes of $\mathbb{P}^2_F$ are known to be those sequences $\{s_i\}_{i\geq 0} = \{1, \ldots\}$ whose difference sequence $\{s'_i\}_{i\geq 0}$ is as described above and in addition satisfies the property:

$$(*) \qquad \text{if } s'_{i+1} < s'_i \text{ then } s'_{j+1} < s'_j \text{ for all } j \geq i.$$

(This is the necessary condition described in [RR, Theorem 4.5]). Below I describe numerical evidence that suggests that if $n = 2$, $p = 2$, then every such sequence occurs for some choice of P.

The generic Hilbert function $GH(d, n)$ of degree d in $\mathbb{P}^n$ is defined by

$$GH(d,n)(i) = \min(\binom{n+i}{n}, d).$$

If the groundfield is algebraically closed this is the Hilbert function of d points in generic position in $\mathbb{P}^n$. If the groundfield is finite, as in our examples, there are only a finite number of points of degree d. Thus "generic" does not have its usual meaning of "something that happens on a dense open set", but is the purely algebraic property of having Hilbert function $GH(d, n)$. However, for $n = 2$ and some d the generic Hilbert function $GH(d, 2)$ seems to occur "most of the time" as we shall see.

4. Examples

To systematically search for points P with specified Hilbert function, I consider P of the form $[1 : \alpha : \alpha^i]$ where α is a primitive $(p^d - 1)^{st}$ root of unity over F_2. Since $F = F_2(\alpha)$ we have that P is of degree d. The element α was chosen primitive to give distinct points P as i ranges over the interval $1 \leq i \leq p^d - 1$. However we can of course have $F = F_2(\alpha)$ without α being primitive (I did a few calculations with α of degree d but not primitive, and it does not seem to matter much whether or not α is primitive), and if d has more than one prime factor we can have $[1 : \alpha_1 : \alpha_2]$ of degree d without either α_1 or α_2 generating F. Thus we are not considering all possible points of degree d. Primitive polynomials over F_2 (i.e. irreducible polynomials of degree d whose roots are primitive $(p^d - 1)^{st}$ roots of unity) of degree ≤ 34 are tabulated in [PW, appendix C]. Given d, and a primitive polynomial f_d of degree d, we can calculate $H = H_{R/I}$in the manner described above, representing α by $X \pmod{f_d}$. Because of various known properties of H, we need only do the calculation for a few low values of i. All the permitted Hilbert functions were observed to occur up to degree 25 (although in several cases, as explained below, I had to use $P's$ of a different form).

In each degree I have used the first example of primitive polynomial given in the table of [PW]. In low degrees these are: x^2+x+1, x^3+x+1, x^4+x+1, x^5+x^2+1, x^6+x+1, x^7+x^3+1, $x^8+x^4+x^3+x^2+1$. I will describe my results first in a few low degrees:
If $d = 6$ the Hilbert functions permitted by (*) together with their difference sequence are:

1 2 3 4 5 6 →	1 1 1 1 1 1 0
1 3 5 6 →	1 2 2 1 0
1 3 6 →	1 2 3 0

If $d = 7$ the possible Hilbert functions are

1 2 3 4 5 6 7 →	1 1 1 1 1 1 1 0
1 3 5 7 →	1 2 2 2 0
1 3 6 7 →	1 2 3 1 0

If $d = 8$ the possible Hilbert functions are

1 2 3 4 5 6 7 8 →	1 1 1 1 1 1 1 1 0
1 3 5 7 8 →	1 2 2 2 1 0
1 3 6 8 →	1 2 3 2 0

In each of degrees 6, 7, 8 the three cases are characterized by $H(1) = 2$, $H(2) = 5$, and $H(2) = 6$ respectively. In degree 6 the three cases first occur respectively as the Hilbert function of $[1 : \alpha : \alpha^i]$ for $i = 1, 2, 4$ and in degrees 7, 8 for $i = 1, 2, 3$. (I started with $i = 1$.) Over the range $1 \leq i \leq 2^d - 1$ for $d = 6$ there are 3, 36, 24 occurrences of each respective case, for $d = 7$ there are respectively 3, 28, 96 occurrences, and for $d = 8$ there are respectively 3, 44, 208 occurrences.

If $d = 9$ the possible Hilbert functions are

1 2 3 4 5 6 7 8 9 →	1 1 1 1 1 1 1 1 1 0
1 3 5 7 9 →	1 2 2 2 2 0
1 3 6 8 9 →	1 2 3 2 1 0
1 3 6 9 →	1 2 3 3

The number of respective cases are 3, 28, 96, 384. The 3 in each case can be explained as follows: these are the cases where $1, \alpha, \alpha^i$ span a two dimensional vector space over F_2. Hence $\alpha^i = a + b\alpha$, for $a, b \in F_2$, not both zero. The three cases are $1, \alpha, 1+\alpha$ (but I don't think there is any easy way to say what value of i yields $1+\alpha$). For each degree there is only one possible Hilbert function with $H(1) = 3$, $H(2) = 5$, and this must occur for $i = 2$, since $[1 : \alpha : \alpha^2]$ clearly lies on a quadric, and $i = 2$ is not the third case with $H(1) = 2$ (so long as $d \geq 3$) since that would mean that $\alpha^2 = \alpha + 1$, which is not possible since α is a root of an irreducible polynomial of degree d. The distribution of the possibilities was observed in general to depend on the choice of primitive polynomial. However there appears to be a trend for the more generic Hilbert functions to occur more frequently, and this trend seems to get stronger as d increases. However I have no idea how to prove this. I do not know if the two occurrences of $H(2) = 28$ and $H(3) = 96$ are a coincidence or not.

I will now describe some of my experience with a few intermediate cases. If $d = 15$ the possible difference sequences are

(1)	1 1 1 1
(2)	1 2 2 2 2 2 2 2
(3)	1 2 3 3 3 2 1
(4)	1 2 3 3 3 3
(5)	1 2 3 4 3 2
(6)	1 2 3 4 4 1
(7)	1 2 3 4 5

These were all found quite quickly with points of the form $P = [1 : \alpha : \alpha^i]$ with α primitive, $i = 1, 2, 3, 10, 4, 6, 36$ respectively. Now $d = 15$ is too large to enumerate all the possibilities. I chose 100 "random" values of i (according to the APL random number generator) from the interval $1 \leq i \leq 30,000$ (= approx. 2^{15}) and from the resulting 50 points $[1 : \alpha^i : \alpha^j]$, where (i, j) is the $(2b-1, 2b)^{th}$ terms of our random sequence $(1 \leq b \leq 50)$. I obtained sequence (3) once, sequence (5) 6 times, sequence (6) 26 times, and the generic sequence (7) 17 times, the remaining sequences did not occur. The trend towards the more generic sequences occurring more often thus continues, although the peak is not with the most generic.

In a similar experiment with $d = 16$ the difference sequences

1 2 3 4 3 2 1
1 2 3 4 4 2
1 2 3 4 5 1

were obtained respectively 1, 13, 36 times. Perhaps, heuristically at least, the greater proportion of the generic case with 16 than with 15 is because 16 is one more than the binomial coefficient $\binom{4+2}{2} = 15$. Genericity cannot first fail in degree 5 if $d = 16$. It must fail in degree 4 or lower, and it seems hard for it to fail in "low" degrees, as will be seen in other cases.

For $d = 14$ (to give an example below the binomial coefficient) the generic case 1 2 3 4 4 occurred 35 times out of 50, and 1 2 3 4 3 1 occurred 15 times.

In these "random" trials points of the form $P = [1 : \alpha^i : \alpha^j]$ were chosen in the hope of being more random than points of the form $[1 : \alpha : \alpha^i]$, although I suspect it doesn't matter much. For some i and j, P might not be of degree d, but this did not happen in any of my trials.

In degree < 24, all the possibilities were quickly found with the points $[1, \alpha, \alpha^i]$. However, with $d = 24$ I did not find the difference sequence 1 2 3 3 3 3 3 3 3 0. (This supports the observation in the preceding paragraph about failure of genericity in low degrees.) This missing case can be found by considering points of the form $P = [1 : \alpha^i : \alpha^{3i}]$. Such a point lies on a cubic, hence (except in the unlikely case that P also lies on a quadric) the Hilbert function must have difference sequence either 1 2 3 3 3 3 3 3 2 1 or 1 2 3 3 3 3 3 3 3 0. With the first degree 24 polynomial from [PW] the first case occurred for $1 \leq i \leq 16$, and the second for $i = 17$. For large random i the second case seems to occur most of the time. A more theoretical approach is given in the next section, which will also give some idea why low values of i do not give very "random" choices for the Hilbert function.

In the case $d = 25$ all the possible Hilbert functions were found quickly for points of the form $P = [1 : \alpha : \alpha^i]$, except

$$1\ 2\ 3\ 4\ 4\ 4\ 4\ 2\ 1 \text{ and } 1\ 2\ 3\ 4\ 5\ 5\ 5.$$

Points of the first type lie on a quartic, and were quickly found by testing points of the type $[1 : \alpha^i : \alpha^{4i}]$ ($i = 5$ worked using the first irreducible polynomial of degree 25 from [PW]). Points of the second type lie on a curve of degree 5, and were quickly found by testing points of the type $[1 : \alpha^i : \alpha^{5i}]$ ($i = 11$ worked).

I stopped looking for all cases with $d = 25$. One of the main hopes in [RR] was to prove that the generic case always occurs. In degrees $26 \leq d \leq 36$ I checked enough "random" examples (by the method used above for $d = 15, 16$) to find the generic case. In the binomial cases $d = \binom{e+2}{2}$ the generic Hilbert function again seemed to occur less frequently than for nearly values of d. For this reason calculations were also done in some higher binomial cases. In more detail my observations are as follows:

In degree $d = 26$ the generic Hilbert function $H = GH(26, 2)$ is characterized by $H(6) = 26$. In 50 random trials $H(6) = 26$ occurred 38 times, and $H(6) = 25$ (two possible Hilbert functions, which I did not try to distinguish) occurred 12 times. In degree $d = 27$ the generic Hilbert function $H = GH(27, 2)$ is characterized by $H(6) = 27$. In 50 random trials $H(6) = 27$ occurred 24 times, $H(6) = 26$ (one possible Hilbert function) occurred 22 times, and $H(6) = 25$ (2 possibilities) occurred 4 times. In degree $d = 28$ the generic Hilbert function $H = GH(28, 2)$ is characterized by $H(6) = 28$. In 50 random trials $H(6) = 28$ occurred 18 times, $H(6) = 27$ (one possible Hilbert function) occurred 25 times, and $H(6) = 26$ (1 possibility) occurred 7 times. In degree $d = 29$ the generic Hilbert function $H = GH(29, 2)$ is characterized by $H(6) = 28$. In 50 random trials $H(6) = 28$ occurred 38 times, $H(6) = 27$ (one possible Hilbert function) occurred 9 times, and $H(6) = 26$ (2 possibilities) occurred 3 times. Thus among degrees 26,27,28,29 the generic case seems to occur least frequently for $d = 28 = \binom{2+6}{2}$, but the generic case is still reasonably common for $d = 28$. Similar results were noted near 36, but with fewer observations.

The table of [PW] only goes up to $d = 34$. For higher degrees I obtained a primitive polynomial from table D of [LN] (which was taken from [W]). In degrees $35, 36, 45, 55, 66$ the generic case was found fairly quickly, occurring in about $1/4$ to $1/2$ of the cases (but with a small random sample).

5. Some theoretical results

Suppose that $d = 3e$, with $e > 2$, and let α be a root of an irreducible polynomial of degree d. The point $P = [1 : \alpha : \alpha^3]$ lies on a cubic but not a quadric, and hence has Hilbert function with difference sequence either

$$1\ 2\ 3\ 3\ 3\ \cdots\ 3\ 0\ (e-1 \text{ threes) or}$$
$$1\ 2\ 3\ 3\ 3\quad 3\ 2\ 1\ (e-2 \text{ threes}).$$

These cases are distinguished by $H(e) = d$ or $H(e) = d - 1$ respectively. Consider the array

$$\begin{array}{cccccc} 1 & \alpha^3 & \alpha^6 & \alpha^9 & \cdots & \alpha^{d-3} \\ \alpha & \alpha^4 & \alpha^7 & \alpha^{10} & \cdots & \alpha^{d-2} \\ \alpha^2 & \alpha^5 & \alpha^8 & \alpha^{11} & \cdots & \alpha^{d-1}. \end{array}$$

The first $e-1$ diagonals $1/\alpha, \alpha^3/\alpha^2, \alpha^4, \alpha^6/, \ldots /\alpha^{d-7}, \alpha^{d-5}, \alpha^{d-3}/$ span a vector space of dimension $H(e-1) = d-3$ (being the monomials of degree $\leq e-1$ in α, α^3), and if we add $\alpha^{d-4}, \alpha^{d-2}, \alpha^d$ to the list of monomials they span a vector space of dimension $H(e)$. From the construction of the array we see that of the $\alpha^i (i < d)$ only α^{d-1} is missing. Thus we have that $H(e) = d$ if and only if the coefficient of x^{d-1} is non-zero in the minimal polynomial for α, and $H(e) = d-1$ if and only if the coefficient of x^{d-1} is zero. This distinguishes the two cases.

Suppose that $d = 4e$, with $e > 3$, and let α be a root of an irreducible polynomial of degree d. The point $P = [1 : \alpha : \alpha^4]$ lies on a quartic but not a cubic, and hence has Hilbert function with difference sequences either

$$1\ 2\ 3\ 4\ 4\ \cdots\ 4\ 4\ 2\ (e-2 \text{ fours) or}$$
$$1\ 2\ 3\ 4\ 4\ \cdots\ 4\ 3\ 2\ 1\ (e-3 \text{ fours}).$$

These cases are distinguished by $H(e) = d-2$ or $H(e) = d-3$ respectively. Consider the array

$$\begin{array}{cccccc} 1 & \alpha^4 & \alpha^8 & \alpha^{12} & \cdots & \alpha^{d-4} \\ \alpha & \alpha^5 & \alpha^9 & \alpha^{13} & \cdots & \alpha^{d-3} \\ \alpha^2 & \alpha^6 & \alpha^{10} & \alpha^{14} & \cdots & \alpha^{d-2} \\ \alpha^3 & \alpha^7 & \alpha^{11} & \alpha^{15} & \cdots & \alpha^{d-1}. \end{array}$$

The first $e-1$ diagonals $1/\alpha, \alpha^4/\alpha^2, \alpha^5, \alpha^8/, \ldots /\alpha^{d-13}, \alpha^{d-10}, \alpha^{d-7}, \alpha^{d-4}/$ span a vector space of dimension $H(e-1) = d-6$, and if we add $\alpha^{d-9}, \alpha^{d-6}, \alpha^{d-3}, \alpha^d$ to the list, the monomials span a vector space of dimension $H(e)$. From the construction of the array we see that of the $\alpha^i (i < d)$ only $\alpha^{d-1}, \alpha^{d-2}$, and α^{d-5} are missing. Thus we have that $H(e) = d-2$ if and only if the coefficients of x^{d-1}, x^{d-2}, and x^{d-5} are not all zero in the minimal polynomial for α, and $H(e) = d-3$ if and only if these coefficients are all zero. This distinguishes the two cases.

In the tables of [PW] can be found lots of polynomials yielding either Hilbert function in each of the cases above.

It is clear that one can continue cranking out ever more complicated " computations" of the Hilbert function of $P = [1 : \alpha : \alpha^i]$, in terms of the coefficients of the minimal polynomial of α. It is also clear from such arrays that the Hilbert function of $P = [1 : \alpha : \alpha^i]$ agrees with the generic Hilbert function in degree j, so long as $ji < d$, and $j \leq i-1$ (since in this range the jth diagonal contains $j+1$ elements, each with exponent less than d).

6. Concluding observations

The main question posed by this note is whether every Hilbert function of the type described in section 3 is the Hilbert function of some reduced closed point in $\mathbb{P}^2_{F_2}$. All permitted Hilbert functions were found in degrees ≤ 25. Given this experimental evidence I would be very surprised if the answer to my question is negative. However, one should be cautious as is shown by the following result from [RR]: Given p, there is a differentiable O-sequence with $H(1) = 3$ which is not the Hilbert function of a reduced subscheme (i.e. a finite union of reduced closed points) in $\mathbb{P}^2_{F_p}$. The smallest example that we found (for $p = 2$) was of degree $d = 51$, and one would be rather hard pressed to find such examples by random searching. The argument in [RR] was to exhaust the lines in $\mathbb{P}^2_{F_2}$. However in the case of closed points satisfying the above restriction I do not see anything to exhaust. Perhaps the closed points found here can be combined to show that every differentiable O-sequence of degree less than 51 is the Hilbert function of a reduced subscheme of $\mathbb{P}^2_{F_2}$.

A more modest problem would be to prove that for each d the generic case always occurs. In [RR, Theorem 2.3] we gave a counting argument which showed that if $d < \binom{n+q}{n}$ then $GH(d,n)$ is the Hilbert function of some reduced closed point in $\mathbb{P}^n_{F_q}$. However, if $n = q = 2$ this yields only $d < 6$, so the computations yield many more generic Hilbert functions than our theorem.

The main additional experimental observations are that the generic Hilbert function can be found easily for random choices of points, for the range of d studied ($d \leq 36$, and $45, 55, 66$). The generic Hilbert function appears to be least frequent (compared with nearby values of d) if d is a binomial $\binom{2+e}{2}$. Hilbert functions "close" to the generic Hilbert function are more easily found than those that are not. Some of the latter seemed unlikely to be found by random searching, and were found using points of special form.

One would expect similar results for $p > 2$, and if $n > 2$ one would hope that every Hilbert function of a Cayley-Bacharach scheme would occur as the Hilbert function of a reduced closed point over F_p, for any prime p.

References

[G] A. Galligo, Examples d'ensembles de points en position uniforme, Proceedings of MEGA-90, Birkhäuser 1991.

[GMR] A. V. Geramita, P. Maroscia, L. Roberts, The Hilbert function of a reduced k-algebra, J. London Math. Soc. (2) 28 (1983), 443-452.

[H] J. Hansen, Points in uniform position and maximum distance separable codes, preprint, Aarhus, 1992.

[K] M. Kreuzer, On the canonical module of a 0-dimensional scheme, preprint 1992.

[LN] R. Lidl and H. Niederreiter, Finite Fields, Encyclopedia of Mathematics and its Applications, Addison-Wesley, Reading, MA, 1983.

[PW] W. Peterson and E. Weldon, Error Correcting Codes, 2nd ed., M.I.T. Press, Cambridge, MA, 1972.

[RR] L. Roberts and M. Roitman, On Hilbert functions of reduced and of integral algebras, J. Pure Appl. Math. 56 (1989), 85-104.

[S] R. P. Stanley, Hilbert functions of graded algebras, Adv. Math. 28 (1978), 57-82.

[W] E. J. Watson, Primitive polynomials (mod 2), Math. Comp. 16 (1962), 368-369.

Minimal polynomials and sparse resultants

Bernd Sturmfels and Jie-Tai Yu*

Abstract. We give a formula for the minimal polynomial of $n+1$ polynomials in n variables. This formula is based on the concept of the sparse resultant.

1991 Mathematics Subject Classification: 12Y05, 13P99, 52B20

0. Introduction

We consider a set of polynomials $\{f_0, f_1, \ldots, f_n\} \subset K[t_1, \ldots, t_n]$ which contains n algebraically independent elements over the ground field K. There exists a unique (up to a factor from K^*) irreducible polynomial $G \in K[x_0, x_1, \ldots, x_n]$ such that $G(f_0, f_1 \ldots, f_n) = 0$. We call G the *minimal polynomial* of $f_0, f_1, \ldots, f_n$ over K. In the case $n = 1$, the following explicit formula expresses the minimal polynomial in terms of a Sylvester resultant.

Theorem 1 (Abhyankar [1, §30], McKay and Wang [6]). *Let $f_0, f_1 \in K[t]$ be non-zero univariate polynomials, having the minimal polynomial $G(x_0, x_1)$. Then*

$$\mathrm{Res}_t\left(f_0(t) - x_0, f_1(t) - x_1\right) = c \cdot [G(x_0, x_1)]^q, \quad c \in K^*,$$

where $q = \left[K(t) : K\big(f_0(t), f_1(t)\big)\right]$, the degree of field extensions.

Abhyankar, McKay and Wang ask for an analogue to Theorem 1 in the general case $n \geq 2$. A partial answer to this question is given in [10], where the part of the Sylvester resultant is played by the Macaulay resultant. One disadvantage of the Macaulay resultant is that it vanishes identically whenever the leading forms of $f_0, f_1, \ldots, f_n$ have a common non-zero root. In such a case the formula in [10] reduces to "0 = 0".

In this note we prove a more refined generalization of Theorem 1 within the setting of sparse elimination theory ([3],[4],[7],[8]). Our new formula (Theorem 2)

* Partially supported by the N.S.F (#DMS 9002056, 925847 and MSRI Berkeley #DMS 9022140) and an Alfred P. Sloan Research Fellowship

contains the formula of [10] as a special case. It works for all sparse systems whose coefficients satisfy a certain genericity condition (Theorem 3). In Example 5 our results are illustrated for a bivariate system. We finally address the question of how to describe the Newton polytope of the minimal polynomial G in terms of the Newton polytopes of $f_0, f_1, \ldots, f_n$. We answer this question for the special case of unmixed systems.

1. On the minimal polynomial of a sparse systems

We recall the definition of the sparse resultant [4],[7]. Consider $n+1$ polynomials

$$f_i = \sum_{j=0}^{m_i} c_{ij} \mathbf{t}^{a_{ij}} \in K[t_1, , \ldots, t_n], \quad i = 0, 1, \ldots, n, \tag{1.1}$$

having *support sets* $\mathcal{A}_i := \{a_{ij}\}_{j=0,\ldots,m_i} \subset \mathbf{N}^n$ and *Newton polytopes* $P_i := \operatorname{conv}(\mathcal{A}_i) \subset \mathbf{R}^n$. We suppose that the coefficients c_{ij} are algebraically independent over $\mathbf{C}$. Given any linear functional $w \in (\mathbf{R}^n)^*$, we write P_i^w for the face of P_i which is supported by w, and $\mathcal{A}_i^w := \mathcal{A}_i \cap P_i^w$. The initial form of f_i with respect to w is denoted $\operatorname{init}_w(f_i)$.

We define the *extended support sets* $\widetilde{\mathcal{A}}_i := \mathcal{A}_i \cup \{\mathbf{0}\}$ and the *extended Newton polytopes* $\widetilde{P}_i := \operatorname{conv}(\widetilde{\mathcal{A}}_i)$. Here $\mathbf{0}$ denotes the origin. After a coordinate change if necessary, we can and will assume that the abelian group $\mathbf{Z}^n$ is generated by $\widetilde{\mathcal{A}}_0 + \widetilde{\mathcal{A}}_1 + \ldots + \widetilde{\mathcal{A}}_n$. Let $x_0, x_1, \ldots, x_n$ be new variables and consider the equations

$$f_0(\mathbf{t}) - x_0 = f_1(\mathbf{t}) - x_1 = \ldots = f_n(\mathbf{t}) - x_n = 0. \tag{1.2}$$

The *sparse resultant* $\mathcal{R}_{\widetilde{\mathcal{A}}_0,\ldots,\widetilde{\mathcal{A}}_n}$ is the unique (up to sign) non-zero irreducible polynomial in $\mathbf{Z}[c_{00}, \ldots, c_{nm_n}, x_0, \ldots, x_n]$, which vanishes under a specialization whenever the corresponding system (1.2) has a zero $\mathbf{t}$ in $(\mathbf{C}^*)^n$. We refer to [7] for a product formula and to [3] for a determinantal formula for the sparse resultant. The main difference to these articles is that here we are dealing with the *extended* supports and *extended* Newton polytopes. See [2],[7] for the definition and basic properties of the *mixed volume* $\mathcal{M}$.

Theorem 2.
(1) *The minimal polynomial of* $f_0, f_1, \ldots, f_n$ *over* $K = \mathbf{C}(c_{00}, \ldots, c_{nm_n})$ *equals*

$$G(x_0, x_1 \ldots, x_n) := \mathcal{R}_{\widetilde{\mathcal{A}}_0, \widetilde{\mathcal{A}}_1, \ldots, \widetilde{\mathcal{A}}_n}(f_0 - x_0, f_1 - x_1, \ldots, f_n - x_n). \tag{1.3}$$

(2) *The degree of* G *in* x_i *equals the mixed volume of extended Newton polytopes*

$$\deg_{x_i}(G) = \mathcal{M}(\widetilde{P}_0, \ldots, \widetilde{P}_{i-1}, \widetilde{P}_{i+1}, \ldots, \widetilde{P}_n). \tag{1.4}$$

We next describe the behavior of the formula (1.3) under a specialization of coefficients. Let $\phi : K \to K'$ be any epimorphism of fields, giving rise to specialized coefficients $c'_{ij} := \phi(c_{ij})$. We also write $f'_0, f'_1, \ldots, f'_n, G'$ for the images of $f_0, f_1, \ldots, f_n, G$ under ϕ. Consider the product of face resultants

$$R(c_{10}, \ldots, c_{nm_n}, x_1, \ldots, x_n) := \prod_w \mathcal{R}_{\widetilde{\mathcal{A}}_1^w, \ldots, \widetilde{\mathcal{A}}_n^w}(\mathrm{init}_w(f_1 - x_1), \ldots, \mathrm{init}_w(f_n - x_n)), \tag{1.5}$$

where w runs over all linear functionals such that $P_1^w + \ldots + P_n^w$ has dimension $n-1$ and P_0^w does not contain the origin.

Theorem 3. *Suppose $\phi(R) = R(c'_{ij}) \neq 0$. Then $f'_1, \ldots, f'_n$ are algebraically independent over K', and the minimal polynomial H of $f'_0, f'_1, \ldots, f'_n$ over K' satisfies*

$$G'(x_0, x_1, \ldots, x_n) = c \cdot H(x_0, x_1, \ldots, x_n)^q, \; c \in K'^* \tag{1.6}$$

where $q = [K'(t_1, \ldots, t_n) : K'(f_0, \ldots, f_n)]$. The degree of H in x_0 equals $\mathcal{M}(\widetilde{P}_1, \ldots, \widetilde{P}_n)/q$.

Remark 4. Often the genericity condition (1.5) will not depend on $x_1, \ldots, x_n$ at all. This happens, for instance, in the unmixed case when all polynomials f_i have the same support $\mathcal{A} = \mathcal{A}_0 = \mathcal{A}_1 = \ldots = \mathcal{A}_n$. In the unmixed case we can write

$$R(c_{10}, \ldots, c_{nm_n}) := \prod_w \mathcal{R}_{\mathcal{A}^w}(\mathrm{init}_w(f_1), \ldots, \mathrm{init}_w(f_n)), \tag{1.5'}$$

where the product is over all facet normals w which do not separate $P = \mathrm{conv}(\mathcal{A})$ from the origin. Also in the following example the expressions (1.5) do not depend on the $x'_k s$.

Example 5. Let $n = 2$ and $\mathcal{A}_0 = \{(1,0),(0,1)\}$, $\mathcal{A}_1 = \{(1,1),(0,1)\}$, $\mathcal{A}_2 = \{(2,0),(1,1),(0,1)\}$. The given polynomials are

$$\begin{aligned} f_0(t_1,t_2) &= a_0 t_1 + a_1 t_2, \\ f_1(t_1,t_2) &= b_0 t_1 t_2 + b_1 t_2, \\ f_2(t_1,t_2) &= c_0 t_1^2 + c_1 t_1 t_2 + c_2 t_2. \end{aligned} \tag{1.7}$$

The minimal polynomial of f_0, f_1, f_2 over $K = \mathbf{C}(a_0, a_1, \ldots, c_2)$ equals

$$\begin{aligned} G(x_0,x_1,x_2) &= \mathcal{R}_{\widetilde{\mathcal{A}}_0,\widetilde{\mathcal{A}}_1,\widetilde{\mathcal{A}}_2}(f_0(t_1,t_2) - x_0, f_1(t_1,t_2) - x_0, f_2(t_1,t_2) - x_0) = \\ &(b_0b_1c_0c_1 - b_0^2c_0c_2)\,x_0^3 + (a_0^2b_0^2c_2 - a_0^2b_0b_1c_1)\,x_0x_2 - a_1b_0c_0c_1\,x_0^2x_1 \\ &+ (a_0^2b_1c_1^2 - a_0^2b_0c_1c_2 + 3a_0a_1b_0c_0c_2 - 3a_0a_1b_1c_0c_1 + 2a_1^2b_1c_0^2)\,x_0x_1 \\ &+ (2a_0a_1^2c_0c_1 - a_0^2a_1c_1^2 - a_1^3c_0^2)\,x_1^2 + (2a_0^2a_1b_0c_1 - 2a_0a_1^2b_0c_0)\,x_1x_2 \\ &+ a_1b_0^2c_0\,x_0^2x_2 - a_0^2a_1b_0^2\,x_2^2 + (a_0^3b_1c_1c_2 - a_0^3b_0c_2^2 - a_0^2a_1b_1c_0c_2)\,x_1 \\ &+ (a_0b_1^2c_0c_1 - a_0b_0b_1c_0c_2 - a_1b_1^2c_0^2)\,x_0^2 + (a_0^3b_0b_1c_2 - a_0^3b_1^2c_1 + a_0^2a_1b_1^2c_0)\,x_2. \end{aligned}$$

The degrees of $G(x_0, x_1, x_2)$ in the variables x_0, x_1, x_2 coincide with the mixed volumes of the extended Newton polytopes. The mixed volumes can be read off from Figure 1:

$$\mathcal{M}(\widetilde{P}_1, \widetilde{P}_2) = 3, \; \mathcal{M}(\widetilde{P}_0, \widetilde{P}_2) = 2, \; \mathcal{M}(\widetilde{P}_0, \widetilde{P}_1) = 2.$$

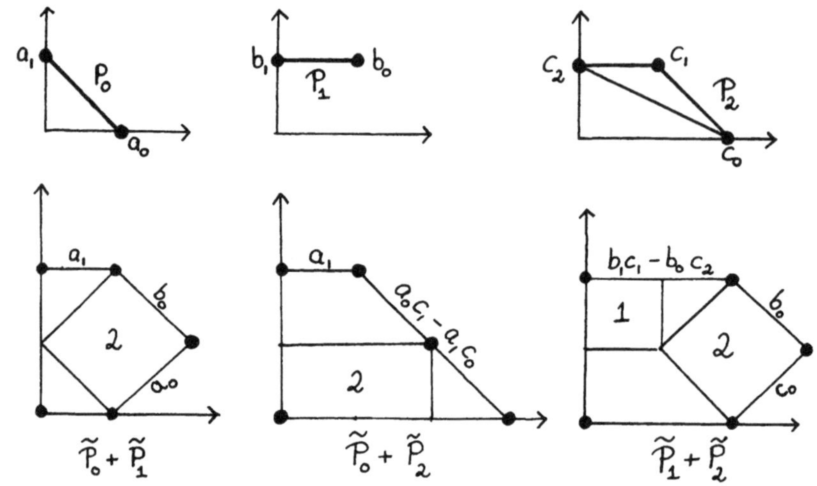

Figure 1. The (extended) Newton polytopes of (1.7)

Let us now see what happens under a specialization of coefficients $\phi : K \to K'$. The special role of f_0 in Theorem 3 can be played by all three polynomials. We form the product (1.5) for all three cases:

(i) $\prod_w \mathcal{R}_{\widetilde{\mathcal{A}}_1^w, \widetilde{\mathcal{A}}_2^w}(\text{init}_w(f_1 - x_1), \text{init}_w(f_2 - x_2)) = b_0c_0(b_1c_1 - b_0c_2)$,

(ii) $\prod_w \mathcal{R}_{\widetilde{\mathcal{A}}_0^w, \widetilde{\mathcal{A}}_2^w}(\text{init}_w(f_0 - x_0), \text{init}_w(f_2 - x_2)) = a_1(a_0c_1 - a_1c_0)$,

(iii) $\prod_w \mathcal{R}_{\widetilde{\mathcal{A}}_0^w, \widetilde{\mathcal{A}}_1^w}(\text{init}_w(f_0 - x_0), \text{init}_w(f_1 - x_1)) = a_0a_1b_0$.

If (i), (ii) or (iii) remains nonzero under some specialization $\phi : K \to K'$, then $G'(x_0, x_1, x_2)$ still gives the minimal polynomial.

For instance, consider the specialization ϕ which maps a_0, a_1, b_0, c_0 to 1 and b_1, c_1, c_2 to 0. Here (iii) remains non-zero, so $\text{degree}_{x_2}(G) = \text{degree}_{x_2}(G') = 2$. Indeed, the minimal polynomial of $f_0' = t_1 + t_2$, $f_1' = t_1t_2$ and $f_2' = t_1^2$ equals $G'(x_0, x_1, x_2) = x_0^2x_2 - x_1^2 - 2x_1x_2 - x_2^2$. □

The proofs of Theorem 2 and Theorem 3 are based on the following lemma. Lemma 6 is a direct consequence of the product formula for the sparse resultant [7, Theorem 1.1].

Lemma 6. *We have the following identity in* $\mathbf{Q}(c_{00},\ldots,c_{nm_n},x_0,\ldots,x_n)$*:*

$$\mathcal{R}_{\widetilde{\mathcal{A}}_0,\widetilde{\mathcal{A}}_1,\ldots,\widetilde{\mathcal{A}}_n}(f_0-x_0,f_1-x_1,\ldots,f_n-x_n) =$$
$$\prod_{\gamma\in V(f_1-x_1,\ldots,f_n-x_n)} (f_0(\gamma)-x_0)\cdot\prod_{w\in S^{n-1}} \mathcal{R}_{\widetilde{\mathcal{A}}_1^w,\ldots,\widetilde{\mathcal{A}}_n^w}(\mathrm{init}_w(f_1-x_1),\ldots,\mathrm{init}_w(f_n-x_n))^{d_w}.$$

where the product is over all facet normals of $\widetilde{P}_1+\ldots+\widetilde{P}_n$. *The exponent* d_w *is a non-negative integer; it is zero if and only if* $\widetilde{P}_0^w$ *contains the origin.*

The zero-dimensional variety $V(f_1-x_1,\ldots,f_n-x_n)$ in Lemma 6 is defined over the algebraic closure of $\mathbf{Q}(c_{10},\ldots,c_{nm_n},x_1,\ldots,x_n)$. By Bernstein's Theorem [2], its cardinality is equal to the mixed volume $\mathcal{M}(\widetilde{P}_1,\ldots,\widetilde{P}_n)$. Note that there is an identity analogous to Lemma 6 for each f_i, not just for $i=0$.

Proof of Theorem 2. The sparse resultant is irreducible as a polynomial in $\mathbf{C}[c_{00},\ldots,c_{nm_n},x_0,\ldots,x_n]$. Gauss' Lemma implies that $G(x_0,x_1,\ldots,x_n)$ is irreducible in $\mathbf{C}(c_{00},\ldots,c_{nm_n})[x_0,\ldots,x_n]$. Obviously $G(f_0,\ldots,f_n)=0$. Therefore, $G(x_0,\ldots,x_k)$ is the minimal polynomial of $f_0,\ldots,f_k$ over $K=\mathbf{C}(c_{00},\ldots,c_{nm_n})$. This proves part (1).

By Lemma 6 and Bernstein's Theorem [2], the degree of the sparse resultant in x_0 is equal to $\mathcal{M}(\widetilde{P}_1,\ldots,\widetilde{P}_n)$. By symmetry we get the same result for the degrees in the other variables $x_1,\ldots,x_n$. This proves part (2). □

Proof of Theorem 3. Suppose that $\phi(R)\neq 0$. This means the right factor on the right hand side of Lemma 6 remains non-zero under ϕ. Consider the affine variety $V(f_1'-x_1,\ldots,f_n'-x_n)$, which is defined over the algebraic closure of $K'(x_1,\ldots,x_n)$. The genericity condition in [2, Theorem B(a)] ensures that $V(f_1'-x_1,\ldots,f_n'-x_n)$ is zero-dimensional of multiplicity $\mathcal{M}(\widetilde{P}_1,\ldots,\widetilde{P}_n)$. This implies that $K'(t_1,\ldots,t_n)$ is an algebraic field extension of $K'(f_1',\ldots,f_n')$, having degree $\mathcal{M}(\widetilde{P}_1,\ldots,\widetilde{P}_n)$. In particular, the polynomials $f_1',\ldots,f_n'$ are algebraically independent over K'.

The points γ in $V(f_1'-x_1,\ldots,f_n'-x_n)$ do not depend on the variable x_0 at all. Therefore $f_0(\gamma)$ cannot be equal to x_0. This implies that the right hand side of the equation in Lemma 6 remains non-zero under the specialization ϕ. Hence $G'(x_0,x_1,\ldots,x_n)$ is non-zero, and its degree in x_0 equals $\mathcal{M}(\widetilde{P}_1,\ldots,\widetilde{P}_n)$.

Let $H(x_0,x_1,\ldots,x_n)$ denote the minimal polynomial of $f_0',f_1',\ldots,f_n'$ over K'. The affine variety defined by H is the Zariski closure of the image of the map

$$\mathbf{t}=(t_1,\ldots,t_n) \quad\mapsto\quad (f_0'(\mathbf{t}),f_1'(\mathbf{t}),\ldots,f_n'(\mathbf{t})). \tag{1.8}$$

Let G'' be any irreducible factor of G', and let $(\alpha_0,\alpha_1,\ldots,\alpha_n)$ be any point on the hypersurface $\{G''=0\}$. By definition of the sparse resultant, the point $(\alpha_0,\alpha_1,\ldots,\alpha_n)$ lies in the Zariski closure of the image of (1.8). Therefore $H(\alpha_0,\alpha_1,\ldots,\alpha_n)=0$, which means that G'' divides H. Since both polynomials are irreducible, they are equal up to a scalar factor. We conclude that $G'=c\cdot H^q$, where

$$q \quad = \quad \frac{deg_{x_0}(G')}{deg_{x_0}(H)} \quad = \quad \frac{[K'(t_1,\dots,t_n) : K'(f_1',\dots,f_n')]}{[K'(f_0',f_1',\dots,f_n') : K'(f_1',\dots,f_n')]}.$$

This completes the proof of Theorem 3. □

Remark 7. An important application of minimal polynomials is the inversion of polynomial automorphisms; see [1, §30],[6],[9],[10],[11]. If $F = (f_1,\dots,f_n) \in (K[t_1,\dots,t_n])^n$ is an automorphism with inverse $G = (g_1,\dots,g_n)$, then g_i coincides with the minimal polynomial of $f_1(x_i = 0),\dots,f_n(x_i = 0)$. Thus Theorem 3 implies new upper bounds for the degree of G in the case when F is sparse.

2. On the Newton polytope of the minimal polynomial

In [6], McKay and Wang proved the following result.

Proposition 8. *Let $f_0, f_1 \in K[t]$ be non-zero univariate polynomials, having the minimal polynomial $G(x_0,x_1)$. Then the Newton polygon of G is a right triangle with perpendiculars on the coordinate axes.*

It is natural to ask whether the analogue of Proposition 8 is true for $n \geq 2$, at least for generic coefficients. In what follows we give an affirmative answer for generic *unmixed* systems. We let $f_0, f_1,\dots,f_n$ be generic polynomials in $K[t_1,\dots,t_n]$ having the same Newton polytope P. Here $K = \mathbf{C}(c_{00},\dots,c_{nm})$ and $G(x_0,x_1,\dots,x_n)$ denotes the generic minimal polynomial as in (1.3). We assume $\dim(\widetilde{P}) = n$, otherwise $f_1,\dots,f_n$ would be algebraically dependent over K. Let e_i denote the standard coordinate vectors in $\mathbf{R}^{n+1}$.

Theorem 9. *The Newton polytope of the minimal polynomial $G(x_0,x_1,\dots,x_n)$ of the generic unmixed system with Newton polytope P equals*

$$\mathcal{N}(G) \;=\; \operatorname{conv}\left\{ n!\,\operatorname{vol}(\widetilde{P})\cdot e_i\,,\; n!\,(\operatorname{vol}(\widetilde{P}) - \operatorname{vol}(P))\cdot e_i\,,\; i = 0,1,\dots,n \right\}. \tag{2.1}$$

Proof. We use the bracket representation of the $\mathcal{A}$-resultant [5] and the constructions in [7, Section 6]. This means we consider $G(x_0,x_1,\dots,x_n)$ as a polynomial in the maximal minors ("brackets") of the matrix

$$\begin{pmatrix} -x_0 & c_{00} & c_{01} & \cdots & c_{0m} \\ -x_1 & c_{10} & c_{11} & \cdots & c_{1m} \\ \vdots & \vdots & \vdots & \ddots & \vdots \\ -x_n & c_{n0} & c_{n1} & \cdots & c_{nm} \end{pmatrix} \tag{2.2}$$

of coefficients of the unmixed system (1.2). We identify the first column of (2.2) with the special point "0" in [7, Section 6]. The $\mathcal{A}$-resultant has degree $r := n!\,\mathrm{vol}(\widetilde{P})$ in brackets. The two antipodal regular subdivisions of $\widetilde{P}$ in [7, Section 6] show that the upper total degree of G in $x_0, x_1, \ldots, x_n$ equals r, while the lower total degree of G in $x_0, x_1, \ldots, x_n$ equals $s := n!\,(\mathrm{vol}(\widetilde{P}) - \mathrm{vol}(P))$. This proves that the Newton polytope $\mathcal{N}(G)$ is contained in the right hand side of (2.1).

To prove the reverse inclusion, it suffices to show that x_0^r and x_0^s appear with non-zero coefficients in G. These coefficients are certain Chow forms, which are denoted $\mathcal{R}_{X \cap P^{n-1}}$ and $\mathcal{R}_Z$ in [7, Section 6]. The value of these Chow forms is non-zero for generic c_{ij} in (2.2). This completes the proof of Theorem 9. □

Remark 10. If the f_i in Theorem 9 have no constant term, then $\mathcal{N}(G)$ is a prism over an n-simplex. If they do have a constant term, then $\mathcal{N}(G)$ is an $(n+1)$-simplex.

Example 11. The conclusion of Proposition 8 and Theorem 9 does not hold for mixed systems. For instance, the generic mixed system

$$f_0(t_1,t_2) = c_{00}t_1^3 + c_{01}t_2^3, \quad f_1(t_1,t_2) = c_{10}t_1^2t_2, \quad f_2(t_1,t_2) = c_{20}t_1t_2^2$$

has the minimal polynomial

$$G(x_0,x_1,x_2) = c_{01}c_{10}^3\,x_2^3 + c_{00}c_{20}^3\,x_1^3 - c_{10}^2c_{20}^2\,x_0x_1x_2.$$

The Newton polytope $\mathcal{N}(G)$ is the triangle with vertices $(0,0,3)$, $(0,3,0)$ and $(1,1,1)$. The vertex $(1,1,1)$ does not lie on any of the coordinate hyperplanes. □

This example raises the following natural question.

Problem 12. For the generic mixed system (1.1), describe the Newton polytope $\mathcal{N}(G)$ of the minimal polynomial $G(x_0, x_1, \ldots, x_n)$ directly in terms of the given Newton polytopes $P_0, P_1, \ldots, P_n$.

Readers familiar with [4] will observe that the following long detour might lead to a solution. By Theorem 2, $\mathcal{N}(G)$ is a coordinate projection of the Newton polytope $\mathcal{N}(\mathcal{R})$ of the sparse resultant $\mathcal{R} = \mathcal{R}_{\widetilde{\mathcal{A}}_0,\ldots,\widetilde{\mathcal{A}}_n}(c_{00}, \ldots, c_{nm_n}, x_0, \ldots, x_n)$. The sparse resultant can be interpreted as an $\mathcal{A}$-*discriminant*, where $\mathcal{A}$ is a certain configuration constructed from the $\widetilde{\mathcal{A}}_i$. One can derive $\mathcal{N}(\mathcal{R})$ from the *secondary polytope* of $\mathcal{A}$, using the machinery of *D-equivalences* in [4, Sect. 3D]. What we seek in Problem 12 is a more direct description.

References

[1] Abhyankar, S.S., Algebraic Geometry for Scientists and Engineers, Amer. Math. Soc., Providence, 1990.

[2] Bernstein, D.N., The number of roots of a system of equations, Functional Analysis and its Applications **9** (1975), 183–185.

[3] Canny J., I. Emiris, An efficient algorithm for the sparse mixed resultant, to appear in Proceedings of AAECC '93, Puerto Rico.

[4] Gel'fand I.M., M.M. Kapranov, A. V. Zelevinsky, Discriminants of polynomials in several variables and triangulations of Newton polytopes, Algebra i Analiz (Leningrad Math. J.) **2** (1990), 1–62.

[5] Kapranov, M., B. Sturmfels, A. Zelevinsky, Chow polytopes and general resultants, Duke Math. J. **67** (1992), 189–218.

[6] McKay, J., and S. Wang, An inversion formula for two polynomials in two variables, J. Pure Appl. Algebra, **40** (1986), 245–257.

[7] Pedersen, P., B. Sturmfels, Product formulas for sparse resultants, Math. Z., to appear.

[8] Sturmfels, B., Sparse elimination theory, in *Computational Algebraic Geometry and Commutative Algebra* (D. Eisenbud and L. Robbiano, Eds.), Proceedings Cortona (June 1991), Cambridge University Press, to appear.

[9] Yu, J.-T., Face polynomials and inversion formula, J. Pure Appl. Algebra, **78** (1992), 212–219.

[10] Yu, J.-T., Resultants and inversion formula for n polynomials in n variables, Manuscript, 1991.

[11] Yu, J.-T., A product formula for minimal polynomials, Manuscript, 1992.

Open problems

collected by E. Ballico

(1) (E. Ballico) Much work was done (with many different motivations) on the cohomological properties of a "generic" finite union of 0-dimensional subschemes "of fixed type" in a given scheme X (see for instance [Hi1], [Hi2] and problem 4 here). Call W this "generic" union. Suppose it is given a subscheme Z of X. It seems that in many different situations, if W consists of a large enough number of subschemes, not only the cohomological properties of $Z \cup W$ are mainly given by the ones of W, but also these can be made very precise (as in several examples with $\dim(W) \leq 1$ and Z union of general points). It is usually very easy to show with explicit bounds that the "cohomology" (i.e., "postulation", "minimal number of generators", "Cohen-Macaulay type", "minimal free resolution") of $W \cup Z$ is not much worst than the one of W. The interesting fact is that sometimes if the "cohomology" of W is "good", than the "cohomology" of $W \cup Z$ is "good" (in the sense that some key linear maps have maximal rank), even if the "cohomology" of Z was "bad". However, every example I can handle (with W disjoint union of general enough points and $\dim(Z) \leq 1$) is checked in a different way (sometimes using ideas in [W1]). Is it possible to find some more general principle at work here unifying at least these scattered examples?

(2) (L. Chiantini) Let $V_{d,g}$ be the Severi variety of plane curves of degree d and geometric genus g and let $V'_{d,g}$ be the open subset corresponding to curves with only nodes as singularities ([S]). There is a natural map (nodal map) $\nu_{d,g} : V'_{d,g} \to (\mathbf{P}^2)^{(\delta)}$, where $\delta = (d-1)(d-2)/2 - g$, which sends every curve to the set of its nodes. This map is known to be birational in many cases ([T]); it is injective when g is small enough, e.g. for values allowed for smooth curves in $\mathbf{P}^n$, $n > 3$ (see [D]).

The image $N_{d,g}$ of $\nu_{d,g}$ is thus stricly related to the Severi variety and also to the variety of linear series of given dimension on abstract curves. Indeed observe that, by Riemann-Roch, the series cut by lines on the normalization X of $C \in V'_{d,g}$ has dimension $r = d-g+i$ with $i = h^0\mathcal{I}_{d-4}$, $\mathcal{I} =$ ideal sheaf of $\nu_{d,g}(C)$. So the pairs (X, L), $L =$ very ample linear series of degree $d \gg 0$ and dimension $r > 3$ on the curve X of genus g, can be "locally parametrized" by the intersection of $N_{d,g}$ with the locus $A^{\delta}_{d-4}(r) = \{$sets of points Z of degree δ in $\mathbf{P}^2$ with $h^0\mathcal{I}_Z(d-4) = g+r-d\}$ (this seems to be the classical approach of Brill and Nother; see also [AC]).

Few things are known in general on the loci $A^a_b(c) \subset (\mathbf{P}^2)^{(a)}$ (see [C] or [BH]), $N_{d,g}$ and their intersection (as far as I know, no sets of equations for $N_{d,g} \subset (\mathbf{P}^2)^{(\delta)}$ are known). A deeper investigation of these loci from a point-theoretical point of view, seems natural by itself and will give important outputs in the theory of special divisors on curves.

(3) (S. Giuffrida, R. Maggioni and A. Ragusa) Let $\mathbf{Q} \cong \mathbf{P}^1 \times \mathbf{P}^1$ (with $\mathbf{P}^1 = \mathbf{P}^1_k$, k an algebraically closed field) be a smooth quadric and let $G(\mathbf{Q}) \cong \mathrm{k}[\mathrm{u}, \mathrm{u}'; \mathrm{v}, \mathrm{v}']$ be the global ring of $\mathbf{Q}$ (see [GM] and [GMR1]). Setting $H^0(a,b) = H^0(\mathbf{Q}, \mathcal{O}_{\mathbf{Q}}(\mathrm{a},\mathrm{b}))$ we have

$$k[u,u';v,v'] \cong \bigoplus_{a,b\in Z} H^0(a,b).$$

Problem 3a). Let $\mathcal{F}$ be a locally free sheaf on $\mathbf{Q}$. Give necessary and sufficient conditions in order that $\mathcal{F}$ be decomposable. Do these conditions depend only on the dimensions of the cohomology groups $H^i(\mathbf{Q}, \mathcal{F}(\mathrm{a},\mathrm{b}))$ $(\mathrm{i} = 0,1;\ \mathrm{a},\mathrm{b} \in \mathbf{Z})$? (Horrocks-type characterization).

In the bigraded ring $G(Q)$ any curve C of Q (=effective divisor), say of type (m,n), is represented by a bihomogeneous equation in the variables $u,u';v,v'$:

$$\sum_{\substack{i=0,\dots,m\\ j=1,\dots,n}} a_{ij}u^{m-1}u'^{i}v^{n-j}v'^{j} = 0.$$

Thus we can associate to C the $(m+1)\times(n+1)$ matrix (a_{ij}), and one can think to study the curves of Q using the associated matrices.

Very few things are known; for instance

- $rk(a_{ij}) = 1$ if and only if C is the union of m $(1,0)$-lines and n $(0,1)$-lines;
- $rk(a_{ij}) \le 2$ if and only if C has the property $R_{m,n}$, i.e., there exist m $(1,0)$-lines $L_1,\dots,L_m$ and n $(0,1)$-lines $L'_1,\dots,L'_n$ such that the complete intersection of $L_1,\dots,L_m$ with $L'_1,\dots,L'_n$ is contained in C.

Problem 3b). To understand the geometrical meaning of the rank of the matrix associated to a curve C of Q. In particular, if C is of type (m,n) with $m \le n$, is true that C has the property $R_{t,n}$ if and only if $rk(a_{ij}) \le m+2-t$?

Let $X \subset \mathbf{Q}$ be a 0-dimensional subscheme, and let $M_X : \mathbf{Z}\times\mathbf{Z} \to \mathbf{Z}$ be its Hilbert matrix (see [GMR1]), i.e., the matrix whose entries are $M_X(i,j) = h^0(i,j) - h^0(\mathbf{Q}, \mathcal{J}_X(\mathrm{i},\mathrm{j}))$, where $\mathcal{J}_X$ is the ideal sheaf of X on $\mathbf{Q}$. Proposition 2.5 and 2.7 of [GMR1] give necessary conditions for the Hilbert matrix of X.

Problem 3c). When is the matrix M the Hilbert matrix of a 0-dimensional subscheme of $\mathbf{Q}$?

If $X, X' \subset \mathbf{Q}$ are two 0-dimensional subschemes which are linked in the complete intersection Z of an $(m,0)$ curve with a $(0,n)$ curve, several questions arise.

Problem 3d). Is the Hilbert matrix of X' just determined by M_X? Alternatively, describe all the possible Hilbert matrices of X' for fixed M_X.

Notice that this question is completely solved when X is arithmetically Cohen Macaulay (ACM for short) (see [GMR1], §4).

Problem 3e). Same question as before for the resolutions, changing the Hilbert matrices with the minimal free resolutions (like (1) below) of the ideal sheaves $\mathcal{J}_X, \mathcal{J}_{X'}$.

Problem 3f). What happens in Problem d) and e) if Z is an ACM subscheme? (Walter-type ACM liaison, see [W2]).

Problem 3g. What happens in Problem d) and e) if Z is ideally a complete intersection (i.e., there exists a sheaf surjection $\mathcal{O}_Q(-m,n)\bigoplus\mathcal{O}_Q(-m',n') \to \mathcal{J}_Z \to 0$)?

Let $X \subset \mathbf{Q}$ be a 0-dimensional subscheme, and let $I(X) \subset G(\mathbf{Q})$ be its saturated bihomogeneous ideal. It is known that X is an ACM if and only if $I(X)$ has a length 2 minimal free resolution. The same is not true for $\mathcal{J}_X$: there exist non-ACM-0-dimensional subschemes of $\mathbf{Q}$ with length 2 minimal free resolution. For instance any ideal ideally complete intersection has a resolution of the following type (Koszul):

$$0 \to \mathcal{O}_{\mathbf{Q}}(-\mathrm{a}-\mathrm{c},-\mathrm{b}-\mathrm{d}) \to \mathcal{O}_{\mathbf{Q}}(-\mathrm{a},-\mathrm{b})\bigoplus\mathcal{O}_{\mathbf{Q}}(-\mathrm{c},-\mathrm{d}) \to \mathcal{J}_{\mathrm{X}} \to 0$$

but if $abcd \neq 0$ X is not ACM. We say that X is a determinantal scheme if there exists a short resolution

$$0 \to \mathcal{F}_1 \to \mathcal{F}_0 \to \mathcal{J}_X \to 0$$

with decomposable $\mathcal{O}_{\mathbf{Q}}$-sheaves $\mathcal{F}_0$, $\mathcal{F}_1$.

Problem 3h). Is any 0-dimensional subscheme of $\mathbf{Q}$ determinantal? Alternatively, characterize them. Notice that a general subscheme with maximal Hilbert matrix is determinantal.

Let now

$$0 \to \mathcal{L}_2 \to \mathcal{L}_1 \xrightarrow{\varphi} \mathcal{L}_0 \xrightarrow{\alpha} \mathcal{J}_X \to 0 \tag{1}$$

be the minimal free resolution of $\mathcal{J}_X$ (see [GMR1] and [GMR2] for details), and let $\mathcal{F}_0$ be a decomposable direct summand of $\mathcal{L}_0$ such that $\alpha' = \alpha\mid_{\mathcal{F}_0}$ gives a sheaf surjection

$$\mathcal{F}_0 \xrightarrow{\alpha'} \mathcal{J}_X \to 0.$$

Problem 3i). When is $\mathrm{Ker}(\alpha')$ decomposable?

This should be a particular way to be determinantal, i.e., determinantal using minimal generators of $I(X)$.

Problem 3j). Let $\mathcal{F}_0 = \bigoplus_{i=1}^{r}\mathcal{O}_{\mathbf{Q}}(-\mathrm{a}_i,-\mathrm{b}_i)$ and $\mathrm{Ker}(\alpha') = \bigoplus_{i=1}^{r-1}\mathcal{O}_{\mathbf{Q}}(-\mathrm{c}_i,-\mathrm{d}_i)$. If (a,b) is the minimal couple such that $(a_i,b_i) \leq (a,b)$ for $i = 1,\dots,r$, is also $(c_i,d_i) \leq (a,b)$ for $i = 1,\dots,r-1$?

Let $X \subset \mathbf{Q}$ be a 0-dimensional subscheme, and let (1) be the minimal free resolution of $\mathcal{J}_X$. Fix a couple $(a,b) \in \mathbf{N} \times \mathbf{N}$, and call $\mathcal{F}_0 = \bigoplus_{i=1}^{r} \mathcal{O}_{\mathbf{Q}}(-\mathrm{a_i}, -\mathrm{b_i})$ (resp. $\mathcal{F}_1 = \bigoplus_{i=1}^{s} \mathcal{O}_{\mathbf{Q}}(-\mathrm{c_i}, -\mathrm{d_i})$) the summand of $\mathcal{L}_0$ (resp. $\mathcal{L}_1$) such that $(a_i, b_i) \leq (a,b)$ (resp. $(c_i, d_i) \leq (a,b)$). Let φ' be the $r \times s$ submatrix of φ consisting of the rows and columns we have chosen.

Problem 3k). Which are the couples (a,b) such that
i) φ' is injective? Or
ii) $\mathrm{Coker}(\varphi') \cong \mathcal{J}_Y$ where $Y \subset \mathbf{Q}$ is a 0-dimensional subscheme? When is $Y = X$?
iii) When do both i) and ii) hold?

Notice that iii) means that Y is determinantal with minimal generators and syzygies.

Let $X \subset Q$ be a 0-dimensional subscheme. X is in *generic position* (on $\mathbf{Q}$) when its Hilbert matrix is maximal; X is in *uniform position* (on $\mathbf{Q}$) when every subscheme of X is in generic position.

Problem 3l). If $X \subset \mathbf{Q}$ is in uniform position, are the graded Betti numbers of the minimal free resolution of $\mathcal{J}_X$ determined by $\Delta^2 M_X$?

If this problem has an affirmative answer, the resolution has the expected graded Betti numbers (see Question 4.8 in [GMR2]): roughly speaking, for each positive entry d_{ij} of $\Delta^2 M_X = (d_{ij})$ there are just d_{ij} first syzygies of degree (i,j); for each negative d_{ij} preceding positive entries there are $-dij$ generators of degree (i,j); the remaining negative entries give the second syzygies.

Problem 3m). Is the question in Problem l) true at least in an open set of $\mathbf{Q}^{\mathrm{N}}$ $(N = \deg X)$? Which conditions on X give rise to such open sets?

Working on the above two problems one can see how useful would be to have a reply to the following question.

Problem 3n). Let $\Sigma \subset H^0(a,b)$ be a linear system of curves of $\mathbf{Q}$. Under which assumption on Σ a generic $(1,0)$-line imposes to Σ independent conditions?

Let $X \subset \mathbf{P}^3$ be a 0-dimensional scheme of degree d, and let $s = \min\{n | H^0(\mathbf{P}^3, \mathcal{J}_{\mathrm{X}}(\mathrm{n})) \neq 0\}$, where $\mathcal{J}_X$ is the ideal sheaf of X in $\mathbf{P}^3$. Suppose that there exist a smooth surface S of degree s containing X.

Problem 3o). If X has the UPP (see [HE]), which are the possible Hilbert functions $HF(X,-)$ of X? There exist an integer $t \geq s$ such that for $n > t$ the first difference sequence $\Delta HF(X,n)$ decreases at least by s (until 0).

For conditions in the case $s = 2$ see [PRR].

Problem 3p). Let $X \subset \mathbf{P}^3$ be a 0-dimensional subscheme with maximal Hilbert function $HF(X,-)$. Which are the bounds - in terms of $\Delta^i HF(X,-)$ - for the graded Betti numbers of the resolutions of all the schemes having the same Hilbert function of X?

Problem 3q). Same question, for a more general Hilbert function, perhaps adding some geometrical assumptions on X.

(4) (B. Harbourne) Let $p_1, \dots, p_n$ be the points of $\mathbf{P}^2_{\mathrm{k}}$, where k is an algebraically closed field, and let $m_1, \dots, m_n$ be positive integers. One would like to know the dimension of the linear system of curves of degree d with assigned base points at each p_i of multiplicity m_i. Of course, the dimension will depend on the disposition of the points and on the numbers d and $m_1, \dots, m_n$. Nevertheless, given d and $m_1, \dots, m_n$, one could hope to determine the dimension if the points are sufficiently general. To pursue this further, it is convenient to change the setting of the task. Denote by S the surface obtained by blowing up the points; one can now consider the essentially equivalent problem of computing $h^0(S,F)$, where $F = de_0 - m_1e_1 - \dots - m_ne_n$, and where e_0 is the divisor class on S of the total transform of a line in the plane, and e_i, $i = 1, \dots, n$, is the exceptional class obtained by blowing up p_i. It is easy to calculate $h^0(S,F) - h^1(S,F)$ from the Riemann-Roch for surfaces; what one needs is to understand $h^1(S,F)$ in the case $h^0(S,F) > 0$. Now it is easy to see that $h^0(S,F)h^1(S,F) > 0$ if we take $F = C + rN$ with r sufficiently large, where C is the class of any curve and N is a reduced irreducible curve of negative self-intersection. We also see that $F \cdot N < 0$ for r sufficiently large. This suggests:

Question 4a). If S is obtained by blowing up sufficiently general points, must $h^0(S,F)h^1(S,F) = 0$ for any numerically effective divisor F?

If the answer to this question is yes, it would be useful (in, say, providing an algorithm for computing $h^0(S,F)$ for arbitrary divisor class F) to know whether a class were numerically effective. For this, one would like to know which classes are the classes of curves of negative self-intersection -1. This raises the question:

Question 4b). If S is obtained by blowing up sufficiently general points, must the class of any reduced irreducible curve of negative self-intersection be an exceptional class?

Taking a moduli-theoretic approach, one can reformulate Question 1 as the following problem:

Problem 4c). For each n, parametrize the locus of points $p_1, \dots, p_n$ such that $h^0(S,F)h^1(S,F) \neq 0$ for some numerically effective divisor F on the blowing up S of the plane at the points $p_1, \dots, p_n$. (It is known that the locus is empty for $n \leq 6$ and nonempty for $n > 8$).

Finally note that an affirmative answer to Question 2 implies that for n sufficiently general blowing up S, the set of self-intersections of reduced irreducible curves on S is bounded below by -1. This recalls the following folklore question. Let S be any smooth complete algebraic surface (not necessarily minimal or rational). Consider the set $D_S := \{C \cdot C | C \in R_S\}$, where R_S is the set of reduced irreducible curves on S.

Question 4d). Is D_S bounded below? (It is easy to see that the answer is yes if mK_S is effective for some $m < 0$).

For very related conjectures and results, see [Hi2] and [dAH].

(5) (M. Kreuzer) For an extensive discussion, detailed explanations, motivations, notations and applications of the following questions, see [Kr] in this volume.

Question 5a). Let $C \subset \mathbf{P}^{\mathrm{d}}$ be a 1-dimensional subscheme. Under what conditions on C is the generic hyperplane section $X := C \cap H$ of C a Cayley-Bacharach scheme?

Question 5b). Let $X \subset \mathbf{P}^{\mathrm{d}}$ be a 0-dimensional subscheme. We say that X *splits cohomologically* if there is a proper decomposition $X = Y \cup Z$ such that $[(\mathbf{I}_{Y/X})_{\sigma X+1} + \mathrm{x}_0\mathrm{R}_{\sigma X}] \cap [(\mathbf{I}_{Z/X})_{\sigma X+1} + \mathrm{x}_0\mathrm{R}_{\sigma X}] = \mathrm{x}_0\mathrm{R}_{\sigma X}$. Here R is the projective coordinate ring of X, $x_0 \in R_1$ a nonzerodivisor, and $\sigma_X := \max\{n : H_X(n) < \deg(X)\}$.

(i) If X does not split cohomologically, does it follow that $H_X(n)H_X(\sigma_X - n) \leq H_X(\sigma_X) + 1$ for $n = 1, \ldots, \sigma_X$?

(ii) If X does not split cohomologically, and if $\{\phi_1, \ldots, \phi_\Delta\}$ is a generic k-basis of $(\omega_R)_{-\sigma_X}$, is it true that $(\omega_R)_0 = R_{\sigma_X}\phi_1 + kx_0^{\sigma_X}\phi_2 + \ldots + kx_0^{\sigma_X}\phi_\Delta$?

Remark: (ii) implies (i).

(6) (L. Roberts) How many O-sequences are there? Recall: O-sequences are those sequences that are the Hilbert functions of a graded ring $R = k[X_1, \ldots, X_n]/I$, with $\deg(X_i) = 1$, I homogeneus, k a field. The dimension of an O-sequence $\{b_i\}_{i\geq 0}$, is the dimension of one (or equivalently every) graded ring with Hilbert function $\{b_i\}$. Dimension zero then means $\sum_{i\geq 0} b_i < \infty$ in which case the degree is $\sum_{i\geq 0} b_i$. In view of [GMR], Theorem 3.3, if k is infinite, it is equivalent to ask how many possible Hilbert functions there are for a set of d (rational closed) points in projective space of arbitrary dimension (although of course $\mathbf{P}^{\mathrm{d}-1}$ suffices). Let a_d be the number of zero dimensional O-sequences of degree d. If an explicit formula for a_d cannot be found, one might ask about the rate of growth of a_d as a function of d. It is obvious that a_d is monotone increasing ($d \geq 2$) and explicit counting of O-sequences suggests that except for a few low cases at the beginning, the ratio a_{d+1}/a_d is monotone decrasing. My question then becomes "Is $\lim_{d\to\infty}(a_{d+1}/a_d) > 1$?" A few cases of a_d (starting with $d = 1$) are $\{1, 1, 2, 3, 5, 8, 12, 18, 27, 40, 57, 82, 116\}$ with ratios a_{d+1}/a_d being $\{1, 2, 1.5, 1.67, 1.5, 1.5, 1.48, 1.425, 1.439, 1.414\}$ (with ratios strictly decreasing until quitting with $a_{26}/a_{25} = 1.305$).

References

[AC] Arbarello, E., Cornalba, M., Footnotes to a paper of Beniamino Segre. Math. Ann. 256 (1981), 341–362.

[BH] Brun, J., Hirschowitz, A.., Le problème de Brill-Noether pour les ideaux de $\mathbf{P}^2$. Ann. Sci. École Norm. Sup. 20 (1987), 171–200.

[C] Chiantini, L., On plane curves with nodes in special position. Rend. Sem. Mat. Univ. Politec. Torino 49 (1991), 29–40.

[D] De Paris, A., On the injectivity of the nodal map, this volume.

[dAH] D'Almeida, J., Hirschowitz, A., Quelques plongements projectifs non speciaux de surfaces rationelles. Math. Z. 211 (1992), 479–483.

[GMR] Geramita, A. V., Maroscia, P., Roberts, L. G., The Hilbert function of a reduced k-algebra. J. London Math. Soc. (2) 28 (1983), 443–452.

[GM] Giuffrida, S., Maggioni, R., On the global ring of a smooth projective surface. Preprint (1992).

[GMR1] Giuffrida, S., Maggioni, R., Ragusa, A., On the postulation of 0-dimensional subschemes on a smooth quadric. Pacific J. Math. 155 (1992), 251–282.

[GMR2] Giuffrida, S., Maggioni, R., Ragusa, A., Resolution of 0-dimensional subschemes on a smooth quadric. Preprint (1992).

[HE] Harris, J., Eisenbud D., Curves in projective space. Les Presses de l'Université de Montréal, Montréal, 1982.

[Hi1] Hirschowitz, A., La méthode d'Horace pour l'interpolation à plusieurs variables. Manuscripta Math. 50 (1985), 337–388.

[Hi2] Hirschowitz, A. Une conjecture pour la cohomologie des diviseurs sur les surfaces rationelles générique. J. reine angew. Math. 397 (1989), 208–213.

[Kr] Kreuzer, M., Some applications of the canonical module of a 0-dimensional scheme, this volume.

[PPR] Paxia, G., Raciti, G., Ragusa, A., Uniform position properties and Hilbert functions for points on a smooth quadric. J. Algebra 149 (1992), 102–121.

[S] Severi, Vorlesungen über algebraische Geometrie. Teubner Leipzig (1921).

[T] Treger, R., Plane curves with nodes. Canad. J. Math. 41:2 (1989), 193–212.

[W1] Walter, C. H., The minimal free resolution of the homogeneous ideal of s general points in P^4, Preprint.

[W2] Walter, C. H., Cohen–Macaulay liaison, Preprint (1992).

List of Participants

A. G. ALEKSANDROV, Institute of Control Sciences. Profsojuznaja Ul., 65, GSP - 7 Moscow, 117806 CSI, Russia.
E-mail address: ALEXANR@ORGMATH.MSK.SU

Edoardo BALLICO, Dipartimento di Matematica, Università di Trento, Via Sommarive 14, 38050 Povo (Trento), Italy.
E-mail address: BALLICO@ITNCISCA.BITNET

Giannina BECCARI, Dipartimento di Matemetica, Università di Torino, Via Carlo Alberto 10, 10123 Torino, Italy.
E-mail address: MASSAZA@ITIPOLI or MASSAZA@POLITO

Antonella BURAGGINA, SISSA, Via Beirut 4, 34100 Trieste, Italy.
E-mail address: BURAGG@ITSSISSA or BURAGG@TSMI19.SISSA.IT

M. Virginia CATALISANO, Dipartimento di Matelmatica, Università di Genova, Via L. B. Alberti 4, 16132 Genova, Italy.
E-mail address: AREZZO@IGECUNIV.BITNET

M. Pia CAVALIERE, Dipartimento di Matematica, Università di Genova, Via L. B. Alberti 4, 16132 Genova, Italy.
E-mail address: CAVALIER@IGECUNIV.BITNET

Karen CHANDLER, Department of Mathematics, University of Chicago, 5734 University Avenue, Chicago, Illinois 60637, USA.
E-mail address: KAC@MATH.UCHICAGO.EDU or CHANDLER@MATH. HARVARD.EDU

Luca CHIANTINI, Facoltà di Ingegneria, Università di Roma "La Sapienza", Piazzale A. Moro 2, 00185 Roma, Italy.
E-mail address: CHIANTIN@ITCASPUR.BITNET

Ciro CILIBERTO, Dipartimento di Matematica, Università di Roma "Tor Vergata", Via O. Raimondo, 00173 Roma, Italy.
E-mail address:CILIBERTO@NAT.UTOVRM.IT

M. Grazia CINQUEGRANI, Dipartimento di Matematica, Università di Catania, Viale A. Doria 6,95125 Catania, Italy.

Francesca CIOFFI, Dipartimento di Matematica e Applicazioni "R. Caccioppoli", Università di Napoli "Federico II", Complesso Universitario di Monte S. Angelo, Via Cintia, 80126 Napoli, Italy.

Maria CONTESSA, Dipartimento di Matematica e Applicazioni, Università di Palermo, Via Archirafi 34, 90123 Palermo, Italy.

Caterina CUMINO, Dipartimento di Matemetica, Università di Torino, Via Carlo Alberto 10, 10123 Torino, Italy.
E-mail address: TEDESCHI@ITOPOLI

Eduard D. DAVIS, Department of Mathematics, SUNYA, 12222 Albany, NY, USA.
E-mail address: EDAVIS@STAT.ALBANY.EDU

Alessandro DE PARIS, Dipartimento di Matematica e Applicazioni "R. Caccioppoli", Università di Napoli "Federico II", Complesso Universitario di Monte S. Angelo, Via Cintia, 80126 Napoli, Italy.

Lora DI FIORE, Dipartimento di Matematica e Applicazioni "R. Caccioppoli", Università di Napoli "Federico II", Complesso Universitario di Monte S. Angelo, Via Cintia, 80126 Napoli, Italy.

Vincenzo DI GENNARO, Dipartimento di Matematica e Applicazioni, Università di Salerno, 84081 Baronissi, Salerno, Italy.

Giuliana DISCEPOLO, Istituto di Cibernetica del C.N.R., Via Togliano 8, 80072 Arco Felice (Napoli), Italy.
E-mail address: GR@IMAGM.NA.CNR.IT

Lucia DORETTI, Dipartimento di Matematica, Università di Siena, Via del Capitano 15, 53100 Siena, Italy.

Stefan EHBAUER, Mathematisches Institut, Universität Bayreuth, Universitätsstr. 30, D-95447 Bayreuth, Germany.
E-mail address: EHBAUER@BTM8X1.MAT.UNI-BAYREUTH.DE

David EISENBUD, Department of Mathematics, University of Brandeis, Boston, USA.
E-mail address: EISENBUD@BINAH.CC.BRANDEIS.EDU

Juan ELIAS, Departament d'Algebra y Geometria, Universidad de Barcelona, Gran Via 585, 08007 Barcelona, Spain.
E-mail address: UBAAGE01@EBCESCA1.BITNET

Philippe ELLIA, Istituto Matematico Università di Ferrara, Via Machiavelli 35, 44100 Ferrara, Italy.
E-mail address: FPE@IFEUNIV.BITNET

Giuliana FATABBI, Dipartimento di Matematica, Università di Perugia, Via Vanvitalle 1, 06100 Perugia, Italy.
E-mail address: ANNALOR@IPGUNIV.BITNET

Federico GAETA, Universidad Complutense de Madrid, Calle Braille 34, 4^0 C, 28034 Madrid, Spain.
E-mail address: GAETA@EMDUCM11.BITNET

Letterio GATTO, Dipartimento. di Matemetica, Politecnico di Torino, Via Carlo Alberto 10, 10123 Torino, Italy.
E-mail address: LGATTO@ITOPOLI.BITNET

Anthony V. GERAMITA, Department of Mathematics and Statistics, Queen's University, Kingston, Ontario, K7L 3N6 Canada.
E-mail address: ANTHONY.V.GERAMITA@QUEENSU.CA

Alessandro GIMIGLIANO, Dipartimento di Matematica, Università di Genova, Via L. B. Alberti 4, 16132 Genova, Italy.
E-mail address: GIMIGLIA@IGECUNIV.BITNET

Salvatore GIUFFRIDA, Dipartimento di Matematica, Università di Catania, Viale A. Doria 6,95125 Catania, Italy.
E-mail address: GIUFFRIDA@MATHCT.BITNET.IT

Johan P. HANSEN, Matematisk Institut, Ny Munkegade, 8000 Aarhus C, Denmark.
E-mail address: MATJPH@MI.AAU.DK

Brian HARBOURNE, Department of Mathematics, University of Nebraska, Lincoln, NE 68588-0323, USA.
E-mail address: HARBOURN@HOSS.UNL.EDU

Mitsuyasu HASHIMOTO, Department of Mathematics, School of Science, University of Nagoya, Chikusa, Ku Nagoya 464-01, Japan.
E-mail address: E43301A@NUCC.CC.NAGOYA-U.AC.JP

L. T. HOA. Institute of Mathematics, P. O. Box 631 Bo Ho, 10000 Hanoi, Vietnam.

Monica IDÁ, Dipartimento di Matematica, Università di Bologna, Piazza di Porta S. Donato 5, 40127 Bologna, Italy.
E-mail address: IDA@DM.UNIBO.IT

Sheldon KATZ, Department of Mathematics, Oklahoma State University, Stillwater, OK 74078, USA.
E-mail address: KATZ@MATH.OKSTATE.EDU

Bettina KREUZER, Fakultät für Mathematik, Universität Regensburg, Postfach 397, D 93053 Regensburg, Germany.
E-mail address: KREUZER@VAX1.RZ.UNI-REGENSBURG.DE

Martin KREUZER, Fakultät für Mathematik, Universität Regensburg, Postfach 397, D 93053 Regensburg, Germany.
E-mail address: KREUZER@VAX1.RZ.UNI-REGENSBURG.DE

Ernst KUNZ, Fakultät für Mathematik, Universität Regensburg, Postfach 397, D 93053 Regensburg, Germany.
E-mail address: KUNZ@VAX1.RZ.UNI-REGENSBURG.DE

Luciano LOMONACO, Dipartimento di Matematica e Applicazioni "R. Caccioppoli", Università di Napoli "Federico II", Complesso Universitario di Monte S. Angelo, Via Cintia, 80126 Napoli, Italy.
E-mail address: LOMONACO@NA.INFN.IT

Anna LORENZINI, Dipartimento di Matematica, Università di Perugia, Via Vanvitalle 1, 06100 Perugia, Italy.
E-mail address: ANNALOR@IPGUNIV.BITNET

Antony MACIOCIA, Department of Mathematics and Statistics, J. Clerk Maxwell Building, The King's Buildings, Mayfield Road, Edinburgh EH9 3JZ, Scotland, UK.
E-mail address: AMA@MATH.EDU.AC.UK

Renato MAGGIONI, Dipartimento di Matematica, Università di Catania, Viale A. Doria 6,95125 Catania, Italy.
E-mail address: MAGGIONI@MATHCT.BITNET.IT

Mirella MANARESI, Dipartimento di Matematica, Università di Bologna, Piazza di Porta S. Donato 5, 40127 Bologna, Italy.
E-mail address: MANARESI@DM.UNIBO.IT

Hideyuki MATSUMURA, Department of Mathematics, School of Science, University of Nagoya, Chikusa, Ku Nagoya 464, Japan.

Emilia MEZZETTI, Dipartimento di Scienze Matematiche, Università di Trieste, Piazzale Europa 1, 34100 Trieste, Italy.
E-mail address: MEZZETTI@UNIV.TRIESTE.IT

Juan MIGLIORE, Department of Mathematics, University of Notre Dame, Notre Dame, 46556 IN, USA.
E-mail address: MIGLIORE@CARTAN.MATH.ND.EDU

Rick MIRANDA, Department of Mathematics, Colorado State University, Fort Collins, CO 80523, USA.
E-mail address: MIRANDA@RIEMANN.MATH.COLOSTATE.EDU

Uwe NAGEL, Universität GH Paderborn, Fachbereich Mathematik Informatik, Warburger Str. 100, D 33098 Paderborn, Germany.
E-mail address: UWEN@UNI-PADERBORN.DE

Hans Joachim NASTOLD, Westfälische Wilhelms Universität, Mathematisches Institut, Einsteinstr. 62, 48149 Münster, Germany.

Paolo MAROSCIA, Dipartimento di Matematica, Università di Roma "La Sapienza", Piazzale A. Moro 2, 00185 Roma, Italy.
E-mail address: MAROSCIA@VAXRM.INFN.IT

Roberto NOTARI, Dipartimento Di Matematica e Applicazioni "R. Caccioppoli", Università di Napoli "Federico II", Complesso Universitario di Monte S. Angelo, Via Cintia, 80126 Napoli, Italy.

Ferruccio ORECCHIA, Dipartimento di Matematica e Applicazioni "R. Caccioppoli", Università di Napoli "Federico II", Complesso Universitario di Monte S. Angelo, Via Cintia, 80126 Napoli, Italy.
E-mail address: ORECCHIA@NA.INFN.IT

Giuseppe PAXIA, Dipartimento di Matematica, Università di Catania, Viale A. Doria 6,95125 Catania, Italy.
E-mail address: PAXIA@MATHCT.CINECA.IT

Grazia RACITI, Dipartimento di Matematica, Università di Catania,,Viale A. Doria 6,95125 Catania, Italy.

Alfio RAGUSA, Dipartimento di Matematica, Università di Catania, Viale A. Doria 6,95125 Catania, Italy.
E-mail address: RAGUSA@MATHCT.BITNET.IT

Isabella RAMELLA, Dipartimento di Matematica e Applicazioni "R. Caccioppoli", Università di Napoli "Federico II", Complesso Universitario di Monte S. Angelo, Via Cintia, 80126 Napoli, Italy.
E-mail address: ORECCHIA@NA.INFN.IT

Irene RASPANTI, Dipartimento di Matematica, Politecnico di Torino, corso Duca degli Abruzzi 24, 10129 Torino, Italy.
E-mail address: RASPANTI@POLCLU.TO.CNR.IT

Lorenzo ROBBIANO, Dipartimento di Matematica, Università di Genova, Via L. B. Alberti 4, 16132 Genova, Italy.
E-mail address: ROBBIANO@IGECUNIV.BITNET

Leslie ROBERTS, Department of Mathematics and Statistics, Queen's University, Kingston, Ontario, K7L 3N6 Canada.
E-mail address: ROBERTS@QUEENSU.CA

Marilina ROSSI, Dipartimento di Matematica, Università di Genova, Via L. B. Alberti 4, 16132 Genova, Italy.
E-mail address: ROSSI@IGECUNIV.BITNET

Paolo SALMON, Dipartimento di Matematica, Università di Bologna, Piazza di Porta S. Donato 5, 40127, Bologna, Italy.

Frank O. SCHREYER, Mathematisches Institut, Universität Bayreuth, Postfach 3008, D 8580 Bayreuth, Germany.
E-mail address: SCHREYER@BTM8X1.MAT.UNI-BAYREUTH.DE

Aron SIMIS, Instituto de Matematica, Universidade Federal de Bahia, Av. Ademar de Barros, Ondina, 40210 Salvador, Brazil.
E-mail address: USERARON@LNCC.BITNET

M. Luisa SPREAFICO, Dipartimento di Matematica, Università di Milano, Via Saldini 50, 20133 Milano, Italy.
E-mail address: SONIA@IMIUCCA.CSI.UNIMI.IT

Rosario STRANO, Dipartimento di Matematica, Università di Catania, Viale A. Doria 6,95125 Catania, Italy.
E-mail address: SSTRANO@MATHCT.CINECA.IT

Bernd STURMFELS, Department of Mathematics, Cornell University, Ithaca, NY 14853, USA.
E-mail address: BERN@MATH.CORNELL.EDU

Giulio TEDESCHI, Dipartimento di Matemetica, Politecnico di Torino, Via Carlo Alberto 10, 10123 Torino, Italy.
E-mail address: TEDESCHI@ITOPOLI.BITNET

Apostolos THOMA, Department of Mathematics, University of Crete, 71409 Iraklion, Crete, Greece.
E-mail address: THOMAP@TALOS.CC.UCH.GR

Ngo Viet TRUNG, Institute of Mathematics, P. O. Box 631 Bo Ho, 10000 Hanoi, Vietnam.

Giuseppe VALLA, Dipartimento di Matematica, Università di Genova, Via L. B. Alberti 4, 16132 Genova, Italy.
E-mail address: VALLA@IGECUNIV.BITNET

Danielle VILLET, Département de Mathématiques, Université de Nice, Avenue Valrose 28, 06034 Nice Cedex, France.
E-mail address: DIAS@VALA.UNICE.FR

Wolfgang VOGEL, Department of Mathematics, Martin Luther Universität, D 06122 Halle, Germany.
E-mail address: VOGEL@MATHEMATIK.UNI-HALLE.DBP.DE

Joachim YAMEOGO. Département de Mathématiques, Université de,Nice, Avenue Valrose 28, 06034 Nice Cedex, France.
E-mail address: YAMEOGO@FRNICE51.BITNET

Jie-Tai YU, Department of Mathematics, University of Notre Dame, Notre Dame, 46556 IN, USA.
E-mail address: YU@CARTAN.MATH.ND.EDU

Bei Fragen zur Produktsicherheit wenden Sie sich bitte an:
If you have any questions regarding product safety,
please contact:

Walter de Gruyter GmbH
Genthiner Straße 13
10785 Berlin
productsafety@degruyterbrill.com